VOLUME 2

INDUSTRIAL FLUID POWER

Fourth Edition

ADVANCED TEXTBOOK ON THE INDUSTRIAL AND MOBILE USE OF
HYDRAULIC AND PNEUMATIC FLUID POWER

Prepared by Charles S. Hedges
Assisted by the Technical Staff of Womack Machine Supply Company

Sponsored by Robert C. Womack
Member Fluid Power Society and Fluid Power Educational Foundation

Published by

Womack Educational Publications

Department of Womack Machine Supply Company

2010 Shea Road   •   Dallas, Texas 75235, U.S.A.   •   Phone: (214) 357-3871

Send mail orders to P. O. Box 35027, Dallas TX 75235

PUBLISHER'S INFORMATION

Title: Industrial Fluid Power — Volume 2
©1988 by Womack Machine Supply Co. — All rights reserved
Second Printing of the Fourth Edition — Printed January 1989
Library of Congress Catalog Card 82-199733 — ISBN No. 0-943719-01-1
Printed in the U.S.A.

ABOUT THE AUTHOR

Mr. Charles S. Hedges is a graduate engineer from the University of Kansas. He has worked in industry as an electrical engineer, Mechanical engineer, and now for the last 25 years in his association with Womack Machine Supply Company has written the Womack textbooks on fluid power, has worked with customers on a variety of fluid power applications, has prepared and published many design data sheets and technical bulletins, has taught fluid power classes in the larger cities of Texas, Oklahoma, and Louisiana, and has trained new salesmen in fluid power applications and circuitry.

WHAT THIS BOOK IS ABOUT

This is the second in a 3-volume series of textbooks on industrial and mobile fluid power. Basic principles, laws, terms, and components were covered in Volume 1. The present book covers some of those topics in greater detail, especially circuits and components. A copy of Volume 1 should be available for reference as the present book is studied. Other Womack books on fluid power, described inside the rear cover, may also be valuable as references.

This is a more advanced book than Volume 1 but only in the sense that certain subjects covered briefly in Volume 1 are expanded into much greater detail. Anyone who comprehends Volume 1 should have little difficulty in also comprehending Volumes 2 and 3. Our method of teaching fluid power is to emphasize the practical, everyday aspects, and to avoid theory and mathematics as much as possible. So this is not a book for scientific study; it is a textbook, a reference, and a guide for those who intend to actually use fluid power.

Thank you for your interest in my books. I have tried to communicate simply and clearly. If I have succeeded and if you have benefitted, then I am glad.

— *Charles S. Hedges*

NEW EDITION

This is the new Fourth Edition. The major updating from the previous edition has been a complete re-write of Chapter 3 with some new circuits to replace those using impulse bleed buttons and impulse limit switches, components which have become difficult to obtain. The book has been enlarged by 16 pages to include additional information in several places and additional design tables in the back of the book. If you have a set of slides for the Third Edition they can be used for this new edition except those covering Chapter 3. A set of 18 slides from #44 through #61 is available for updating your set of slides for this new edition.

HOW TO ORDER WOMACK TEXTBOOKS

Whether you are an individual, a company, library, or book store you can order textbooks in any quantity from the address on the title page. See description of book titles on inside back cover of this book. We discourage book stores from carrying our books in stock because they are revised and updated from time to time and may get out of date on a book shelf. We carry a large stock of all our books in our Dallas warehouse and can ship your order in from 1 to 3 days after we receive it.

Companies, schools, and book stores can order by letter or phone for net 30-day invoicing. Or, place your order on our Telefax No. (214) 350-9322. Individuals please send check with order, or we will ship C.O.D. by UPS.

We would like to send you our latest price list because prices do change. When we publish a revised or enlarged edition we raise the price to cover the additional cost. Thank you again for your purchase of our books.

Chapter Contents–Volume 2

See also the general index on Page 270, and design data tables on Pages 248 through 269.

1

DESIGNING WITH
Fluid Power Cylinders

FIGURE 1-1. Typical Double-Acting Cylinder.

The function of a cylinder in a fluid power system is to convert energy in the fluid stream into an equivalent amount of mechanical energy. Its power is delivered in a straight-line, push-pull motion. Fluid circuits and applications in the present book are designed for cylinder output. Rotary power output, as delivered by hydraulic motors, air motors, rotary actuators, and hydrostatic transmissions is covered in Volume 3 of this textbook series.

The present book covers advanced circuit design using cylinder output. The reader is referred to Volume 1 for basic information on cylinders including methods of calculating power output, speed, standard mountings and seals, and other topics. Additional information on cylinder installation, operation, and maintenance will be found in another Womack book "Fluid Power in Plant & Field".

Graphic Symbols. Figure 1-2 illustrates the standard ANSI (American National Standards Institute) graphic symbols for use in circuit diagrams. Six of the more often used types are shown. The standard

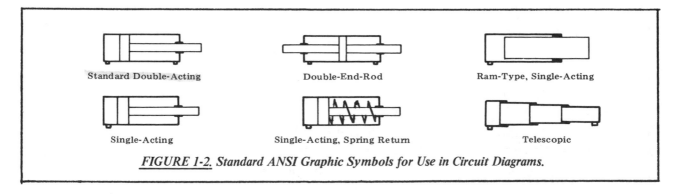

Standard Double-Acting — Double-End-Rod — Ram-Type, Single-Acting

Single-Acting — Single-Acting, Spring Return — Telescopic

FIGURE 1-2. Standard ANSI Graphic Symbols for Use in Circuit Diagrams.

double-acting cylinder with piston rod out one end, and pictured on the preceding page, is used in the majority of applications. It develops force in both directions of piston travel. The double-end-rod type is a variation of the standard cylinder but having a piston rod extending out both end caps. It is occasionally used where it is necessary to have equal area on both sides of the piston, or where one of the rod extensions is to be used for mounting a cam for actuation of a limit switch, or for mounting a stroke limiting stop. The single-acting cylinder develops force in one direction, and is retracted by internal or external spring or by reactive force from the load. The single-acting ram is a construction often used on high tonnage press cylinders. The telescoping cylinder is built in both single-acting and double-acting types. Its purpose is to provide a long stroke with a relatively short collapsed length

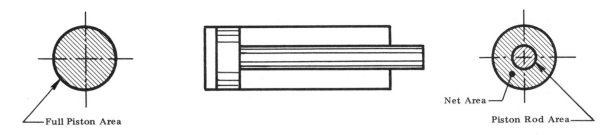

FIGURE 1-3. Significant Areas in a Double-Acting Cylinder, Single-End-Rod Type.

FORCE PRODUCED BY A CYLINDER

Figure 1-3. A standard double-acting cylinder has three significant internal areas. The full piston area when exposed to fluid pressure, produces force to extend the piston rod. The amount of this force, in pounds, is calculated by multiplying piston square inch area times gauge pressure, in PSI.

The "net" area on the front side of the piston is less than full piston area because part of the piston surface is covered by the rod. Net area is calculated by subtracting rod area from full piston area. Because net area is always less than piston area, cylinder force for rod retraction is always less than can be developed for extension when working at the same pressure.

The "rod area" becomes significant when the cylinder is used in a regenerative mode for rapid advance as described in a later chapter in this book.

Cylinder Force Against a Load. Figure 1-4.
The force which a cylinder can exert against a load is determined by making two calculations. First, extension force is calculated according to piston area and PSI pressure against it. Then, the opposing force on the opposite side of the piston is calculated the same way. Net force against a load is the difference between the two.

Caution! It is incorrect, on a single-end-rod cylinder to calculate cylinder net force as piston area times ΔP across the piston. This is true only for double-end-rod cylinders which have equal areas on both sides of the piston.

Example: In Figure 1-4, the extension force is 95 PSI x 50 sq.in. = 4750 lbs. The opposing

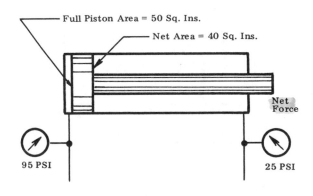

FIGURE 1-4. On a single-end-rod cylinder, two calculations are necessary to determine the net force a cylinder can exert against a load.

force on the rod side is 25 PSI x 40 sq. in. = 1000 lbs. Therefore, the net force which the cylinder can exert against a load in its extension direction is 4750 – 1000 = 3750 lbs. In making cylinder force calculations we sometimes assume that the opposite side of the piston is at atmospheric pressure, and that the counter-force is zero. On some kinds of loads this can lead to serious error.

TYPES OF CYLINDER LOADS

A cylinder may be required to move any one of several kinds of loads — gravity (non-friction) loads, friction loads on a horizontal plane, friction loads on an inclined plane, rolling (low friction) loads, stall-out loads with or without free traverse up to the point of maximum load, loads which do not move parallel to the cylinder axis, or combinations of these loads. In this section we will show how to calculate various kinds of cylinder loads to determine the force required from the cylinder. This will lead, in later chapters, to a study of the pressure level which must be produced at the pump or air compressor to overcome losses in the fluid circuit as well as to move the load.

If several of the above load conditions exist at the same time during the cylinder stroke, the force to meet each condition must be separately calculated, then combined. If two or more conditions occur during the stroke but at different times, the cylinder force must be sized for the condition which requires the greater force. With a massive load, if rapid acceleration is required, additional pressure may be required, and this will be treated in later chapters.

The packing friction in most standard cylinders may amount to about 5% of the maximum force for which the cylinder is rated. The 5% (or 10% to be safe) extra pressure is consumed while the cylinder is in motion but it is also lost to the load even when the cylinder is stalled. Flow losses through the cylinder ports will also consume a small amount of pressure while the cylinder is moving, but these porting losses will disappear when the cylinder stalls. The pump must supply these friction and flow losses, and the methods for estimating them will be described as this study progresses.

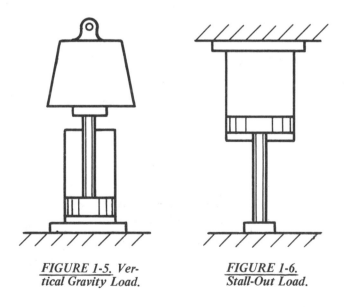

FIGURE 1-5. Vertical Gravity Load.

FIGURE 1-6. Stall-Out Load.

Vertical Gravity Load. Figure 1-5. This is essentially a non-friction load, and full cylinder force must be produced throughout the stroke. The force to balance the load can be calculated easily by the usual method: PSIG x piston square inch area. A small additional force will be needed to accelerate the load and to supply flow and friction losses.

Stall-Out Load. Figure 1-6. Cylinders used on bonding, laminating, coining presses and similar applications, working either vertically or horizontally, usually run in "free traverse" for most of the stroke. When the press platen contacts the work the cylinder stalls with little or no further movement. Very little pressure is required on the cylinder piston during free traverse, and the pump or air compressor supplies only the circuit losses due to friction and fluid flow. At stall, all of the flow losses disappear. The pressure required at the cylinder port is calculated in the usual way, allowing about 5% for packing friction losses even though there is no movement through the packings.

Horizontal Friction Loads. Figure 1-7. Cylinders may be used for pulling or pushing high friction sliding loads such as machine slides, lathe tailstocks, milling machine or grinder tables.

The force for moving a load horizontally against friction resistance is usually less than the force required to lift it, and is calculated by multiplying load weight times the coefficient of static friction between the two sliding surfaces. The coefficient for steel on steel is usually about 0.2 to 0.4 depending on how well the surfaces are lubricated, and may be as high as 0.8 if there is no lubrication. For lubricated surfaces a cylinder force equal to about 1/2 the weight of the load should be sufficient.

Static friction coefficients for other materials can be obtained from a machine handbook.

In Figure 1-7, assuming a friction coefficient of 0.5, a cylinder would have to exert a force of 4500 x 0.5 = 2250 lbs. to move the load.

Reactive Force. Figure 1-8. On any cylinder application, be sure to provide sufficient extra force to overcome any external active force which is trying to move the load in the opposite direction.

Rolling Loads. Figure 1-9. Loads which roll on low friction ball, needle, or roller bearings, can be kept in motion on a horizontal plane with very little force after breaking them away from a

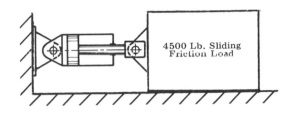

FIGURE 1-7. Horizontal Friction Load.

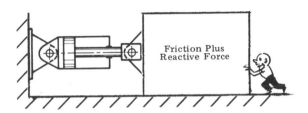

FIGURE 1-8. Live Reactive Force.

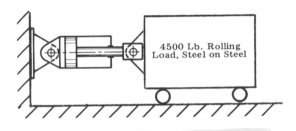

FIGURE 1-9. Rolling (Low Friction) Load.

standstill. Usually, a force as little as 1/10th the load weight will break them away and even less force is required to keep them rolling at a constant speed. Massive loads may require special handling. Additional force may be required to accelerate them rapidly, and some means may be required to stop them without damaging shock to themselves or to the cylinder. Deceleration circuits are covered elsewhere in later chapters of this book.

Cast iron or steel wheels running on steel rails is a special case of a rolling load. The wheels make a small depression in the rails and the wheel is also slightly flattened. The wheel is always forced to roll up out of this depression, and to keep it rolling, a certain amount of horizontal force must be continually applied. This force, from the cylinder, must be sufficient to raise the entire load weight out of the small depression. The following formula may be used to estimate the force required.

F (cyl. force) = W (weight, in lbs., on wheels) x C (coefficient) ÷ R (wheel radius, inches)

The coefficient, C, for iron or steel wheels on steel rails is approximately 0.02. For other coefficients and further information on rolling resistance, see Volume 3 of this textbook series.

In Figure 1-9, the force to keep a 4500 lb. load, on steel wheels of 6'' radius, and coefficient of 0.02, rolling is: F= 4500 x 0.02 ÷ 6 = 15 lbs. Of course up to 450 lbs. might be required to start, and perhaps an additional force, above the 15 lbs., might be required for axle and bearing friction.

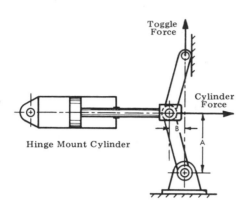

FIGURE 1-10. Toggle Lever Load.

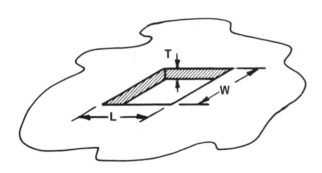

FIGURE 1-11. Shearing Area Must be Calculated.

Toggle Loads. Figure 1-10. A toggle lever system is useful on operations like coining and marking which require exact depth control and extremely high force through a very short distance. Leverage developed in the toggle can greatly amplify the force from a cylinder, producing high tonnage over a distance of a fraction of an inch.

In this figure cylinder force is horizontal and toggle force is taken off vertically. Support bearings at each end of the toggle levers must be precision fitted, and sufficiently heavy to carry the full amplified toggle force.

The cylinder moves in free traverse through most of its stroke, then picks up the load very near the end of its stroke. Toggle force can be calculated with this formula:

$$T = F \times A \div 2B \quad \text{in which:}$$

T = force, in lbs., developed on the toggle leg.
F = force, in lbs., produced by the cylinder.
A = length of toggle arm, in inches.
B = distance, in inches, to full toggle closure.

Note: Dimension A on Figure 1-10 is not actually the toggle lever length, but can be used for lever length in making calculations. The error will be small because the lever is nearly vertical.

Punching, Notching, Shearing. Figure 1-11. An approximation can be made of the cylinder force needed to punch a hole, slot, or notch using the following method:

Find the cross section, in inches, of the section to be sheared. In Figure 1-11 it is the distance around the hole (the perimeter) multiplied by material thickness.

A (area to be sheared) = (L + W + L + W) x T   *(All dimensions must be in inches)*

After finding the number of square inches of material to be sheared, multiply this times the shear strength of the material. This gives the cylinder force needed to make the cut, assuming the punch is flat and shears the entire perimenter simultaneously. To reduce the force needed, die makers sometimes bevel the punch so it enters the shear area gradually, although this may distort either the material or

TABLE 1-1. SHEAR STRENGTH OF COMMON METALS

Sheet Material	Shear Strength, Range, PSI	Sheet Material	Shear Strength, Range, PSI
Steel, low carbon	45,000 to 75,000	Aluminum, half-hard	8,000 to 34,000
Steel, medium carbon	60,000 to 135,000	Brass, half-hard	28,000 to 48,000
Steel, high carbon	65,000 to 160,000	Dural (aluminum)	22,000 to 83,000
Stainless steel 18-8	65,000 to 95,000	Copper, rolled	22,000 to 29,000

the slug. If several holes are to be punched on one stroke of the cylinder, punch heights may be staggered to reduce the maximum force required. The cylinder is sized for the largest cut to be made.

For approximate shear strength of various materials, use the table at the foot of the preceding page. Shear strength can vary widely according to the degree to which the the material has been work hardened. For more accurate shear values, consult the supplier of the material. Values for other materials may be found in a machinery handbook. For metals (only), the shear strength can be assumed to be about 75% of rated tensile strength. Shear strength is expressed in PSI (lbs. per square inch).

Example: Find the cylinder force needed to shear a rectangular hole measuring 1/2" x 5/8" and is 3/16" thick. The material is steel with a shear strength of 60,000 PSI.

Solution: First, convert fractional dimensions to decimals: Hole measurements are .500 x .625 x .1875" thick. Area to be sheared is: (.500 + .625 + .500 + .625) x .1875 = .422 square inches. Shearing force is 60,000 x .422 = 25,300 lbs., or 12.65 U.S. tons.

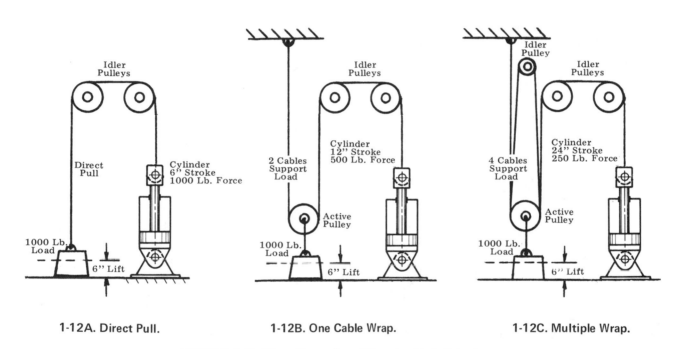

1-12A. Direct Pull.　　1-12B. One Cable Wrap.　　1-12C. Multiple Wrap.

FIGURE 1-12. Three Examples of Hoisting With Cylinders.

Hoisting With Cylinders. Figure 1-12. Cylinders may be used for lifting gates and doors through cables. The lifting force of a cylinder can be increased through multiple wraps of the cable over idler pulleys. Three examples are shown in this figure.

Direct Pull. Figure 1-12A. One cable supports the load and the ratio between load lift height and cylinder stroke is 1:1. To raise a 1000 lb. load to a height of 6 inches, the cylinder must have a 6-inch stroke and sufficient net area on its rod end to produce 1000 lbs. of force. Cable tension is 1000 lbs., and the cable must be sized accordingly.

One Wrap. Figure 1-12B. Since two cables support the load, the tension in each cable is 500 lbs., and a smaller cable diameter can be used. The cylinder to load ratio is now 2:1. Its stroke must be twice as long but it must produce only 500 lbs. pull. It can be a less expensive model with a piston area only half as large as required for a direct lift at the same hydraulic pressure.

Multiple Wrap. Figure 1-12C. The use of additional pulleys and additional cable wraps further changes the cylinder to load ratio. For example, with four cables supporting the load, the cylinder to load ratio is 4:1. Cable tension is reduced to 250 lbs., allowing the use of a much smaller cable. The cylinder must have a 24-inch stroke but is required to produce only 250 lbs.

The cylinder to load ratio is directly related to the number of cables supporting the load. Cylinder force decreases and the stroke increases in direct proportion to the number of cables.

An interesting point to note on the three examples of Figure 1-12 is that whether the cylinder has a large bore and short stroke or a small bore with a longer stroke, the lifting capacity, lifting speed, hydraulic system PSI and GPM required remains the same provided the cylinder piston area is chosen to operate at the same PSI in each case. The same hydraulic power unit, with the same electric motor would give similar results except for higher mechanical losses with a greater number of pulleys.

On a multiple wrap system, the cylinder is lighter and less expensive, more HP can be purchased (in the fluid system) for each dollar invested, and the cable, while longer, can be smaller in diameter, more flexible, lighter in weight, and less expensive than in a direct pull system. However, additional pulley friction may require slightly more power input. The additional input may amount to 18% on a 2-cable system or 33% on a 4-cable system (4 cables supporting the load). In addition, an extra 5 to 7% may be required for each idler pulley in the system.

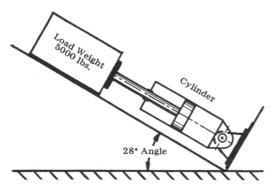

FIGURE 1-13. Combination Load. Cylinder Working a Load on an Inclined Plane.

Sliding Load on an Inclined Plane. Figure 1-13. Cylinder and load axes are in alignment and the cylinder is delivering a straight push to the load. But the load is moving upward at an angle with the horizontal. This is a combination of the gravity load of Figure 1-5 and the friction load of Figure 1-7.

At a 0 degree angle with the horizontal, the entire cylinder load is friction with the sliding surface, and none of it is gravity lift. At 90 degrees to the horizontal, all of the load is gravity lift and none of it is friction. At intermediate angles, part of the cylinder load is gravity lift and part of it is friction. The load against the cylinder is the sum of these two calculations.

TABLE 1-2. TRIGONOMETRIC SINES AND COSINES											
Angle, Degrees	Sine (sin)	Cosine (cos)	Angle, Degrees	Sine (sin)	Cosine (cos)	Angle, Degrees	Sine (sin)	Cosine (cos)	Angle, Degrees	Sine (sin)	Cosine (cos)
2	.035	.999	24	.407	.914	46	.719	.695	68	.927	.375
4	.070	.998	26	.438	.899	48	.743	.669	70	.940	.342
6	.105	.995	28	.469	.883	50	.766	.643	72	.951	.309
8	.139	.990	30	.500	.866	52	.788	.616	74	.961	.276
10	.174	.985	32	.530	.848	54	.809	.588	76	.970	.242
12	.208	.978	34	.559	.829	56	.829	.559	78	.978	.208
14	.242	.970	36	.588	.809	58	.848	.530	80	.985	.174
16	.276	.961	38	.616	.788	60	.866	.500	82	.990	.139
18	.309	.951	40	.643	.766	62	.883	.469	84	.995	.105
20	.342	.940	42	.669	.743	64	.899	.438	86	.998	.070
22	.375	.927	44	.695	.719	66	.914	.407	88	.999	.035

At any angle, the friction load is calculated by multiplying total load weight times the cosine (cos) of the angle which the sliding surface makes with the horizontal. The table of cosines on the preceding page may be used, or for values not in this table consult a mathematical handbook. The gravity lift load is calculated by multiplying total load weight times the sine (sin) of the same angle. When the results of these two calculations are added, this is the force the cylinder must produce.

Example: Calculate the cylinder force required to move a load as in Figure 1-13 which is traveling at a 28-degree angle with the horizontal. The load weight is 5000 lbs., and the coefficient of friction with the sliding surface is 0.7.

Solution, Step 1: First, find the friction load:
   WF (load weight creating friction) = 5000 (total weight) x .883 (cos 28°) = 4415 lbs.

Step 2: Next, calculate the gravity lift load:
   WG (weight to be lifted against gravity) = 5000 (total weight) x .469 (sin 28°) = 2345 lbs.

Step 3: Calculate cylinder force to overcome the friction weight calculated in Step 1:
   FF (friction force) = 4415 x 0.7 (coefficient of friction) = 3091 lbs.

Step 4: Add the two loads to find the total force which the cylinder must produce:
   TF (total force) = 3091 (friction) + 2345 (gravity lift) = 5436 lbs.

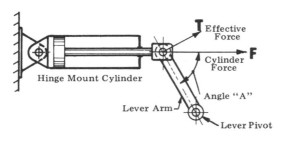

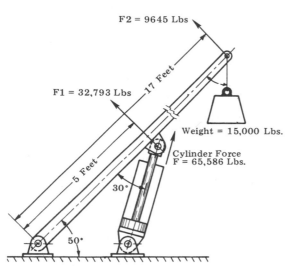

FIGURE 1-14. Hinged Cylinder Producing Torque on a Hinged Lever.

FIGURE 1-15. Hinged Lever Operating a Crane.

Cylinder Working a Rotating Lever. Figure 1-14. A cylinder working a hinged lever can exert its maximum force on the lever only when the lever axis and cylinder axis are at right angles. When Angle A (Figure 1-14) is greater or less than a right angle, only part of the cylinder force is effective on the lever. The cylinder effective force is found by multiplying the full cylinder force times the sine (sin) of the least angle between cylinder and lever axes.

Example: Find the effective force exerted by a 3-inch bore cylinder against a lever when the cylinder is operating at 3000 PSI and when its axis is at an angle of 55 degrees with the lever axis.

Solution: Step 1: First, find the full force developed by the cylinder:
FF (full force) = 7.07 (piston area) x 3000 PSI = 21,210 lbs.

Step 2: Next, find the effective force at 55°:
EF (effective force) = 21,210 x .819 (sin 55°) = 17,371 lbs.

Since maximum cylinder force is delivered in the right angle position, the hinge points for cylinder and lever should be located, if possible, so the right angle point falls close to the lever position which requires the greatest torque.

Cylinders on Cranes. Figure 1-15. The cylinder must be sized for the beam angle and load condition which will require the greatest force. Since the angles change as the cylinder strokes, it may be difficult to determine the working angle and load condition at which maximum force will be required. A rough model can be constructed with cardboard arms and thumbtack hinges from which the beam angle requiring the greatest force can be determined. Exact calculations can then be made for this condition. Calculations will involve sines and cosines of the various angles between cylinder, load, and beam. They will also involve direct and inverse proportion.

Example: In Figure 1-15 the maximum load to be handled is 15,000 lbs. Find the cylinder force required when the beam is in the position shown, at a 50o angle with the horizontal.

Solution: Step 1: First find the force F2 at right angles to the beam which must be present to support the 15,000 lb. hanging load.

F2 = 15,000 x .643 (cos 50°) = 9645 lbs.

Step 2: Next, find the force F1, also at right angles to the beam, which must be produced by the cylinder to support the 15,000 lb. load. This is calculated by proportion. F1 will be greater than F2 in the same ratio that arm length 17 feet is greater than arm length 5 feet.
Arm length ratio of 17 ÷ 5 = 3.4. Therefore, F1 = 9645 x 3.4 = 32,793 lbs.

Step 3: Finally, calculate the cylinder force, at an angle of 30° to the beam, which will produce a force of 32,793 lbs. at its rod hinge point at right angles to the beam:
F (cylinder force) = F1 ÷ sin 30° = 32,793 ÷ .500 = 65,586 lbs.

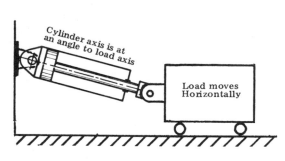

FIGURE 1-16. Low Friction Load.

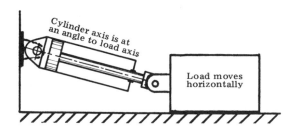

FIGURE 1-17. High Friction Load.

Cylinder Working at an Angle to the Direction of Load Travel. If the cylinder and load axes are not in alignment, only a part of the cylinder force will be effective in the direction of load movement. The greater the angle of mis-alignment, the less the useful force delivered to the load.

Low Friction Load. Figure 1-16. If a relatively small part of the load is friction and most of it is from inertia or overcoming an active opposing force, useful force from the cylinder is calculated by multiplying full cylinder force times the cosine of the mis-alignment angle. Or, if the amount of useful force which will be needed against the load is known, the cylinder force to produce it is calculated by dividing the useful force by the cosine of the angle of mis-alignment.

Example: If the cylinder can produce 5000 lbs. of force and if it is working at a 20° mis-alignment angle, how much of this force is effective as useful force in the direction of load travel? *Solution:* 5000 x .940 (cosine 20°) = 4700 lbs.

High Friction Load. Figure 1-17. If most of the load is friction, and if the angle of the cylinder axis is downward, there is an additional problem. The force of the cylinder actually creates additional friction which further adds to the cylinder load. The mathematical solution involves algebra and trigo-

TABLE 1-3. MULTIPLIERS FOR ANGLE FRICTION PROBLEMS (SEE TEXT)

Angle w/ Horizontal	Coefficients of Friction																
	0.10	0.15	0.20	0.25	0.30	0.35	0.40	0.45	0.50	0.55	0.60	0.65	0.70	0.75	0.80	0.85	0.90
10°	.103	.156	.210	.266	.321	.380	.436	.496	.557	.619	.682	.746	.810	.878	.951	1.05	1.09
15°	.107	.162	.219	.278	.337	.401	.463	.529	.598	.668	.741	.815	.891	.972	1.12	1.19	1.23
20°	.111	.169	.230	.293	.358	.428	.498	.572	.650	.733	.817	.907	.998	1.10	1.21	1.38	1.43
25°	.116	.178	.243	.313	.384	.462	.542	.628	.719	.817	.920	1.03	1.14	1.27	1.42	1.66	1.71
30°	.123	.190	.261	.338	.418	.508	.600	.701	.811	.931	1.06	1.20	1.35	1.53	1.74	2.10	2.17
35°	.132	.205	.284	.371	.463	.567	.677	.801	.939	1.09	1.27	1.46	1.68	1.93	2.26	2.88	2.97
40°	.143	.224	.314	.415	.523	.648	.784	.943	1.12	1.34	1.58	1.87	2.21	2.64	3.26	4.66	4.81
45°	.158	.250	.353	.473	.605	.763	.942	1.15	1.41	1.73	2.13	2.63	3.29	4.26	5.95	12.5	12.9
50°	.178	.285	.408	.556	.724	.938	1.19	1.51	1.93	2.48	3.28	4.47	6.59	11.1	34.7	----	----
55°	.204	.334	.488	.681	.914	1.22	1.63	2.19	3.04	4.50	7.36	16.0	----	----	----	----	----
60°	.243	.408	.613	.887	1.25	1.78	2.59	4.14	7.45	----	----	----	----	----	----	----	----

nometry, and is not within the scope of this book. The reader is referred to a good textbook on engineering mechanics. However, in Table 1-3, the mathematical solutions are presented in chart form and are quite easy to use. Figures in the body of the table are multipliers to be multiplied times the weight of the load. This will give the total required cylinder force under the stated conditions for both the friction load and for the mis-alignment angle between cylinder axis and the horizontal.

Example: If the load weight is 15,000 lbs. and the coefficient of friction is 0.40, what force is required from the cylinder when it is operating at a 20° angle with the horizontal?

Solution: From Table 1-3, the multiplier for 0.40 coefficient of friction and an angle of 20° is .498. Cylinder force required is, therefore: 15,000 x .498 = 7470 lbs.

The reader is referred to standard books on engineering mechanics for solutions to other geometrical arrangements such as a cylinder pushing or pulling at an upward angle, or the load and cylinder moving on non-horizontal planes.

Cable Tensioning With Cylinders. Figure 1-18. This is a simplified method of calculating the cylinder force for tensioning a horizontal cable which is supporting a hanging weight with a specified maximum amount of "sag". The method is based on trigonometry but can be used by anyone without a knowledge of trigonometry by following the simple steps below. The chart on the next page makes the solution of these problems easy.

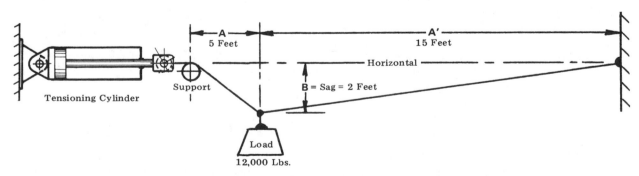

FIGURE 1-18. Pulling Tension on a Cable With a Cylinder.

15

TABLE 1-4. RATIO/MULTIPLIER CHART

Use this for solving cable tensioning problems as explained in the text

Ratio A ÷ B	Multiplier	Ratio A ÷ B	Multiplier	Ratio A ÷ B	Multiplier	Ratio A ÷ B	Multiplier	Ratio A ÷ B	Multiplier
60.00	60.01	21.00	21.02	7.00	7.07	3.60	3.74	2.10	2.33
55.00	55.01	20.00	20.03	6.50	6.58	3.50	3.64	2.05	2.28
50.00	50.01	19.00	19.03	6.25	6.33	3.40	3.54	2.00	2.24
45.00	45.01	18.00	18.03	6.00	6.08	3.30	3.45	1.95	2.19
40.00	40.01	17.00	17.03	5.70	5.79	3.20	3.35	1.90	2.15
35.00	35.01	16.00	16.03	5.40	5.49	3.10	3.26	1.85	2.10
30.00	30.02	15.00	15.03	5.20	5.30	3.00	3.16	1.80	2.06
29.00	29.02	14.00	14.03	4.90	5.00	2.90	3.07	1.75	2.02
28.00	28.02	13.00	13.03	4.70	4.81	2.80	2.97	1.65	1.93
27.00	27.02	12.00	12.04	4.50	4.62	2.70	2.88	1.60	1.89
26.00	26.02	11.00	11.04	4.30	4.42	2.60	2.79	1.50	1.80
25.00	25.02	10.00	10.04	4.15	4.27	2.50	2.69	1.45	1.76
24.00	24.02	9.00	9.05	4.00	4.12	2.40	2.60	1.40	1.72
23.00	23.02	8.00	8.05	3.80	3.93	2.30	2.51	1.35	1.68
22.00	22.02	7.50	7.57	3.70	3.83	2.20	2.42	1.30	1.64

This chart is for use in solving cylinder cable tensioning problems as described in the text on the preceding page.

Figures in the "Ratio A ÷ B" columns are trigonometric tangents of the angle θ between the cable leg and the vertical in Figure 1-18.

Figures in the "Multiplier" columns are trigonometric secants for the same angles.

Interpolation may be used to find intermediate values in both columns. For ratios beyond the range of this chart, find your ratio in the "tangent" columns of a standard trigonometric table, and for the multiplier, use the "secant" listing for the same angle.

Method of Solution: Assume the weight of the cable itself is small compared to the weight of the hanging load. If the weight were suspended in the exact center of the cable, each leg of the cable would support exactly half the load weight. But when suspended nearer one end, that leg of the cable must support more than half the load weight.

The first step in the solution is to calculate the percentage of the load weight supported by each leg of the cable. This will be inversely proportional to the distance from the load to the end support as related to the total span distance between supports. In Figure 1-18, 75% or 9000 lbs. is supported by the leg nearest the cylinder and 25% or 3000 lbs. by the other leg..

After finding the part of the weight supported by the cable leg nearest the cylinder, calculate the ratio of the distance A to the distance B (A ÷ B). Use this ratio in Table 1-3 to find the corresponding multiplier. Then, use this multiplier to multiply times the part of the weight which is creating the tension. This is the force which the cylinder must exert to raise the cable and weight to the position of maximum sag.

Example: Using Figure 1-18, find the cylinder force to pull the cable to raise a 12,000 lb. hanging weight to a position of no more than 2 foot sag. The weight is suspended 5 feet from the left support and the total cable span between supports is 20 feet.

Solution: Distance A is 25% of the total 20-foot span and Distance A' is 75% of the total span. Therefore, the cable leg nearest the cylinder will support 75% of the weight or 9000 lbs.

Next, calculate the ratio of A ÷ B: 5 ft. ÷ 2 ft. = 2.5 ratio. Enter the table above and find the multiplier 2.69 corresponding to a 2.5 ratio. Multiply 9000 x 2.69 = 24,210 lbs. This is the force which must be exerted by the cylinder.

FINDING THE CYLINDER STROKE

On those applications where the cylinder is directly pushing or pulling the load, the required length of stroke will be obvious. But on some applications, such as the rotation of a hinged lever, the required cylinder stroke depends on the length of the lever and the angle through which it must be moved. On those applications the cylinder stroke must either be calculated or must be measured from a scale layout.

A cylinder which operates any hinged device must be free to swing with the stroke. This means it must be hinge or trunnion mounted and must have a hinged end on its piston rod. Fluid connections must be made to both fluid ports through hoses.

Note: The slot in the rod hinge must be sufficiently deep so the lever does not hit the bottom of the slot before the piston reaches the end of its stroke. This might cause a permanent bend in the rod.

TABLE 1-5 — CHORD FACTOR CHART

Use this for solving cylinder stroke as explained in the text.

Angle "A" Degrees	Chord Factor	Angle "A" Degrees	Chord Factor	Angle "A" Degrees	Chord Factor	Angle "A" Degrees	Chord Factor
5	0.087	45	0.765	85	1.351	125	1.774
10	0.174	50	0.845	90	1.414	130	1.813
15	0.261	55	0.923	95	1.475	135	1.848
20	0.347	60	1.000	100	1.532	140	1.879
25	0.433	65	1.075	105	1.587	145	1.907
30	0.518	70	1.147	110	1.638	150	1.932
35	0.601	75	1.217	115	1.687	155	1.953
40	0.684	80	1.286	120	1.732	160	1.970

Finding Cylinder Stroke by the Chord Factor Method. Figure 1-19. If the cylinder is mounted so it rotates the lever to an equal angle each side of the line which is perpendicular to the cylinder axis as in this figure, the length of stroke can readily be calculated by multiplying pin-to-pin lever length times the chord factor taken from Table 1-5.

Example: Find the cylinder stroke needed to swing a 14" (pin-to-pin) lever through a 105° arc when mounted as in Figure 1-19.

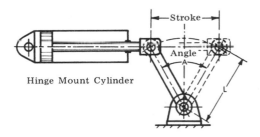

Hinge Mount Cylinder

FIGURE 1-19. Cylinder Axis is Perpendicular to Mid Position of Lever.

Solution: S (stroke) = 14" (lever length) x 1.587 (chord factor taken from Table 1-5) = 22.2 inches.

Calculation by Sines. The example above can be solved by trigonometry using the right angles which the cylinder axis makes with the vertical line marking one-half the total cylinder stroke. The half stroke of the cylinder is lever length times the sine of the angle through which the lever moves in reaching the mid-point of its stroke. Total stroke will be twice this distance.

Solution: S (total stroke) = 2 x 14 x .793 (sin ½105°) = 22.2 inches. If the listing of sines in Table 1-2 is not adequate, consult a complete sine table in a mathematical or machinery handbook.

Calculation by the Law of Squares. Figure 1-20. The cylinder stroke for *any* geometrical arrangement of lever and cylinder can be found by the law of squares. This law states that in any *right angle* triangle, the square of the hypotenuse is equal to the sum of the squares of the two shorter sides.

Example: Figure 1-20. In this right triangle, the two sides are 3 feet and 4 feet long. Solve for the hypotenuse by the law of squares:

H (hypotenuse) = $\sqrt{3^2 + 4^2}$ = $\sqrt{9 + 16}$ = 5 feet.

However, before any mathematical solution can be started, the locations of the cylinder hinge and the lever starting and ending positions must be established in relation to each other. On a new design, where it is permissible to mount the cylin-

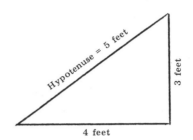

FIGURE 1-20. *Illustration of the Law of Squares.*

der hinge point wherever it is most convenient, it may be necessary to approximately locate a hinge point and make a preliminary calculation. If this location does not put the cylinder rod hinge in a position where it can be coupled to the lever, it will be necessary to tentatively move the hinge location and re-calculate. Several trials may be necessary to establish the best position for the cylinder.

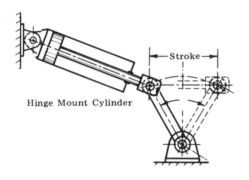

FIGURE 1-21. *This cylinder operates at a random angle with respect to the lever. Stroke can be found with the law of squares.*

Example. Figure 1-21. This is a pictorial diagram. Assigned dimensions and the right angle triangles involved in the solution are shown in the next diagram directly below.

Solution. Figure 1-22. The method of solution is to solve these right triangles for XZ, the retracted length (pin-to-pin) of the cylinder, and XW the extended length. When subtracted from each other this will be the cylinder stroke required to move the end of the lever from Point Z to Point W.

XZ = $\sqrt{20^2 + 80^2}$ = 82.5″ retracted length.

XW = $\sqrt{20^2 + 120^2}$ = 121.7″ extended length.

S (cylinder stroke) = 121.7 – 82.5 = 39.2″

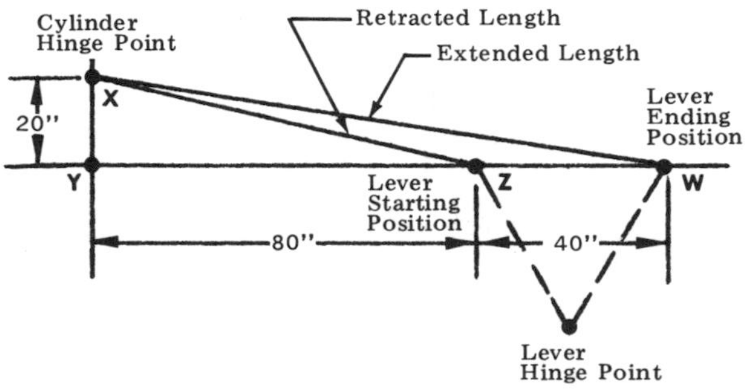

Cylinder Stroke by the Layout Method. On some applications it may be more practical, at least for experimental determination of best cylinder mounting position, simply to take a large sheet of wrapping paper and draw an exact layout to full or reduced scale. The lever should be shown both in starting and ending positions. The cylinder which has been tentatively selected should be laid out on another piece of paper or a sheet of cardboard. Its pin-to-pin hinge points should be laid out accurately. Thumbtacks can be used as pivots for moving parts. A ruler, tape, or metal scale can be used to measure distances from the cylinder hinge point to the

FIGURE 1-22. *This is a vector diagram of angles involved in the cylinder and lever problem in the diagram directly above.*

starting and ending positions of lever movement. By subtracting these two measurements, the minimum cylinder stroke can be determined. If there are positive stops on the lever rotation, the cylinder can have extra stroke which is not used.

After experimentally locating the best cylinder hinge point it can be verified with one of the mathematical solutions.

In order to get an exact pin-to-pin cylinder dimension to satisfy a desired geometrical layout, industrial cylinders can be ordered with a piston rod extending further from the front end of the cylinder than normal catalog standard.

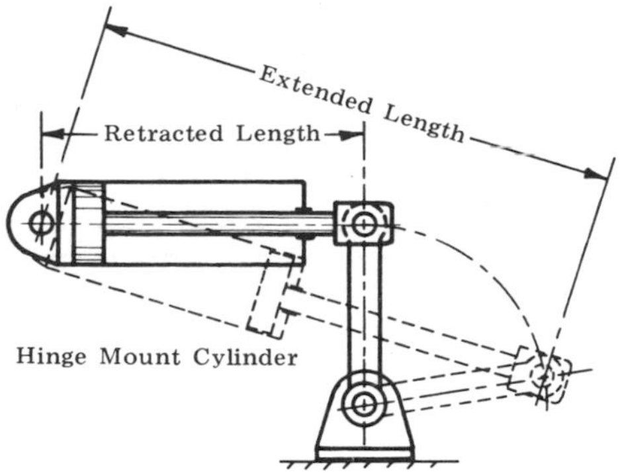

COLUMN STRENGTH OF PISTON ROD

The piston rod of a cylinder may buckle in column failure, when extended, if its diameter is too small for its length and for the load against it. Manufacturers of industrial cylinders offer a range of rod diameters from which to choose, and it is the responsibility of the purchaser, when ordering, to specify a rod diameter sufficiently large to withstand column failure on his application.

The resistance of a piston rod to buckling depends on several factors: the mounting style; how well it is supported; size and weight of the cylinder; side load on the rod, etc.

FIGURE 1-23. Cylinder stroke can be determined by making a scale layout and measuring retracted and extended distances.

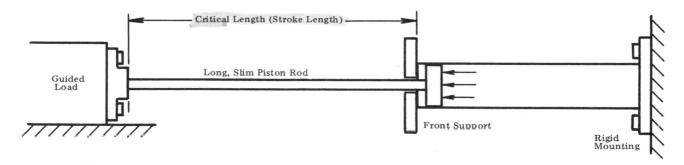

FIGURE 1-24. *This type of mounting has a maximum resistance against column failure of the piston rod.*

Best Mounting. Figure 1-24. A cylinder mounted in this way is least likely to suffer a bent rod from column failure. Its back end is rigidly mounted; its front end is supported against lateral movement; and its rod is rigidly anchored in a guided mechanism. The minimum diameter of its piston rod can be determined from Table 1-6 on the next page. The "critical length" to use in the table is the actual unsupported length of piston rod. This is approximately equal to the cylinder stroke in most cases.

Example: Select the minimum rod diameter for a cylinder with 60-inch stroke operating against a load of 40,000 lbs., if it is mounted as shown in Figure 1-24.

Selection: Use Table 1-6. In the horizontal line opposite 40,000 and in the column headed 60 inches, the table shows a minimum diameter of 2-3/8''. Use the next larger rod diameter listed in the cylinder manufacturers catalog.

Weakest Mounting. Figure 1-25. A cylinder mounted in this way will be much more likely to suffer a column failure. Its back end is hinged, perhaps to a movable member; its front end is not supported; the end of its rod is hinged into a member which may be movable. The suggested minimum rod diam-

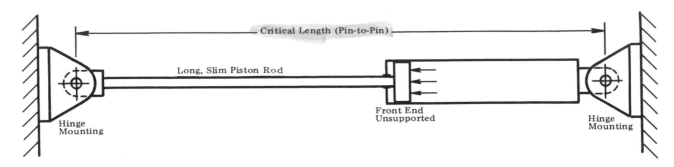

FIGURE 1-25. *Hinged mountings are the ones most likely to develop column failure in the piston rod.*

TABLE 1-6 — MINIMUM PISTON ROD DIAMETERS

Figures in the body of this chart are suggested minimum piston rod diameters, in inches

Load, Pounds	Critical Length, Inches (See Mounting Illustrations)							
	10"	20"	40"	60"	70"	80"	100"	120"
1500	- - - -	- - - -	13/16	1-1/16	- - - -	- - - -	- - - -	- - - -
3000	- - - -	11/16	15/16	1-3/16	1-3/8	1-1/2	- - - -	- - - -
6000	13/16	7/8	1-1/8	1-3/8	1-9/16	1-5/8	1-7/8	- - - -
10,000	1	1-1/8	1-5/16	1-9/16	1-5/8	1-7/8	2-1/8	2-3/8
20,000	1-3/8	1-7/16	1-5/8	1-7/8	2	2-1/8	2-7/16	2-3/4
40,000	2	2	2-1/8	2-3/8	2-1/2	2-5/8	2-7/8	3-1/4
80,000	2-3/4	2-3/4	2-7/8	3	3	3-1/4	3-1/2	3-3/4
150,000	3-3/4	3-3/4	3-7/8	4	4	4-1/8	4-3/8	4-1/2
300,000	5-3/8	5-3/8	5-3/8	5-1/2	5-1/2	5-1/2	5-3/4	6

eter can be found from Table 1-6. The "critical length" to use in the table is the pin-to-pin distance between mountings. Usually this will be about twice the actual stroke length.

Example: Use the same cylinder as in the preceding example, with 60-inch stroke working against a load of 40,000 lbs., and assume it is mounted as in Figure 1-25.

Selection: Use Table 1-6. The "critical length" to use in the table is 120" which is approximately twice the stroke. If pin-to-pin distance is known, use this for the critical length. In Table 1-6, in the horizontal line opposite 40,000 and in the column headed 120 inches, the table shows a minimum diamter of 3-1/4". Use the next larger rod diameter listed in the cylinder manufacturers catalog.

Other Mountings. For cylinders mounted less securely than in Figure 1-24 but more securely than in Figure 1-25, an intermediate rod diameter between the two extremes would be indicated.

Column strength is usually a factor only on hydraulic cylinders operating at high pressure, when they are at or near the end of their stroke, and only while they are pushing. It would normally not be a factor on air cylinders except on extremely long stroke with very small bore cylinders.

Rod failure is not a factor during cylinder retraction. The smallest rod diameter (standard) offered in each bore size is sufficiently large to avoid any chance of structural failure provided the cylinder is operated within the manufacturers pressure rating.

Caution! Since there are several factors which contribute to column failure, Table 1-6 should be used primarily for estimating. On doubtful applications, be sure to check your rod diameter selection and your application with the cylinder manufacturer before purchasing.

FORCE REQUIRED TO ACCELERATE A LOAD

Formulae and rules for calculating many kinds of cylinder loads have been given in the preceding pages. These formulae have given cylinder force to overcome load resistance including friction while the load is in steady motion, and have not included additional power which must be provided to accelerate a load from a standstill up to a steady speed. On most applications the additional power for acceleration is very small and is usually neglected because there is a sufficient reserve to supply this small amount for a brief acceleration period. But certain applications may require a significant additional amount of power and this extra power must be provided if performance is to be satisfactory.

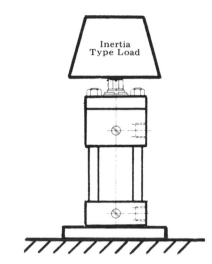

Inertia Type Load

FIGURE 1-26. Overcoming Inertia of a massive cylinder load.

If the load is heavy it has high mass and high inertia to resist acceleration. Additional power must be provided to accelerate from a standstill to a final velocity in a specified time interval. The amount of additional power is proportional to the mass of the load and inversely proportional to the time allowed for acceleration. While the load is in motion, this extra power is carried as kinetic energy. When the load is stopped, the extra power comes back into the system and produces shock and heat. A means of gradual deceleration may be necessary to allow the power to be absorbed gradually by the time the load comes to a complete stop.

Information on this page is to show how to calculate the *extra* force (and pressure) required to rapidly accelerate a heavily loaded cylinder. For massive rotating loads, see information in Design Data section on Pages 252 and 253.

Calculating Inertia Loads. Figure 1-26. Before calculating the extra PSI needed to rapidly accelerate this vertically moving load, the normal PSI needed to raise it at a constant rate of speed must be calculated by the usual formula: *PSI = Load Weight (lbs) ÷ Piston Area (sq. inches)*. Allowance should also be made for friction in ways or guides if this is significant.

Next, use the following formula to calculate the extra force needed to accelerate the load mass to final velocity in a specified time:

$$F = (V \times W) \div (g \times t) \; Lbs., \; in \; which:$$

F is the accelerating force, in pounds, that will be needed.
V is the final velocity, in feet per second, starting from a standstill.
W is the weight of the load, in pounds.
g is acceleration of gravity to convert load weight into mass; always 32.16.
t is the time, in seconds, during which acceleration takes place.

Finally, calculate the additional PSI needed for acceleration by using the square inch area of the cylinder piston. A = piston area in square inches; other symbols are the same as for the formula above. Use this formula for PSI calculation:

$$PSI = (V \times W) \div (A \times g \times t) \; (See \; example \; on \; next \; page)$$

Example of Inertia Calculation. (Use the formulae on the preceding page). A vertically-acting cylinder is lifting a steady load of 35,000 lbs. The cylinder bore is 4'' (piston area = 12.57 square inches). Starting from a standstill, the load must be accelerated to a final velocity of 12 feet per second in a time of 2 seconds. Calculate both the PSI required to keep the load in motion once it has reached final velocity and calculate the additional PSI which must be available from the pump during the initial 2 seconds accelerating period.

PSI for steady movement: *35,000 lbs. (load weight) ÷ 12.57 (piston area) = 2784 PSI.*

Additional PSI for acceleration: *(12 x 35,000) ÷ (12.57 x 32.16 x 2) = 520 PSI.*

Total PSI required from pump: *2724 + 520 = 3304 PSI.*

STRETCH IN MECHANICAL MEMBERS

Figure 1-27. The method described here can be used to find the strain (elongation in tension) of mechanical members such as cylinder barrels or tie rods, bolts holding a pipe flange to the body of a component, or stretch in bolts holding a flange union together.

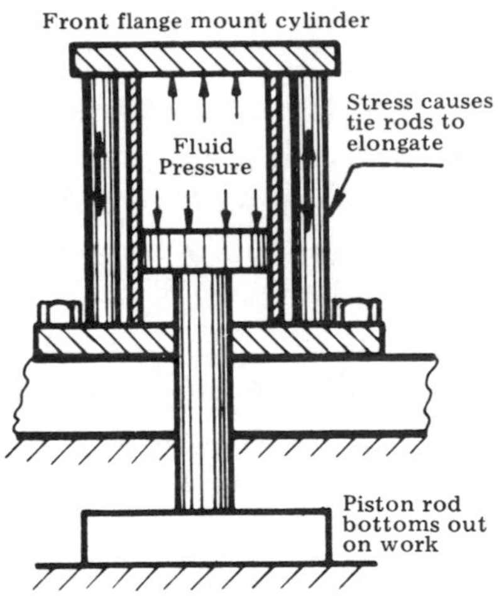

Front flange mount cylinder

Stress causes tie rods to elongate

Fluid Pressure

Piston rod bottoms out on work

FIGURE 1-27. Tie rods elongate under fluid pressure on cylinders mounted in this manner.

The data required for making a calculation are: (1), the length, in inches, of the member being strained, (2), the cross-sectional area of the strained member, in square inches, (3), the total applied force, in pounds, and (4), the modulus of elasticity, E, of the material in the member. This can be found from a machine handbook, or for common materials, refer to the table at the foot of this page.

We are primarily interested in steel which has an E = 30,000,000 for any kind of steel. High carbon steel stretches the same amount as low carbon steel but, of course, will resist a higher force before fracturing.

The formula for making a strain calculation is:

Elongation (in inches) = [L x F] ÷ [A x E], in which:

L = length, in inches; F = applied force, in lbs.; A = cross section area, in square inches: E is the modulus of elasticity for the stretched material (from table).

Example: Find the stretch in the four 1-inch tie rods of a 6-inch bore cylinder at 3000 PSI. The rods are 27'' long. Note: Please refer to Page 245 for more information on cylinder mounting.

Solution: First, get all the facts together. Compute total applied force in the usual manner for a 6'' bore cylinder operating at 3000 PSI (or refer to table on Page 249) = 84,822 lbs. Modulus of elasticity for steel (from table) = 30,000,000. Cross sectional area being stretched is 4 tie rods each with a sectional area of 0.785 square inches = 3.14 square inches. Then use the formula:

Stretch = [27 x 84822] ÷ [3.14 x 30,000,000] = 0.0243''.

Modulus of Elasticity for Common Metals (in Tension)	
Material	*Modulus of Elasticity, E*
Aluminum	10,000,000
Brass	15,000,000
Cast iron, gray	12,000,000
Copper	17,000,000
Magnesium	6,500,000
Monel	26,000,000
Nickel	30,000,000
Phosphor bronze	16,000,000
Steel, carbon	*30,000,000*
Steel, stainless	28,000,000

2

Direction, Force, & Speed Control

OF AIR CYLINDERS

INTRODUCTION TO AIR CIRCUITRY

Decision: Air or Hydraulics? A designer considering the use of fluid power for a certain job must decide between air power or hydraulic power. A third choice, that of vacuum may also be considered. But vacuum is not usually competitive with the other media, and some of the jobs it can do cannot be done as well or at all with either of the others. Vacuum applications are covered in Volume 1 and will not be repeated in this volume. Some of the characteristics of air and hydraulic systems that may influence the designer's choice are these:

(a). Power Level. This may well be the major factor in deciding between air and hydraulics. Generally speaking, air systems operate at low power, in the range of 1/4 to 1 horsepower, while hydraulic systems usually operate from 1 horsepower up. This is a general observation, as occasionally there may be a high power air system or a very low power hydraulic system. In these exceptional cases there were probably other factors that affected the final choice of medium.

A 100 horsepower air compressor may feed 200 or more unrelated branch circuits of 1/2 horsepower or less. But a 100 horsepower hydraulic system usually operates only one machine, although this machine may include several related branches.

(b). Noise Level. An air system can usually be designed to operate at a lower noise level than an equivalent hydraulic system. For one thing it is completely silent between cycles or when holding a steady force. For lowest sound level, the compressor should be located remotely from the area in which silence must be maintained. The air exhaust from cylinders and valves can be piped away from the immediate area, eliminating the exhaust hiss so characteristic of compressed air. If this is not possible, all exhaust air can be passed through mufflers before discharging it to atmosphere.

The primary source of noise in a hydraulic system is the pump and/or hydraulic motor. To minimize this noise a pump should be selected of a type which naturally produces a low noise level, then it should be operated at below catalog rated maximum RPM and pressure.

Electric motors, although not as noisy as pumps, also add to the overall noise level. Low RPM electric motors produce less noise than those which operate at higher RPM.

(c). Cleanliness. Compressed air is cleaner to use, especially if it is unlubricated. A leak will not cause a mess around the machine. A hydraulic system may never leak if it is carefully built, but the potential is always there, around cylinder or pump shaft seals, under reservoirs, and at plumbing joints.

When working near food products, or near other materials which could be damaged by contact with oil, compressed air is preferred if it will do the job.

(d). Speed. Compressed air usually produces faster movement in lightweight mechanisms operating at low power. Extremely fast actuation is possible by using oversize piping and valves, and by supplying a much higher pressure than required simply to balance the load resistance.

Air circuitry may be simpler because an additional circuit can usually be added to the air power supply which is already in existence, and the cylinder can be stalled indefinitely without having to unload a hydraulic pump.

(e). Rigidity. There are certain applications which require a high degree of rigidity in the fluid system. These applications are unsuited for compressed air and can only be done with hydraulics. Examples are: lifts or elevators where loading or unloading may be done at an intermediate point in the cylinder stroke; feeding of a metal cutting tool; slow movement of a machine slide or other mechanism which has a large area of sliding friction; any press application using two or more cylinders on the same platen, particularly with uneven loads on the two cylinders; any system where a close control of cylinder position must be maintained; all these, and others, should always be assigned to hydraulics even though the horsepower level may be low.

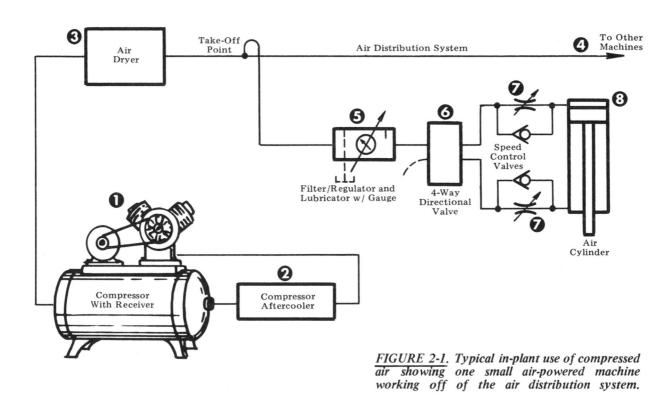

FIGURE 2-1. Typical in-plant use of compressed air showing one small air-powered machine working off of the air distribution system.

Elements of an In-Plant Compressed Air System. Figure 2-1. One large air compressor usually supplies air power, through distribution piping, to many small machines. In this diagram, one typical branch circuit supplying one machine is illustrated.

Item 1, Figure 2-1. Air Compressor. Pressure in a shop air supply is usually in the range of 60 to 120 PSI. Two-stage piston-type compressors are commonly used. The electric motor drive on small compressors, say 50 HP and less, is stopped when pressure in the receiver tank reaches the desired level. Large compressors are allowed to remain running but their air flow is vented to atmosphere when the cut-off pressure is reached.

Item 2, Figure 2-1. Compressor Aftercooler. Larger compressors, 25 to 50 HP and up, should have an aftercooler installed between the compressor and the receiver tank. The purpose of an aftercooler is to precipitate water from the compressed air to minimize the water problem further downstream. An air cooled aftercooler may be used with small compressors but a water cooled model is required on large compressors. An aftercooler adds to the original cost of the installation but saves enough money to pay for itself in a few years by eliminating water from the system.

Item 3, Figure 2-1. Air Dryer. While an aftercooler precipitates most of the air line water and is sufficient for many systems, a dryer is usually required for total elimination of water. It may be installed in the main distribution line or in a branch line. Since a dryer is not effective on incoming air temperatures greater than 110° F, an aftercooler must also be used on many systems to reduce air temperature to an acceptable level for a dryer.

Two types of air dryers are in common use, the refrigeration type and the chemical type. They are described in Volume 1, Chapter 6.

Item 4, Figure 2-1. Air Distribution System. Air to each branch circuit is tapped off of the top of the distribution piping to prevent water from following the air down to the machine. General information plus a table of recommended distribution pipe sizes is given in Volume 1, Chapter 1.

Item 5, Figure 2-1. Filter-Regulator-Lubricator. A trio unit including these three items should be interposed between the raw air from the distribution system and the air-powered machine. Each machine should have its own trio unit; more than one machine should not be served by one trio unit. It should be physically placed as close as possible to the inlet of the 4-way valve.

The symbol shown at Position 5 includes the filter placed first, followed by an adjustable pressure regulator, usually with a pressure gauge, then a lubricator. These units are individually described in Volume 1, Chapter 6.

Item 6, Figure 2-1. Four-Way Valve. The directional control valve is usually a 5-port, dual exhaust type, with a choice of actuators including manual, solenoid, cam, pilot, etc. They are described and their symbols are shown in Volume 1, Chapter 4.

Item 7, Figure 2-1. Speed Control Valves. Flow control valves, including both a needle valve and a built-in check valve for free return flow, are installed in lines connecting 4-way valve to cylinder. Brass valves are preferred over steel because of the water problem in many air systems. Two valves are required for control of compressed air, and are usually connected so they meter outgoing air rather than inlet air to the cylinder.

Item 8, Figure 2-1. Air Cylinder. The air cylinder must be sized to produce sufficient force to overcome the load when operated on a pressure about 25% less than will be delivered by the pressure regulator. The additional 25% pressure is not needed by the load but is used to push air through the flow losses in piping and valving. See Volume 1, Chapter 2 for sizing of air cylinders.

Graphic Symbols Frequently Used in Air Circuits. Schematic circuit diagrams in the Womack fluid power books are those approved by the NFPA (National Fluid Power Association) and the ANSI (American National Standards Institute), and by most foreign standardization agencies.

The following is a summary of the more commonly used symbols in this book for compressed air diagrams. These symbols are an important communication link between manufacturers and users of fluid power, and they override any language barrier. We encourage every student of fluid power to learn to recognize and to draw these symbols. If he understands their meaning it will help him not only in learning basic ideas expressed in this book but will help him understand the operation of machines he may be called on to service. Most of these components are described in greater detail in Volume 1.

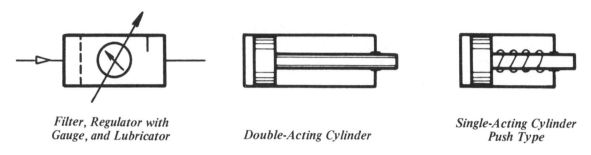

Filter, Regulator with Gauge, and Lubricator

Double-Acting Cylinder

Single-Acting Cylinder Push Type

Figure 2-2. Some Component Symbols Used in this Book.

Filter-Regulator w/Gauge-Lubricator. One simplified composite symbol shows all four units. The dotted vertical line indicates a filter, the long arrow an adjustable regulator, the short arrow the gauge, and the short vertical line the lubricator.

For simplicity in presentation of basic ideas these components are usually omitted from diagrams in this book, although their use is vital to proper operation of cylinders and valves. Refer to Volume 1 for individual description of each component.

Double-Acting Cylinder. The bore and stroke of each cylinder is sometimes noted on the circuit diagram to help identify various cylinders. It is not customary to specify mounting style nor type of seals unless this information is essential to circuit operation. Ram-type or big-rod cylinders which are often used in hydraulic circuits are seldom used on compressed air.

Single-Acting Cylinder. The illustration indicates a push-type cylinder, spring loaded toward the retracted position. Other single-acting cylinders can be drawn with appropriate symbols according to their function. Other types would include pull-type cylinders and those cylinders having no springs, with return by gravity or by load reaction.

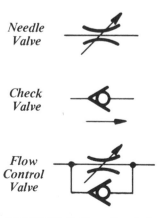

Needle Valve

Check Valve

Flow Control Valve

FIGURE 2-3. More Symbols.

Needle Valve. This symbol may be used to indicate any kind of throttling device which presents equal restriction in both directions of flow. In air circuitry it would usually indicate a needle valve, but could also represent any kind of 2-way valve meant for throttling. The arrow indicates adjustability. If drawn without the arrow, it would represent any kind of non-adjustable orifice or line restriction.

Check Valve. Air flow in one direction, reverse flow blocked. The body is marked for the free flow direction. A compressed air check valve for operation on shop pressure of 150 PSI will usually have a cracking pressure of 3 PSI when passing rated flow.

Flow Control Valve. An adjustable needle valve with built-in check valve for free reverse flow. Built as a composite valve for economy and convenience in plumbing. Most manufacturers stamp an arrow on the body to indicate *controlled flow* direction, since this is the significant feature of the valve. Other manufacturers may use the opposite designation, and the valve will be so marked. If stamped IN and OUT, controlled flow will be with air entering the IN port. If in doubt, blow air through the ports to determine free flow direction.

Manual Shut-Off

Shut-Off Valve. For simple 2-way shut-off valves this simplified symbol is easier to draw than the full symbol shown below, and is acceptable practice. It is used in preference to the needle valve symbol above when the primary purpose is complete shut-off rather than throttling.

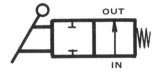

*2-Way,
Normally Open*

Two-Way Valves. According to the rules given in Volume 1 for construction of valve symbols, external circuit connections on these two symbols would be made to the block on the right, furthest from the actuator on the left end. The first valve, being spring returned to its wide open position, is a "normally open" type. The second one, being spring returned to its closed position, is a "normally closed" type. These symbols, showing all porting positions in relation to the type of actuator, are a more accurate and complete way of describing the operation of a 2-way valve than with the simplified symbol shown above.

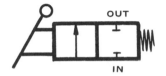

*2-Way,
Normally Closed*

Three-Way Valves. Diverter, selector, and directional control 3-way valves can all be represented with this basic symbol, although the arrowheads might have to be reversed in some cases.

The symbol shown here is for a normally closed type. If the actuator were moved to the opposite end, the symbol would represent a normally open type.

3-Way, 2-Position

Four-Way, Two-Position Valves. For directional control of double-acting air cylinders. The example shown here is a single-solenoid valve with a spring to return the spool to its normal position after the solenoid has been de-energized. Other common actuators are manual lever, cam, pedal, treadle, and pilot.

4-Way, 2-Position

Two-Position, 4-Way Valves With Double Actuator. The same actuator can be used on both ends, or different actuators on each end. Examples: Double solenoid, double pilot, manual with spring return or pilot return, solenoid with manual or pilot return, pedal with pilot return, etc.

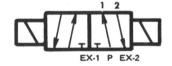

*4-Way, 2-Position, With
Double Actuator*

Four-Way, Three-Position Valves. These types are seldom used on compressed air because an air cylinder cannot be accurately stopped nor rigidly held in position except against positive stops at each end of its stroke. However, for unusual circuitry, 3-position directional valves are occasionally used, and the closed center spool shown here is the more popular. Alternate spools sometimes used on compressed air are shown on Page 32.

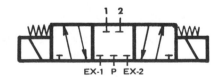

*4-Way, 3-Position, With
Closed Center Spool*

FIGURE 2-4. *Directional Valve Symbols.*

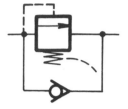

*Sequence Valve
(With Return Flow Check)*

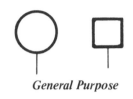

General Purpose

FIGURE 2-5. *More Symbols.*

Sequence Valve. Used with or without a return flow check valve depending on its location in the circuit; whether there will a return flow at any time. Its symbol resembles a relief valve except its spring chamber must always be vented to atmosphere, and this will identify it as a sequence valve on a diagram. Atmosphere venting is vital to its performance; however, on some *air* circuits a relief valve can be substituted if a sequence valve is not available. Information is on Page 62.

Other Symbols. The symbols listed in this section take care of most circuitry in this book. A few others, like the shuttle valve, quick exhaust valve, and pressure regulator are used occasionally and their symbols will be shown where they appear.

Occasionally a component may be shown for which no suitable symbol can be found. In this case a circle, square, or triangle may be used, and an appropriate note should be added to the drawing to explain the nature of the item.

DIRECTIONAL CONTROL OF SINGLE-ACTING CYLINDERS

Single-acting cylinders are those built to deliver power in only one direction. Push-type cylinders deliver power on their extension stroke. Pull-type cylinders deliver power on retraction. The piston in a single-acting cylinder is usually returned to home position by an internal spring after air pressure has been removed. Or, it may be returned by weight of the load or by external springs on the mechanism, installed by the user. The reader may wish to review single-acting cylinders in Volume 1, Chapter 2.

Double-acting cylinders are preferred for most applications; single-acting models are used primarily for light loads requiring only a short stroke. A good application is for clamping of a workpiece in a machine; or for the use of several small air clamps instead of a larger cylinder for distribution of clamping force over a large area.

The only significant advantage of single-acting cylinders is that only one hose connection is required between cylinder and control valve. However, they can be operated with inexpensive 3-way valves which may sometimes be less expensive than a 4-way valve to operate a double-acting cylinder. They have many important disadvantages, and for the reasons listed here are not generally recommended except for air clamping.

1. They have less power than a double-acting cylinder of comparable size because the internal return spring must be compressed during the power stroke. Thus, their power decreases toward the end of their stroke, and the ultimate force at stall may be as much as 25% less than that of a double-acting cylinder of the same bore and stroke.

2. Their ability to retract a load is quite limited. The return spring is of minimum size, just strong enough to retract the piston if there is no load connected to it. If made stronger, it would take away some of the force on the power stroke. A single-acting cylinder should not be used, for example, for punching holes. Its return force, being quite limited, could not withdraw the punch from the hole.

3. They offer no price advantage, and sometimes cost more. While the cost of a rod seal is saved, the extra cost of a return spring assembly plus the extra barrel length needed to accommodate the return spring make their cost higher than that of a comparable double-acting cylinder.

4. They are limited to short strokes because of the return spring. If designed for a long stroke, the cylinder would have to be unreasonably long to get a good spring characteristic. If designed with a steep spring rate, to keep cylinder length within reason, the entire cylinder force might be expended in compressing the spring by the time the piston traveled to the end of its stroke.

5. Because of their greater length, due to the return spring, they require a longer mounting space than a double-acting cylinder of the same bore and stroke.

Single-Acting Cylinder Controlled With 2-Way Valves. Figure 2-6. We will begin the study of directional control with the simplest type of control valve. At least two simple shut-off valves are required to control a single-acting cylinder — one to supply pressure to extend its piston, the other to vent the piston to atmosphere so it can retract.

When air pressure is connected to the cylinder by opening Valve 2, the piston moves upward provided exhaust Valve 1 is closed. During piston travel, air in front of it is pushed out to atmosphere.

To permit the internal spring to retract the piston, air pressure behind the piston must be vented to atmosphere by opening Valve 1. At this time, Valve 2 must be closed to prevent escape of compressed air to atmosphere.

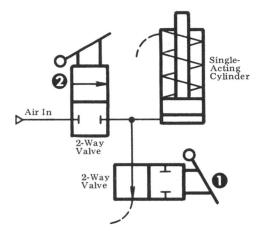

FIGURE 2-6. Single-Acting Cylinder Controlled With 2-Way Valves

Single-Acting Cylinder Controlled With a 3-Way Valve. Figure 2-7. Normally, one 3-way valve can be used in place of two 2-way valves, for convenience and cost. A 3-way valve performs both pressure and exhaust functions at the same time.

If a 3-way valve is not available, a 4-way valve with one of its cylinder ports plugged may be used instead. In fact, some valve manufacturers offer only 4-way valves because a 3-way valve costs about as much to manufacture and will not serve as wide a range of applications.

Please refer to Volume 1, Chapters 3 and 4, for a more complete explanation of the action of a 3-way valve on a single-acting cylinder.

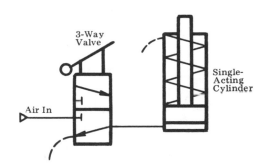

FIGURE 2-7. Single-Acting Cylinder Controlled With a 3-Way Valve

Air Clamping. Figure 2-8. An air clamp is a cylinder usually a single-acting type, which is used specifically for holding a workpiece in a jig or fixture while various drilling, machining, grinding, or forming operations are performed on it. It is, in effect, an air operated vise. Single-acting cylinders of small to medium bore have ample force for most ordinary clamping. Double-acting air cylinders or hydraulic cylinders can be used where greater clamping force is required.

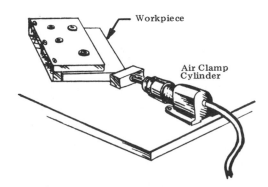

FIGURE 2-8. Example of Air Clamping.

An air clamping application is considered as one in which the cylinder moves in free travel until it contacts the workpiece. Then it stalls and builds up clamping force.

Multiple Clamps. Figure 2-9. If the workpiece is large or flexible such as sheet metal, plastic, card-board, etc., several small clamps should be used rather than one large one, to distribute clamping

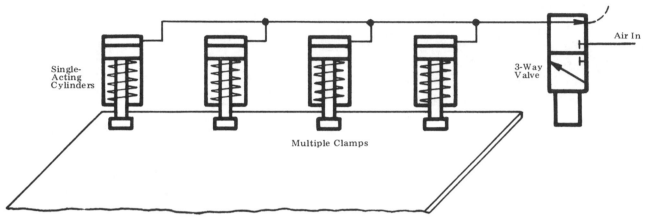

Single-
Acting
Cylinders

3-Way
Valve

Air In

Multiple Clamps

*FIGURE 2-9. When clamping large, thin, or flexible workpieces, use several small air clamps
to distribute clamping force over a wide area.*

force over a wider area. It is usually more convenient to use small bore, short stroke, spring return cylinders. Only one air feed line is required, and it may be run from cylinder to cylinder by any convenient routing. However, the line diameter should be sufficiently large to carry return air with low flow resistance so the clamps will release quickly. On very long runs, it may be necessary to install a quick exhaust valve at each clamping cylinder to speed the release action.

Clamp and Work Action. Two cylinders are often used in combination — one in a free-running condition to rapidly clamp a workpiece in a jig or fixture, to close a mold, or to index a table or tool head; the other to furnish power to the working tool. Circuits involving these two cylinders in a pre-determined sequence or program are called "clamp and work" circuits. Many examples are shown later in this chapter using button bleeder, solenoid, and sequence valve action.

DIRECTIONAL CONTROL OF DOUBLE-ACTING CYLINDERS

Double-acting cylinders are built to deliver power in both directions of travel, although the force (on a single-end-rod type) will be about 10% less on retraction because of the "net" area being smaller than the full piston area. They are preferred over single-acting cylinders on most applications.

Machines using air cylinders should be designed, if practical, to allow the cylinder to operate from a stall in one direction to a stall in the opposite direction, either against its own end cap or against external positive stops. Full air pressure should remain against the piston while in either stall position. An air cylinder cannot be accurately stopped at a designated point in mid-stroke, and when stopped, cannot be held firmly in mid-stroke because of the compressible nature of air. If the air supply was cut off while the piston was moving, the piston would not immediately stop. It would continue to travel a short distance until a balance was reached between air pressure on one side and load resistance on the opposite side of the piston. If while stopped in mid-stroke, the air line pressure should change, or the load resistance should change, the piston would start by itself and move to a new balance position. So, mid-stroke stopping of an air cylinder is to be avoided, if practical. If a mid-stroke stop must be used and must be at a precise point, hydraulics or air-over-oil operation will give better results. See air-over-oil circuitry in Chapter 12.

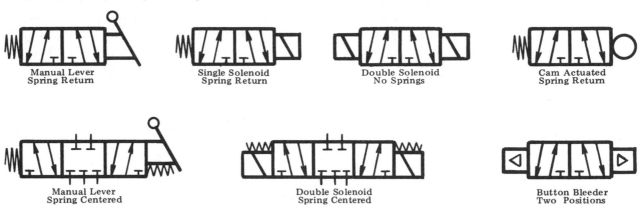

FIGURE 2-10. Typical ANSI Graphic Symbols of 4-Way Valves for Use on Circuit Diagrams.

ANSI Graphic Symbols for Circuit Diagrams. Figure 2-10. Standard 4-way valves exhaust on both ends of their spools. In hydraulic valves, these two internal exhausts are internally combined, either by cross drilling or by coring in the body casting, and they are brought out as one "tank" port. On the other hand, most air valves have both internal exhausts brought out on individual ports. This kind of valve is called a "dual exhaust" type. One of these ports carries exhaust air to atmosphere from the rod end of the cylinder while the piston is extending. The other port does the same for the blind end while the piston is retracting. By placing variable restrictors (needle valves) in these ports, speed in both directions of piston travel can be adjusted independently. In this section of the book, which covers air cylinders, all 4-way valves are shown as the dual exhaust type.

In Figure 2-10 only a few representative valve types are shown. For a more complete listing of available 4-way valve types, please refer to Volume 1, Chapter 4.

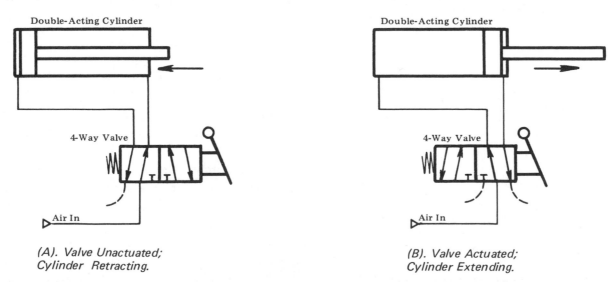

(A). Valve Unactuated;
Cylinder Retracting.

(B). Valve Actuated;
Cylinder Extending.

FIGURE 2-11. Directional Control of Double-Acting Cylinder With 2-Position, 4-Way Valves.

Four-Way, 2-Position Valves. Figure 2-11. The preferred way of controlling a double-acting air cylinder is with a 4-way valve which has no neutral position. On the left side of this figure the valve is shown in its unactuated or normal position. Porting through the valve causes the cylinder piston to move to and remain in its retracted position with full pressure on the piston. On the right side of the diagram the valve is shown in its actuated or shifted position. Porting through the valve causes the cylinder piston to extend. When the piston is at the end of its stroke, it will remain stalled against the cylinder end cap under full pressure, until the valve is again shifted to its normal position. Since the valve has no neutral position, the piston cannot be stopped at any intermediate point in its stroke. The student is referred to Volume 1, Chapter 4 for further information on 4-way valve control.

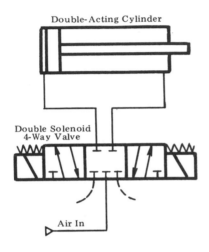

FIGURE 2-12. 4-Way, 3-Position Valve With Closed Center Porting; Double Solenoid Type.

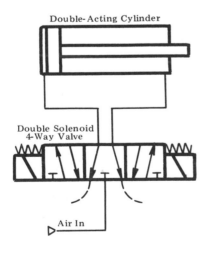

FIGURE 2-13. 4-Way, 3-Position Valve With Float Center Porting; Double Solenoid Type.

Four-Way, 3-Position Valves. Figures 2-12 and 2-13. If the application is one of those rare ones where it may be necessary to stop the cylinder piston at some point in mid stroke, a 4-way valve with center neutral position should be used. Neutral is always in center; the two outside positions give normal 4-way action for extending and retracting the cylinder piston. Although there are quite a number of neutral porting arrangements for hydraulic valves, for air valves the two center neutral porting arrangements shown in these two figures are the only ones commonly used.

Figure 2-12 shows closed center porting in which all ports are blocked from each other and from exhaust. This spool is used in about 95% of all applications in which a 3-position valve is required.

Figure 2-13 shows float center porting in which the pressure inlet port is blocked but all other ports are open to exhaust, hence to each other. This spool is used in those circuits where the cylinder piston rod must be free so it can be re-positioned by external mechanical forces during parts of the cycle when the cylinder is not under pressure.

On rare applications other neutral porting arrangements may be required such as a "regenerative" spool in which both exhaust ports are blocked and inlet pressure is admitted to both cylinder ports.

Stable Air Circuit. Figure 2-14. Where necessary to have a mid-travel stop, the piston will have greater stability while stopped if held between two columns of relatively high, active air pressure, supplied through two 3-way valves. Neutral, or piston stop position is with both 3-way valves de-energized. This puts full air line pressure on both sides of the cylinder piston. Since these sides have equal areas, the cylinder cannot produce a net force in either direction. While this arrangement does not have the stability of a hy-

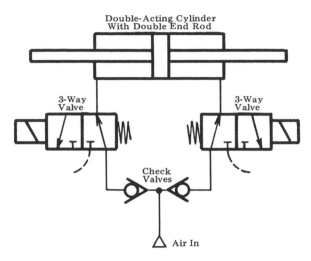

FIGURE 2-14. Air Circuit With Improved Stability.

draulic or an air-over-oil circuit, it is a definite improvement in stability over a standard circuit using one 4-way valve.

The double-end-rod cylinder must have both piston rods the same diameter so it will be in balance when stopped. It can be jogged to the left or to the right by momentarily energizing one solenoid or the other

A typical application for this circuit is where the cylinder has its home position at some mid point in its stroke, and must be slightly re-positioned from time to time. For example, the cylinder may be tilting or shifting a roll of material to keep it feeding straight into the machine.

If the valves are solenoid type they can be treated in the electrical circuit as if they were opposite solenoids on a double solenoid valve.

Even on applications where a mid-travel stop is not required, but where an air cylinder is to be operated at a very slow speed, this circuit has a better chance of operating smoothly without chatter than does a conventional circuit.

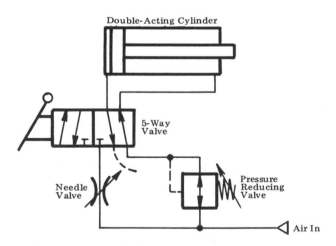

Fig. 2-15(A). Valve in Normal Position,
Cylinder Retracts.

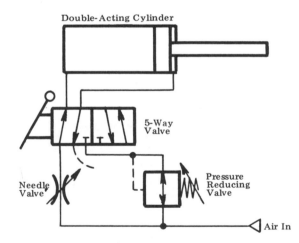

Fig. 2-15(B). Valve in Shifted Position,
Cylinder Extends.

FIGURE 2-15. Air Cylinder Operation With a 5-Way Directional Control Valve.

Five-Way Valve Control. Figures 2-15(A) and (B). A 5-way valve is a special variation of a 4-way valve, with pressure and exhaust ports interchanged. This gives it two independent inlet ports and one common exhaust port. Two levels of inlet pressure can be used, one for extension the other for retraction of the piston. In Part (A) of this figure, the valve is in its normal or unactuated position and the cylinder piston has retracted with reduced pressure from the pressure regulator. In Part (B), the valve is in its shifted position and the cylinder piston has extended under full air line pressure, its speed being controlled with the needle valve. A speed control valve is not needed for the return stroke, the regulator itself acting to limit speed. Adjust it to the lowest pressure which will give acceptable speed.

PRESSURE CONTROL IN AIR CIRCUITS

The Need for Pressure Control. In an industrial air system the pressure applied to air cylinders should be controlled within close limits, and a constant pressure maintained to give repeatibility of performance, stroke after stroke. If air pressure is allowed to vary, cylinder speed and force will also vary a proportional amount.

It is not practical to adjust the air compressor to maintain a constant line pressure. It must be allowed to work between a low (cut-in) pressure and a high (cut-out) pressure. For example, a typical shop air system might be adjusted to start the compressor when distribution pressure dropped to 110 PSI, then to stop the compressor when the pressure built up to 140 PSI.

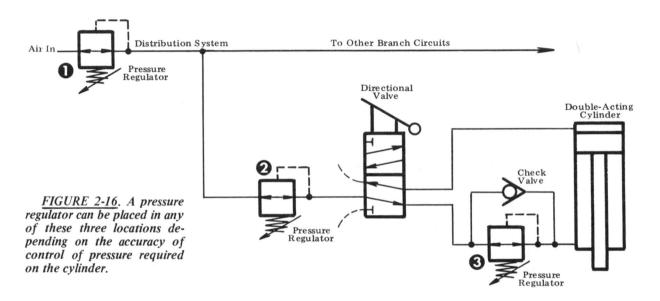

FIGURE 2-16. A pressure regulator can be placed in any of these three locations depending on the accuracy of control of pressure required on the cylinder.

Air Pressure Regulator. Figure 2-16. Pressure to an air cylinder can be kept constant by installing a pressure regulator in the machine circuit and adjusting it to a pressure less than the cut-in pressure of the air compressor. The regulator will accept an inlet pressure of any level within its range and will deliver a constant outlet pressure. It is basically an air pressure reducing valve, the counterpart of a hydraulic pressure reducing valve.

A pressure regulator is not a pressure relief valve. Relief valves are seldom used with compressed air because the loss of air to atmosphere while they are relieving would represent a major loss of power from the system. Figure 2-16 shows the official ANSI graphic symbol to be used on circuit diagrams.

There are three possible locations where a pressure regulator can be installed. These are designated (1), (2), and (3) in the diagram. If installed at (1) near the compressor it could serve many machines connected to the air supply. If air pressure does not have to be maintained accurately this may be a good location, but if installed here it should be preceded by a filter. Ordinarily this is not a good location because each machine may require a different level of pressure for best performance. Also it could not maintain close control of pressure level applied to the cylinders because of flow resistances of lines and valving between its outlet and the cylinder. No matter how accurate the regulator might be, there would be pressure loss through the connecting lines as soon as flow started, causing a drop in pressure on the cylinder piston. The greater the volume of air flow, the higher this pressure drop would be, and it would be impossible to maintain close control of the pressure level on the cylinder.

Position (2) for the regulator is preferred for most applications providing the connecting lines between

4-way valve and cylinder are relatively short. In spite of the pressure loss through the 4-way valve, a good regulator can maintain the pressure on the cylinder piston at a sufficiently constant level for most application. However, the regulator should be installed as closely as practical to the 4-way valve inlet, being preceded by a filter and followed by a lubricator.

Position (3) for the regulator gives a more accurate control of pressure applied to the cylinder piston where accurate maintenance of pressure is desirable and/or where the connecting lines between valve and cylinder are unusually long and would create a substantial pressure loss between valve and cylinder. The regulator should be installed as near as practical to the cylinder port, and should be by-passed with a check valve for free reverse flow around the regulator. A second regulator may be placed in the other cylinder line if needed. The regulator in Position (3) can sometimes serve as a speed control, eliminating a separate flow control valve.

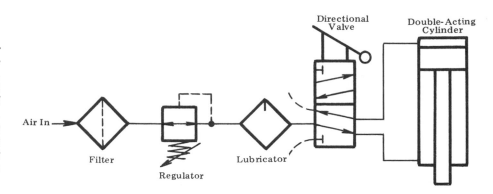

FIGURE 2-17. Air processing equipment includes a filter, regulator, and lubricator. Most machines operated from a compressed air line should include all three units.

See Figure 2-2 for the simplified symbol which is ordinarily used to represent an assembly of these three units as a trio.

Processing the Air. Figure 2-17. Air tapped off the plant distribution system is not suitable for direct use in precision components such as air tools, directional valves, and cylinders. Every machine operating from the distribution system should be preceded by a trio which includes a filter installed first to take out dirt, pipe scale, and water, followed by a pressure regulator for levelling off the pressure to a constant value, and finally a lubricator which adds a fine mist of oil to the air stream for downstream lubrication of components. Units should always be connected in the sequence described.

For sake of simplicity the air diagrams in this book do not show this trio of units but it is understood that they are required on every industrial compressed air circuit. A more complete description of each of these units will be found in Volume 1, Chapter 6.

Power Loss Through a Pressure Regulator. A flow of any fluid whether compressed air or hydraulics, carries a potential and kinetic level of energy which is in direct proportion to its pressure above atmospheric. If any of this fluid is permitted to discharge to a lower pressure or to atmospheric pressure, without performing mechanical work through a cylinder, fluid motor, or rotary actuator, the fluid energy represented by this difference in pressure level will convert into heat in the fluid. The heat energy which is produced will be in direct proportion to the pressure difference and, in the case of a hydraulic system, can be calculated with the fluid horsepower formula. We know this is true in the case of hydraulic oil discharging across a relief valve or flowing through a reducing or flow control valve, but it is also true of air reduced from a higher to a lower pressure by means of a pressure regulator. For this reason the air in the compressor tank should not be pumped to a pressure higher than necessary. Air leaks should be promptly repaired.

To accurately determine how much power is wasted when air is reduced to a lower pressure is a rather involved mathematical procedure. However it can be easily estimated to an accuracy which is sufficient for most industrial compressed air work. The method is to find the compressor HP needed to

Load Balance PSI + Flow Loss PSI = Inlet PSI

pump air to the higher pressure. Then, to find the compressor horsepower to pump the same amount of air to the lower pressure, and to subtract this from the first calculation. This will give a reasonable estimate of horsepower which goes into waste when the air pressure is bled down, as in a pressure regulator, without doing mechanical work.

This method is described in greater detail in the data charts at the end of this book. Tables are included for estimating compressor horsepower. This method can also be used for calculating the wasted compressor horsepower because of pressure drop across a flow control valve.

THE EFFECT OF AIR PRESSURE ON AIR CYLINDER SPEED

Air Flowing Through a Pipe. When there is air under pressure in a pipe, but if there is no flow, there is no loss of pressure from one end to the other. This is according to Pascal's Law (See Volume 1, Chapter 1). But as soon as air starts to flow through the pipe there will be a loss of pressure through the pipe. The amount of this loss depends on the volume of air flowing, the size and kind of pipe, and bends and restrictions in the pipe. Air pressure supplied to an air cylinder circuit must be high enough not only for the load but to make up all pressure losses which develop as a result of the flow.

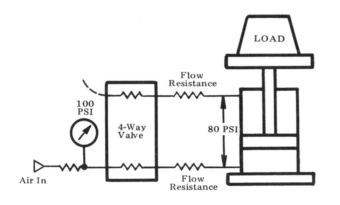

FIGURE 2-18. The supply pressure must be greater than load resistance to make up for flow losses in the valves and plumbing.

Cylinder Speed. Figure 2-18. The speed at which the piston of an air cylinder moves is directly affected by the level of pressure available to the circuit from the pressure regulator. On a properly designed air circuit most of the pressure is used by the cylinder piston to produce a force equal to the load resistance, but additional pressure must be available to move the air through the flow resistance of the circuit. The cylinder speed depends on how much additional pressure is available.

Example: in Figure 2-18, the weight load is such that 80 PSI ΔP (pressure difference) across the cylinder ports is required to equal the load resistance. The cylinder is connected through valves and plumbing to a source of 100 PSI air pressure (from a regulator). Flow resistances in the circuit are shown thus: ⋀⋀⋀ which is a symbol borrowed from electrical circuitry.

As air flows through this circuit some of its pressure will be lost in flow resistance. As air flow increases, pressure losses will also increase. Since 80 PSI is required for the load, this leaves only 20 PSI which can be consumed in flow losses. As the cylinder starts moving, and as it accelerates, flow losses appear in various parts of the circuit. Air flow (and cylinder speed) can increase only to the condition where the flow losses reach 20 PSI. Air flow (and cylinder speed) can increase no further because flow losses greater than 20 PSI would rob some of the pressure which the cylinder must have against the load, and the cylinder speed (and air flow) would have to decrease to recover 80 PSI to equal load resistance. In all air circuits, the sum of the pressure needed for load balance plus all flow losses in the circuit must exactly equal inlet pressure.

An air cylinder will accelerate, usually very quickly, up to a speed (and air flow) where flow losses plus load pressure equals inlet pressure. To increase cylinder maximum speed, either the inlet pressure must be raised or the flow losses reduced by increasing the size of valves and/or plumbing.

In the diagrams to follow, an air cylinder and 4-way valve are operating from a plant distribution system of 170 PSI. A pressure regulator, mounted very close to the 4-way valve, reduces this pressure, and the circuit action described starts at the outlet port of the regulator. Air flows through the 4-way valve, through the plumbing to the cylinder. Return air flows through return plumbing to the 4-way valve, through the valve and to atmosphere. Through all of these passages there is flow resistance.

The cylinder is opposing a load of 1000 lbs., which includes not only load weight but all mechanical friction in the cylinder. The cylinder bore is 4 inches (area = 12.57 square inches). Therefore, a pressure of 80 PSI is required to exactly equal load resistance. The following diagrams show the effect on cylinder speed by changing the pressure at the inlet of the 4-way valve.

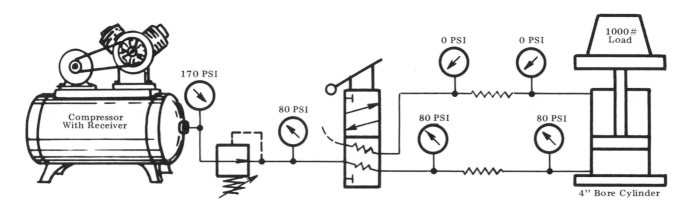

FIGURE 2-19. *In this example the regulator has been adjusted to give exact pressure required by the load but no more. The cylinder is unable to move.*

Figure 2-19. In this first diagram the pressure regulator has been adjusted to deliver 80 PSI to the inlet of the 4-way valve. On a 4" bore cylinder the 80 PSI will produce a force of 1000 lbs., which is exactly the force needed to equal load resistance. There is no surplus of pressure which can be lost in circuit flow resistance. Therefore, the cylinder cannot move. Any air flow, however small, would create a loss of pressure through the flow resistance of the valve and plumbing, and there would no longer be 80 PSI available for the cylinder to move the load.

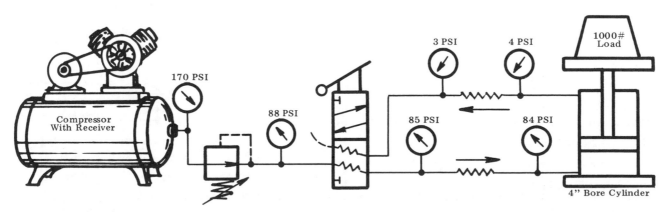

FIGURE 2-20. *The inlet pressure has been raised to 8 PSI more than that required against the load and there is a small surplus of pressure to supply circuit flow losses up to 8 PSI. The cylinder will move slowly.*

Figure 2-20. The pressure regulator has been adjusted upward, to deliver 88 PSI to the circuit. This gives a surplus of 8 PSI above the 80 PSI required for the load. Air can now start to flow through the

37

circuit. The cylinder will start to move and will accelerate to the speed (and air flow) where circuit flow losses reach 8 PSI. This is the balance condtion for this inlet pressure and the cylinder can go no faster.

Please notice the pressure gauge readings in all these diagrams. They show the progressive pressure losses around the circuit from valve inlet to atmosphere. Flow losses in the diagrams are assumed values for the purpose of illustration. Actual values of these losses will vary according to valve and line size, line length, number of bends and fittings, and possibly to other factors.

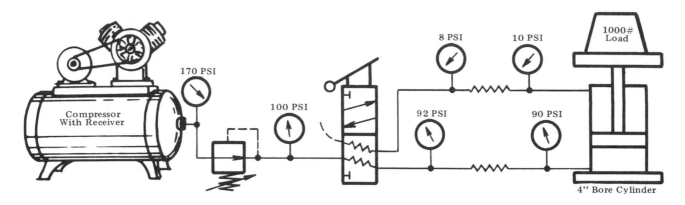

FIGURE 2-21. *The pressure regulator has been adjusted to a pressure about 25% greater than required by the load. On most applications this will give sufficient speed.*

Normal Design. Figure 2-21. The pressure regulator setting has again been raised, this time to 100 PSI. There is now a surplus of 20 PSI available for circuit losses. Therefore, air flow (and cylinder speed) will increase, seeking a new balance condition. Cylinder speed will level off when flow losses reach 20 PSI. At this point load pressure (80 PSI) + flow losses (20 PSI) = 100 PSI (inlet pressure). Each time inlet pressure is increased, cylinder speed will increase until a new balance point is reached. When that point has been reached it can go no faster. Cylinder speed will change if load resistance is changed until a new balance condition between speed and inlet pressure has been reached.

Supplying the circuit with about 25% more pressure than required against the load should give ample speed for normal applications. If it does not, the circuit is improperly designed, with undersize valving

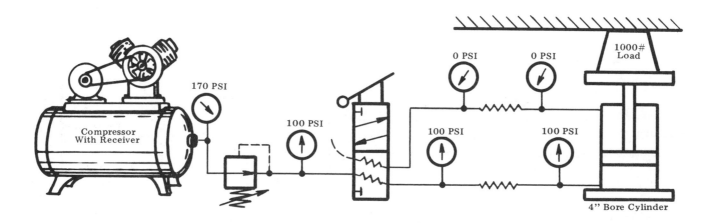

FIGURE 2-22. *If the cylinder stalls, all circuit flow losses will disappear and full inlet pressure will appear across the cylinder piston.*

or undersize plumbing. If speed is greater than desired, it can be reduced with flow control valves installed in the cylinder ports or in the 4-way valve exhaust ports.

On applications designed for exceptionally high speed, inlet pressure should be increased above the 25% rule-of-thumb for normal applications, even as high as 100% more than required against the load.

Pressure at Stall. Figure 2-22. Should the cylinder reach a stall condition against its own end cap, against an external stop, or against a load which it cannot move because of insufficient air pressure, air flow through the circuit would cease, and all flow losses through valving and plumbing would disappear, leaving full inlet pressure against the cylinder piston.

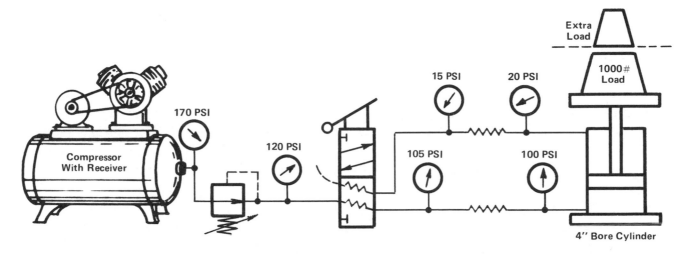

FIGURE 2-23. If the cylinder should pick up additional load during its stroke, the surplus pressure for flow losses will become less and the cylinder is forced to reduce its speed.

Load Change During Stroke. Figure 2-23. If the cylinder, while traveling at a certain rate of speed which is a balance between load resistance, circuit flow losses, and inlet pressure, should pick up an additional load during its stroke, a greater proportion of the inlet pressure would be required for the load. This would leave less surplus pressure for making up flow losses. Cylinder speed would have to decrease until the sum of load resistance plus flow losses was exactly equal to inlet pressure.

In Figure 2-23, for example, the normal load resistance requires 80 PSI of the inlet pressure, leaving a surplus of 40 PSI for flow resistance. If an additional load should be picked up which required an additional 25 PSI, the surplus pressure available for flow losses would decrease to 15 PSI. Air flow (and cylinder speed) would automatically decrease to the condition where flow losses were only 15 PSI.

Calculation of Cylinder Speed. It is very difficult and usually impractical to calculate the cylinder speed to be expected when first designing a circuit. The rules-of-thumb in the box on the next page for selecting cylinder bore size and inlet pressure usually work just as well as highly involved mathematical calculations and are simple to use. When using these rules, select plumbing size and valve port size with the same flow area as the cylinder ports, or the same as the combined port area if more than one cylinder operates through common valving and plumbing.

Design Recommendations for Inlet Pressure. When operating from an air distribution system there is always a loss of power from the system when distribution pressure must be reduced for operation of an air cylinder. But the power loss is the same whether pressure is reduced with a regulator or is absorbed with flow control valves.

Generally, the best procedure when starting up an air operated machine is first to adjust the output

RULE-OF-THUMB — CYLINDER SIZE vs INLET PRESSURE

The cylinder size, and system pressure, should be able to provide about 25% more force than needed to equal load resistance. This will give sufficient surplus pressure, on normal applications, for ample speed. If the speed should be too fast, it can be reduced with flow control valves. If speed is to be unusually fast, the cylinder, and system pressure, should be able to provide about twice the force needed to just equal load resistance.

pressure on the regulator to a pressure sufficiently reduced so it can be maintained as a constant level regardless of pressure fluctuations in the distribution system. Then, use flow control valves in the cylinder lines or in the 4-way valve exhaust ports to absorb surplus pressure not needed for the desired cylinder speed. One exception to this procedure (see Figure 2-22) would be if cylinder force when stalled against a load would tend to distort or damage the load. In this case, reduce inlet pressure from the regulator and open up the flow control valves. This will produce the same cylinder speed and the same circuit efficiency, but the stall-out force will be reduced. See special circuits in the following section.

SPECIAL CIRCUITS FOR PRESSURE CONTROL

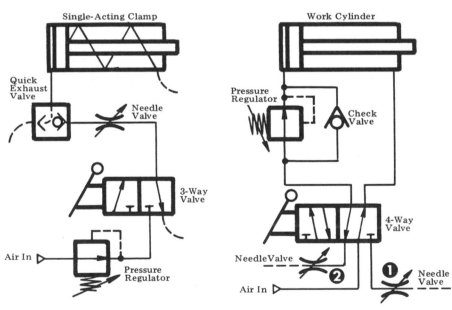

Part (A). Single-Acting Air Clamp.   *Part (B). Double-Acting Air Clamp.*

FIGURE 2-24. Methods of Clamping a Delicate Workpiece.

Delicate Clamping. Figure 2-24. An air cylinder, either single-acting or double-acting, which is used for clamping a workpiece in a fixture while being worked on, is called an air clamp. When clamping a delicate workpiece there are two forces which must be controlled to prevent damage to the workpiece. One is impact due to velocity of the clamp as it closes. The other is the "squeeze" which develops after the clamp stalls against the workpiece. Impact can be reduced with a flow control or needle valve, squeeze with a pressure regulator. Both of these components will be required.

Single-Acting Clamp. Figure 2-24, Part (A). This is only one of several circuits which could be used for clamp control. Usually a 3-way valve is used for directional control, and the pressure regulator can be installed at the inlet to this valve and adjusted to a level which will not damage the workpiece from excessive stall-out squeeze. A needle valve for speed control can be installed in the line between 3-way valve and clamp, and adjusted to a clamping speed which will not produce excessive impact. For quick release of the clamp a quick exhaust valve can be installed at the clamp port.

Double-Acting Clamp. Figure 2-24, Part (B). A 4-way valve, usually one with dual exhaust ports, will normally be used for directional control. Squeeze on the workpiece may be limited with a pressure regulator installed in the line leading to the blind end of the clamp cylinder. It should be by-passed with a check valve for free flow of air around the regulator when the clamp is retracting. Impact can be controlled with a needle valve, 1, installed in the 4-way valve exhaust port which is discharging air on the forward stroke of the clamp. A second needle valve, 2, may be installed in the other exhaust port to limit speed on the return stroke if necessary.

Circuit Adjustment of Figure 2-24. The pressure regulator should always be adjusted first because it affects speed as well as the squeeze. Set needle valve, 1, wide open, and adjust the pressure regulator to a pressure as low as will give sufficient squeeze for the application. Then adjust needle valve, 1, to as high a speed as will not produce excessive impact. Finally, adjust needle valve, 2, for the desired return speed.

Saving Air With Reduced Pressure. Figure 2-25. By reducing the pressure on the idling stroke of an air cylinder (usually the return stroke), the overall air consumption can be reduced. This increases circuit efficiency and saves energy. On applications where the cylinder cycles frequently or for long periods of time, a great deal of energy can sometimes be saved in this way. Air consumption is reduced because when the cylinder stalls at the end of its stroke, the volume of air which continues to flow into the cylinder is less at lower pressure. Of course air saving is possible only on those applications where the cylinder is running free either on the forward or the return part of its cycle.

In Figure 2-25 the cylinder runs free as it retracts. A regulator installed at the rod port reduces return pressure and acts as a speed control. It should be adjusted to the lowest pressure which will give acceptable return speed. No other speed control is needed for the return stroke. The regulator should be by-passed with a check valve for free air flow around the regulator when the cylinder is extending. Speed can be regulated on the extension stroke with a needle valve placed in the 4-way valve exhaust port.

Five-way valves can also be used in air saving circuits, and have the advantage of eliminating all check valves from the circuit. See Figure 2-15.

Table 2-1 shows the percentage of air which can be saved on the return stroke (only) by reducing pressure from values across the top of the chart to lower values along left side.

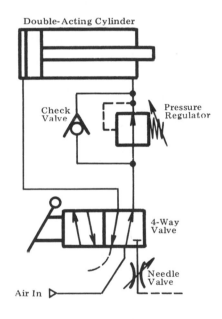

Double-Acting Cylinder

FIGURE 2-25. Regulator reduces air consumption on the return stroke and serves as a speed control as well.

Reduced Pressure, PSI	System Pressure, PSI, on Power Stroke						
	120	110	100	90	80	70	60
80	18%	15%	10%	5%	- - -	- - -	- - -
70	24%	21%	16%	12%	7%	- - -	- - -
60	30%	26%	23%	19%	14%	7%	- - -
50	36%	34%	30%	26%	22%	16%	10%

TABLE 2-1. — Air Savings With Reduced Pressure on the Return Stroke of a Cylinder

SPEED CONTROL OF AIR CYLINDERS

The travel speed of an air cylinder is rather unpredictable in the design stage because it depends on several factors which are unknown or are difficult to measure. Even if the precise load resistance is known, pressure loss through valves and plumbing due to flow is unknown. The usual practice is to choose the bore size of the cylinder and the operating pressure to give a force about 25% greater than the load resistance. This allows 25% of the inlet pressure to be consumed in flow losses, and if the valves and plumbing have been chosen to equal the port size of the cylinder, this should give more than enough allowance for flow losses for normal cylinder speed. If the speed should turn out to be excessive, it can be reduced with flow control valves. To obtain unusually fast travel speed, the cylinder force should be made twice as great, or greater, relative to the load resistance.

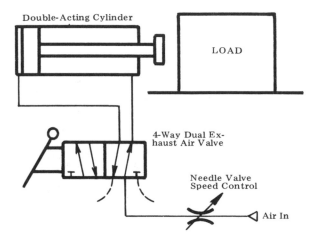

FIGURE 2-26. A needle valve in the inlet line cannot give individual adjustment of speed in each direction of travel.

Needle Valve Speed Control. Figure 2-26. The easiest way to control air cylinder speed would be to place a needle valve in the air inlet line. However, this is not a popular method because speed in each direction may vary widely according to the load. For example, a cylinder carrying a heavy load while extending but no load while retracting, would likely travel too slowly going out and too fast going back in. Since a greater volume of air is required while moving against a load than while running free, most air circuits should have an individual speed control valve for each direction of travel.

Dual Exhaust Speed Control. Figure 2-27. Usually the most convenient way of controlling air cylinder speed is to install a needle valve in each exhaust port of a dual exhaust 4-way directional control valve. This permits individual control of speed in each direction of cylinder travel. Special needle

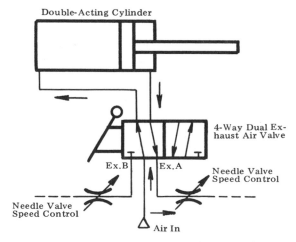

(A). 4-Way Valve Shifted; Cylinder Extends.

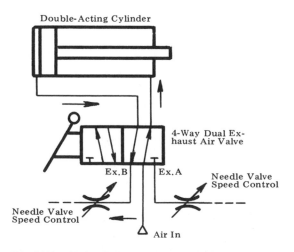

(B). 4-Way Valve Released; Cylinder Retracts.

FIGURE 2-27. Cylinder speed is controlled with needle valves installed in dual exhaust ports of 4-way valve.

valves are available from several manufacturers which have male pipe threads on their inlet. These can be screwed into the valve exhaust ports without requiring extra plumbing

Dual exhaust control is a form of meter-out speed control in which cylinder speed can be reduced by restricting the exhaust flow from the advancing side of the cylinder piston.

In Part (A) of Figure 2-27, the 4-way valve is in its shifted position and the piston is advancing. Exhaust air comes out exhaust Port A, and the extension speed can be regulated with the needle valve installed in Port A. In Part (B), the valve is in its un-actuated state and the cylinder rod is retracting. Exhaust air now comes out exhaust Port B, and retraction speed can be regulated with the needle valve screwed into Port B without affecting cylinder extension speed.

Flow Control Valves for Speed Control. Figure 2-28. A flow control valve includes both a needle valve for metering and a check valve for free passage of air in the reverse or non-metering direction of flow. The needle valve is adjustable and some models have a calibrated knob to allow the operator to return to a reference setting. The check valve orifice is usually smaller than the size rating of the valve and does offer a certain amount of flow resistance in addition to its 3 PSI cracking pressure.

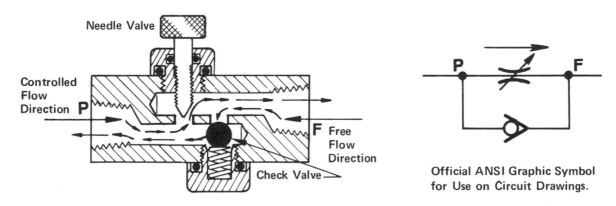

FIGURE 2-28 Flow Control Valve for Speed Control of Air and Hydraulic Cylinders.

Flow control valves are offered in sizes from 1/8" NPT through 3/4" NPT, with brass, steel, or stainless steel bodies. Brass is preferred for compressed air as it is more compatible with water which may be in the air line. Steel is preferred for hydraulics because of its higher pressure rating and because copper products, including brass, should be kept out of hydraulic oil systems as far as practical. This is because of possible chemical incompatibility with some petroleum base oils. Stainless steel must sometimes be used when handling media corrosive to brass or steel.

Direction of metered flow is stamped on the valve body. Some manufacturers stamp an arrow on the body to indicate direction of controlled flow; other manufacturers use the arrow to indicate free flow direction. Before connecting the valve read the instructions metal stamped on the side of the body.

For more detailed information on flow control valves, please refer to Volume 1, Chapter 3.

Meter-Out Cylinder Speed Control. Figure 2-29A. Two flow control valves are used, one in each of the lines connecting the cylinder to its 4-way control valve. They are installed to control the flow pushed out of the cylinder by the advancing (or retracting) piston. Preferred location is as close as practical to the cylinder ports, to reduce the volume of air trapped between the flow controls and cylinder. When installed in this way, Valve 1 regulates extension speed of the piston by restricting the

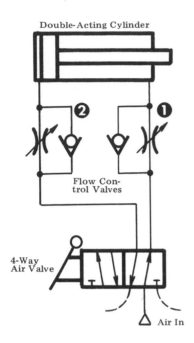

FIGURE 2-29A.
Meter-Out Speed Control.

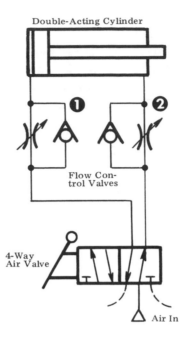

FIGURE 2-29B.
Meter-In Speed Control.

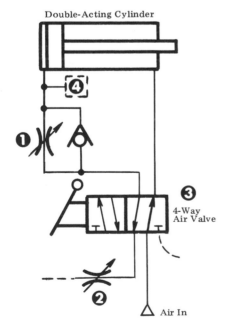

FIGURE 2-30.

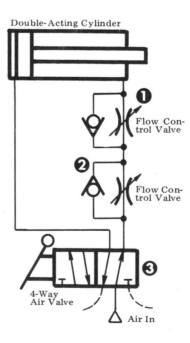

FIGURE 2-31.

Flow control (or needle) valves may be placed in various parts of the circuit for speed control

rate of exhaust flow from the rod end of the cylinder. Valve 2 regulates piston return speed by restricting rate of exhaust flow from the blind end of the cylinder. Note that needle valves alone (without by-pass check valves) will not give satisfactory operation. The built-in check valves prevent interaction between the needle valves, so when one needle valve is adjusted it will not also affect cylinder speed in the opposite direction.

Meter-In Speed Control. Figure 2-29B. The two flow control valves can be turned around so their controlled flow direction is *toward* the cylinder. When connected in this way they meter the rate of air flow into the cylinder while allowing exhaust air to pass out without restriction. In this diagram, Valve 1 regulates forward speed and Valve 2 regulates return speed.

Meter-Out/Meter-In Air Cylinder Speed Control. Figures 2-30 and 2-31. A combination of meter-out and meter-in speed control methods may sometimes be used on the same cylinder where it may be more convenient to install one flow control valve in the cylinder line for meter-in control in one direction of piston travel, and a needle valve in one exhaust for meter-out speed control in the other direction as shown in Figure 2-30. Or, both flow control valves may be placed in the same cylinder line but facing in opposite directions as shown in Figure 2-31. This gives meter-in control in one direction, meter-out control in the other direction.

In Figure 2-30, flow control Valve 1 is connected for meter-in speed control while the piston is advancing. This type of control may be desirable in the working direction if a pressure switch or sequence valve, Item 4, is used to detect a sharp rise in pressure behind the cylinder piston when it stalls against the load. Valve 2 is a simple needle valve installed in the appropriate exhaust port of the 4-way valve to give meter-out control during retraction of the cylinder piston.

In Figure 2-31, flow control Valve 1 is connected for meter-out control while the piston is advancing and Valve 2 for meter-in control while it is retracting. By reversing both valves the metering functions would be reversed.

Cylinder Speed — Factors to be Considered on the Original Design. An air cylinder operates from an infinite supply of air — that which is stored in the receiver tank. Provided the combination of piston area and air pressure is greater than load resistance, the only factor which prevents the cylinder from reaching an infinite velocity is the flow resistance in lines, through 4-way valve porting, and of course in flow control valves. It is a difficult matter to predict just how much additional pressure above that required to balance the load will be required to supply flow resistance losses. We must rely on past experience and rules-of-thumb to come up with a pressure level high enough to give the speed we want. The matter of oversizing the pressure level has been covered earlier in this section. It is far better to have more than enough pressure to give the desired speed, then to reduce cylinder speed with flow control valves than to be short on pressure and have insufficient speed. The following factors should be considered on an original design:

(a). System Supply Pressure. A knowledge of the pressure level available to the machine from the air distribution system is important. For consistent results the pressure regulator on the machine must be set no higher than the lowest pressure available from the air line during those periods when the machine will be in operation. Hopefully, most of this pressure can be used against the load with a minimum amount consumed in flow losses.

(b). Piping and Valve Size. All flow resistance in piping and valving will reduce maximum cylinder speed. To obtain high speed the circuit flow resistance must be held to a minimum.

A rule-of-thumb states that for moderate cylinder speed, the flow areas through piping and valving should be at least equal to flow area through the cylinder ports. For higher speed, piping and valving flow areas should be even larger. Usually, for cylinders up through 3″ bore, a flow area corresponding to 1/4 or 3/8 diameter is sufficient. Of course if two or more cylinders receive air through the same valve, the flow area should be correspondingly larger.

(c). After estimating the pressure available to balance the load (with pressure for flow losses deducted), the cylinder bore should be selected to provide sufficient load force.

How to Increase the Speed of an Air Cylinder. On an existing system in which the cylinder speed is too slow even with the speed control valves wide open, the speed can be increased either by raising the air pressure or by reducing the flow losses. Specific areas where changes can be made include the following: (Note: These remedies do not apply to hydraulic cylinders.)

(a). If possible, increase air pressure by adjusting the pressure regulator which serves the particular branch. It may be necessary in some cases to raise the cut-off pressure of the air compressor before the regulator can be set to a higher pressure. Increasing the pressure will always produce an increase in cylinder speed.

(b). If an exhaust muffler is used, remove and discard it. If exhaust hiss is objectionable, install a muffler of much greater flow capacity than the original muffler. This will reduce back pressure and will noticeably increase cylinder speed.

(c). Remove speed control valves completely from the system. Their internal orifices are smaller than line size, and even when they are wide open they still restrict the air flow. The internal "free flow" check valve also has a small orifice and adds restriction to the returning air. If necessary to have speed control valves, replace the existing ones with larger ones of one to two pipe sizes larger, with reducing bushings to connect into the lines.

(d). Enlarging the size of plumbing lines will increase cylinder speed. Start with the lines connecting between cylinder and 4-way valve. If this does not sufficiently increase speed, enlarge the inlet air line clear back to the distribution piping. Replace the existing trio unit (filter/regulator/lubricator) with the next larger size. Shorten plumbing runs if possible. Eliminate unnecessary bends. Use larger hoses to eliminate the restriction through hose fittings.

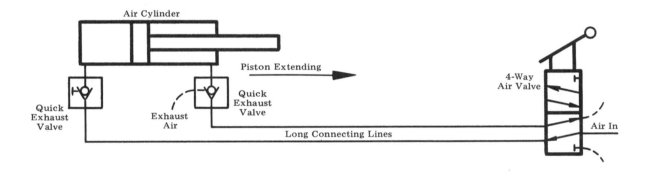

FIGURE 2-32. Quick exhaust valves will increase the speed of an air cylinder.

(e). Quick Exhaust Valve. Figure 2-32. Install a quick exhaust valve directly at the cylinder port which is exhausting the air for the direction in which the speed is to be increased. A 1/4" size valve will usually be sufficient.

(f). If the previous remedies fail to produce a sufficient increase in speed, replace the 4-way control valve with one of higher flow rating.

(g). After all of the above remedies have been exhausted, replace the cylinder with one having a larger piston area.

SLOW FEED CIRCUITS FOR AIR CYLINDERS

Air cylinders are essentially power, not slow feed devices. In general they should not be used to move metal cutting tools at a slow feed rate. An exception is for "crowding" a drill bit into the work. In this case the feed rate is not a function of the cylinder, and the work resistance stabilizes the cylinder movement. Hydraulic cylinder or motor feed must be used on milling machine tables, grinder tables, shapers, and planers.

Air cylinders perform best when allowed to travel at moderate to high speeds. But where air power must be used at moderate to slow speeds, the following suggestions may improve performance-

Speed Control Method. Figure 2-33. Experience shows that an air cylinder, operated at low speed, is less likely to chatter if meter-out speed control is used as shown in Figures 2-27 and 2-29A, rather than the meter-in control of Figure 2-29B. Meterering the fluid out of the cylinder keeps a back pressure on the advancing side of the piston. The higher the back pressure, the greater the stability.

In order to have high back pressure for stability, the cylinder should have a bore large enough to give more force than needed against the load. The more the cylinder is overpowered with respect to the load, the greater will be the back pressure against the piston when meter-out speed control is used.

For example in this figure, if inlet pressure to the blind end is 100 PSI, and if the pressure to overcome the load is only 30 PSI, this gives 70 PSI back pressure for stability. If the load should actually require 80 PSI, and if inlet pressure were 100 PSI, this leaves only 20 PSI to satisfy circuit flow losses and provide back pressure. In this case the cylinder would have a greater tendency to move erratically.

On existing systems, if the inlet pressure can be raised and the speed control valve closed down, this will give a higher back pressure under the same load and speed conditions. On new designs, the air cylinder can be oversized by 200 to 300% over the load resistance to provide a higher back pressure and more stability when moving the load.

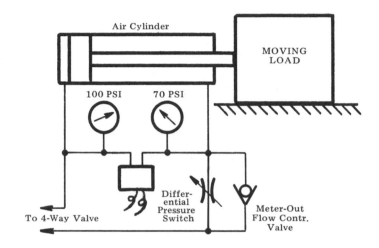

FIGURE 2-33. A high back pressure against the forward face of the piston increases its stability at slow speeds.

While *meter-out* speed control, as described above, does give greater cylinder stability, it presents a problem when a pressure switch or sequence valve is installed as in Figure 2-30 to detect a pressure rise when the load builds up in front of the piston. When the piston stalls there will be only a small rise in pressure as upstream flow losses disappear. Adjustment of the trip setting on the pressure switch may be quite critical. *Meter-in* speed control will give a sharp rise in pressure when the piston stalls but does not have as great stability while the cylinder piston is moving.

A standard pressure switch, for example, teed into the cylinder blind port will not work well with *meter-out* speed control because, unless adjusted very critically, it may prematurely activate before the piston reaches stall, or it may not activate after stall is reached.

Pressure Differential Switching. As shown in Figure 2-33, the use of a pressure differential switch permits *meter-out* speed control to be used, and at the same time will produce a sharp rise in pressure when the cylinder stalls, at which time all downstream pressure on the cylinder rod port disappears, and all flow losses upstream disappear. Full system pressure is then available to trip the pressure switch.

Note: Pressure switch activation for retracting a cylinder is only suitable for circuits where the cylinder is free-moving through its stroke, then stalls against a workpiece; it does not work well for applications where the cylinder is moving at or near full load throughout its stroke.

Air-Over-Oil Metering System. Figure 2-34. This is a combination circuit using air for motive power and using an oil loop between the cylinder and a reserve oil tank into which oil is pushed by the

47

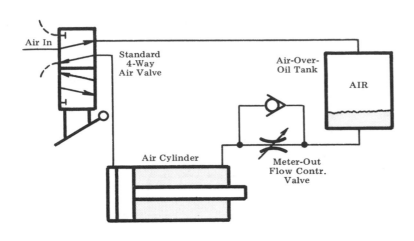

FIGURE 2-34. *Conventional single-tank, air-over-oil system.*

advancing piston on its forward power stroke. Cylinder speed in the forward direction is regulated by a flow control valve placed in a meter-out mode in the oil loop.

When the 4-way valve is shifted to retract the piston, air pressure applied to the top of the reserve oil tank forces oil back into the rod end of the cylinder. Return speed can be regulated, if necessary, with a needle valve placed in the appropriate exhaust port of the 4-way valve. However, return speed control may not be necessary, as the viscosity of the oil may provide enough flow resistance to keep the speed to an acceptable level.

To be successful, the piston seals in the cylinder must be virtually leaktight to prevent leakage of the oil to atmosphere through the 4-way valve. For oil metering in both directions, forward and back, a double tank system is required. Details of this and other air-over-oil circuits are covered in Chapter 12.

While not as stable as a straight hydraulic circuit, the air-over-oil system does reduce the tendency of the cylinder piston, on some applications, to move erratically.

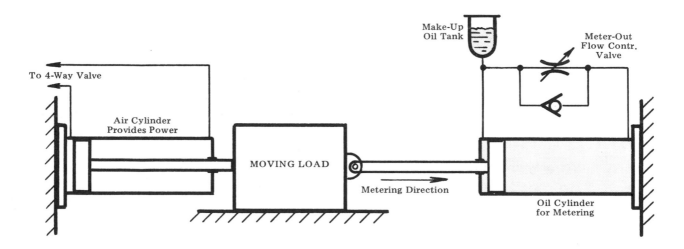

FIGURE 2-35. *An air-powered system is shown, with oil metering for smooth travel at slow speeds.*

Metering Cylinder. Figure 2-35. An improved method for oil metering the speed of an air cylinder is shown. This method can sometimes be added to an existing air system which does not operate smoothly at low speeds. The arrangement has several important advantages over a standard air-over-oil circuit as shown in the preceding figure:

(1). The compressed air does not come into intimate contact with the metering oil. Therefore, there is no oil mist discharged into the atmosphere. In an air-over-oil system some of the oil vaporizes into the air on every cycle and is carried out the exhaust ports of the 4-way valve.

(2). Since oil is not discharged into the atmosphere, the oil loop seldom has to be replenished. This reduces cost of maintenance.

(3). If desired, the oil metering cylinder can be selected with a very large volume. The larger flow through the flow control valve permits more accurate control of feed rate.

(4). All components are standard catalog fluid power items. No specially constructed pressure tank or special cylinder is required.

(5). Only the small make-up oil tank has to be mounted at a higher elevation than the other components. There is no problem of trapped air in the oil loop.

The air cylinder provides the power for moving the load; the oil cylinder, of any bore but having the same stroke, is used solely for speed control of the air cylinder and the load. The placement of the oil cylinder depends on the physical layout of the machine. For simplicity of illustration it is shown with axis in line with the cylinder axis. However, it can be placed in any position where the moving load can actuate it. Its bore can be smaller or larger than the air cylinder bore. The larger its bore the better the speed control action.

During the metering stroke, the extension stroke in this figure, oil is pushed from blind end to rod end of the oil cylinder, passing through the metering needle valve. Surplus oil which cannot enter the rod end is stored in the make-up tank. On the retraction stroke of the air cylinder (the extension stroke of the oil cylinder) the surplus oil again enters the blind end of the oil cylinder. On this part of the cycle oil can pass from rod end to blind end of the oil cylinder through the check valve, and speed is un-metered.

The volume of the make-up tank must be greater than the volume displaced by the piston rod of the oil cylinder. An air line filter with plastic bowl and bowl guard makes a convenient make-up tank.

The oil used for metering should have very low viscosity and must be compatible with seals and with the plastic bowl of the make-up tank. ATF transmission fluid, Type A or Type F is an excellent fluid to use and is available in small quantities.

The oil cylinder in this figure is connected to meter when its piston is pushed in. To make it meter when its piston is pushed out, move the make-up tank to the blind end port and turn the flow control valve around.

Note: Self-contained "hydraulic checking cylinders" can be purchased with built-in flow control valve and reservoir. They can be installed and adjusted so there is some free travel before the slow metering action starts. However, they are available only in relatively short metering strokes.

3
Several Air Cylinders...
...OPERATING ON ONE MACHINE

A machine using several cylinders may have them arranged either for individual control by an operator, or may have them connected to operate either in "random" or in "sequence". An example of random operation is several cylinders connected in parallel to one control valve. It makes little difference which one moves faster or which one completes its stroke first.

"Sequential operation" has come to mean a step-by-step progressive operation of several cylinders, usually one at a time, in a certain programmed order, cycle after cycle. The usual definition of sequencing implies that the programmed operation of the cylinders occurs automatically to the end of the machine cycle, after being started by an operator. Each cylinder, when it completes its stroke, supplies a signal to start the next cylinder. All action, after cycle start, must occur automatically and repetitively in the same order with no attention from an operator except for starting the cycle. The order in which the various cylinders extend and retract is known as the machine "program".

Example of Sequential Operation: An operator manually loads a workpiece into a machine. He presses a button or shifts a valve to start the cycle. A cylinder moves in to clamp the workpiece firmly in a jig or fixture. After the clamping is completed, another cylinder moves in to perform a machining, stamping, shearing, or forming operation, then withdraws. The first cylinder retracts to

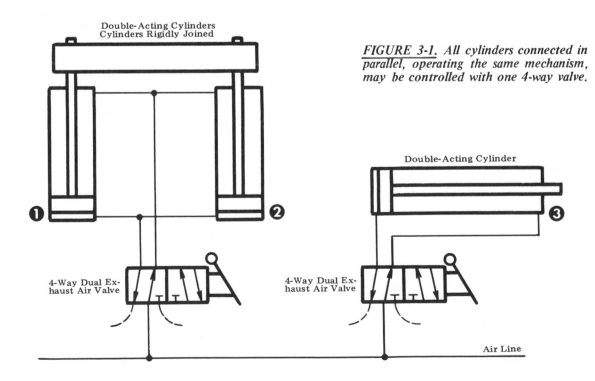

FIGURE 3-1. All cylinders connected in parallel, operating the same mechanism, may be controlled with one 4-way valve.

50

unclamp the piece. Finally, a third cylinder extends to eject the piece, then it, too, retracts. The machine has completed its cycle and stops. In this machine the operator loads and unloads workpieces, starts the machine, and the cylinders go through their "program" with no further attention.

In this section we will show examples of automatic sequencing using solenoid air valves controlled with limit switches, pressure switches, and timers. We will also show non-electrical operation with button bleeder, pilot-operated. sequence, and cam valves.

SEVERAL CYLINDERS ON ONE MACHINE

Please remember that in this section we are dealing with *air* cylinders. The principles stated and the circuits shown may or may not be applicable to hydraulics. Since air is compressible, the action in a compressed air circuit may be different from the action in a similar hydraulic circuit.

Parallel Cylinders. Figure 3-1. All cylinders which power the same mechanism and which are connected in parallel, as Cylinders 1 and 2 in this diagram, may be operated from one 4-way directional valve. The valve should be sized according to the number and size of cylinders receiving air through it, and according to travel speed desired. Rules for oversizing for speed have been covered in the preceding chapter.

Recommended practice is to use, if practical, one larger cylinder rather than several smaller cylinders, although multiple cylinders are sometimes used to distribute force more evenly over a larger area, or where space for a larger cylinder is not available. *Caution!* Two air cylinders tend to travel at different speeds according to the load carried by each. The structure which anchors them together must be sufficiently strong to force the two cylinder to travel in step without distorting or causing a bind in the structure. Two air cylinders, carrying different loads, can only be synchronized to the same speed with special circuitry or components. See air/oil applications in Chapter 12.

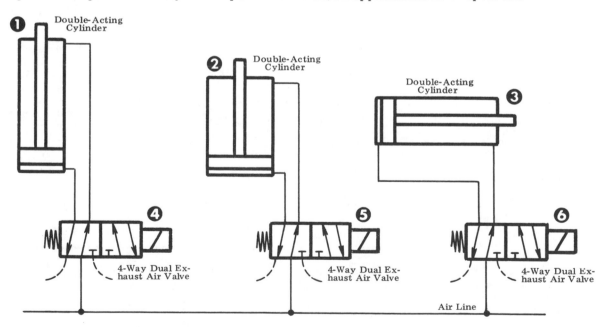

FIGURE 3-2. Each cylinder on the machine, if it operates independently, must have its own 4-way control valve.

Individual Cylinders. Figure 3-2. Each cylinder operating a separate mechanism must have its own 4-way control valve. It is easy to see why this is necessary if each cylinder starts its extension or retraction stroke at different times in the cycle. But, less obviously, it also applies to cylinders which operate separate mechanisms even though they may always be started at the same instant in the cycle.

Using a separate 4-way valve for each cylinder, and feeding them from an adequate distribution system, will completely eliminate any interaction between them at any point in the stroke where the load on either cylinder changes, or where one cylinder may reach the end of its stroke while the other cylinder continues to move. This will be demonstrated in the next series of diagrams.

In Figure 3-2 if Valves 4 and 5, for example, are solenoid, button bleeder, or pilot operated, and if they must be actuated at the same time in the cycle, their control circuits can be tied together, as through relay contacts if solenoid operated. If one valve must be used for two cylinders which operate separate mechanisms, as for manual control, the interaction between cylinders can be minimized, although not entirely eliminated, by choosing a 4-way valve oversize for the application, thus reducing pressure loss through it which is the reason for the interaction.

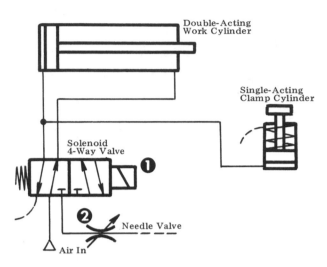

FIGURE 3-3. In this clamp and work circuit there may be unwanted interaction between the two cylinders.

If Cylinders 1 and 2 are fast moving, free running cylinders such as air clamps, interaction between them is usually of little consequence, and both may be operated from one 4-way valve even though they may be clamping on different parts of the machine.

Clamp and Work Action. Figure 3-3. This involves two cylinders. The clamp cylinder is usually of small bore and short stroke and is used like an air vise, to hold a workpiece in a jig or fixture. The work cylinder, usually much larger, provides power for the tool which does the work. This circuit demonstrates interaction which may occur between the two cylinders as a result of being supplied with air through the same 4-way valve.

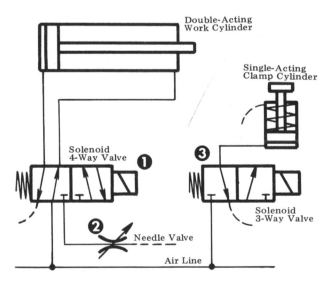

FIGURE 3-4. Interaction between the two cylinders is eliminated by operating each one from a separate directional control valve. Compare with Figure 3-3.

Valve 1 is a dual exhaust 4-way directional control valve controlling both cylinders. When it is shifted, both cylinders start to extend.

The small air clamp snaps in very quickly and stalls against the workpiece before the work cylinder has had time to move very far. However, while the work cylinder is moving in free travel, pressure behind its piston cannot build up to full value, and the clamp cannot come up to its full clamping force until the working tool contacts the work and pressure behind the work cylinder piston rises to its full value.

This means the clamping force may not be sufficient at the critical moment when the work cylinder starts. Furthermore, if the work cylinder should start to chatter, or if the working tool should break through the work, clamping force would momentarily be lost and the workpiece might be released prematurely. This effect can possibly be minimized by selecting a 4-way valve oversize for the application.

Needle Valve 2 is a speed limiting control which can be added in the appropriate valve exhaust port to limit speed of the work cylinder while its piston is advancing.

Elimination of Cylinder Interaction. Figure 3-4. By using a separate 3-way (or 4-way) valve to control the air clamp, all interaction between the two cylinders is eliminated. The clamp can immediately come to full clamping force, and can maintain full force throughout the cycle no matter what happens to the work cylinder circuit. The directional valves, 1 and 3, are shown as solenoid valves. The coils can be connected in parallel and operated in unison from the same switch or relay contacts as if they were a single valve. Or, if desired, their coils can be operated independently. For example, Valve 3 can be left energized after the work has been completed and until the work cylinder has retracted. It will continue to hold the workpiece securely while the work cylinder is withdrawing from the work.

SEQUENCING WITH BUTTON BLEEDER VALVES

Valve Description. Figure 3-5. The button bleeder valve is one of the family of 4-way directional valves available from most valve manufacturers. Its spool is shifted by air pressure obtained from the valve's own pressure inlet. Full air pressure is maintained on both ends of the main spool. The spool shifts when one of the bleed buttons is pressed momentarily, venting that end cap to atmosphere. Pressure on the opposite end of the spool shifts the spool toward the button which has been pressed, The bleed buttons, one on each end, are furnished as part of the main valve. They are small 2-way, normally closed, pushbutton valves. They are screwed into the end caps and may be left there or may be removed and mounted a short distance away, connected into the valve end caps by short lengths of hose. Either of them can be used as an operating pushbutton or can be actuated by a moving cam on the machine.

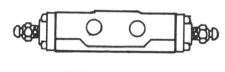

FIGURE 3-5.
Button Bleeder Valve.

The student may want to review the complete description of this valve in Volume 1. The main valve is usually a 4-way type, with two positions, and with no return or centering springs. It is usable only on compressed air and actually depends on air compressibility for its operation. From some sources it may be available with internal return or centering springs, but the springs reduce its sensitivity, making it less reliable and slower to shift, and these models are not recommended.

Each bleed button vents to atmosphere when actuated, through a small unthreaded exhaust hole. It cannot be used in external circuitry where a threaded exhaust valve is required as shown in some circuits to follow. If detached and mounted remotely from the main body, connecting hoses should be kept as short as possible. The longer the connecting hoses, the slower the shifting response. On long lines shifting becomes unreliable and the valve may not shift at all. For best results the separation distance should be limited to about 5 feet. Connecting hoses of 1/4'' size seem to be the optimum diameter, and usually work better than either larger or smaller hoses.

<u>Limitations.</u> Button bleeder valves cannot be used on vacuum nor on hydraulics. They work best on air pressure of 80 PSI and higher. Their main value is on applications where electricity is not permitted. Spool response becomes more sluggish as length of bleed line is increased. Simple circuits for their use will be found in Volume 1. They are available in sizes from 1/8 through 1½'' NPT.

Bleed Button Graphic Symbol. Figure 3-7. The bleed button itself is a 2-way, normally closed, spring returned valve vented to atmosphere through a small unthreaded hole. The standard ANSI graphic symbol for that kind of valve can be used, but for simplicity and clarity we have selected the picture symbol of Figure 3-7 to be used on diagrams to follow in this chapter.

FIGURE 3-7.
Bleed Button.

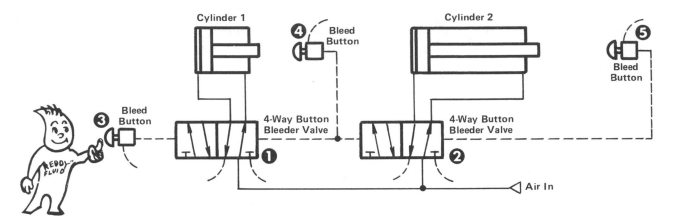

FIGURE 3-8. Standard Two-Cylinder Button Bleeder Circuit.

> *Cyl. 1 Advances*
> *Cyl. 1 Retracts &*
> *   Cyl. 2 Advances*
> *Cyl. 2 Retracts*

Example Circuit No. 1. Figure 3-8. The two cylinders are to be programmed to operate automatically through one cycle every time an operator starts a cycle by pressing bleed Button 3. The program or sequence is shown in the box.

Circuit operation of Figure 3-8 is as follows: The operator presses and releases bleed Button 3. This causes the spool in the main Valve 1 to shift and the cylinder to start forward. When this cylinder reaches and actuates bleed Button 4, both Valves 1 and 2 are shifted. This starts Cylinder 1 toward home and starts Cylinder 2 forward. When Cylinder 2 reaches and actuates bleed Button 5, Valve 2 spool reverses, causing Cylinder 2 to retract. Both cylinders return to home position and stall under pressure. This ends the cycle. The next cycle must be started by the operator with bleed Button 3.

In this and all button bleeder circuits the valve spool should be shown in its un-actuated state. When the bleed button is actuated, the valve symbol is visualized as being pushed in a direction away from the end being actuated. This puts the other valve symbol block into alignment with the circuit connections. This is standard symbology even though in the physical state the spool may actually move in a direction toward the end of the valve where the actuator is located.

Example No. 2. Figure 3-9. The program between the cylinders is different from that in the preceding circuit, and is shown in the box. Cylinder 1, after advancing and actuating Valve 4, must remain extended until Cylinder 2 completes its forward stroke. Then both cylinders will start retraction at the same time.

This sequence of operation is more difficult because after Cylinder 1 advances and actuates Valve 4 it remains standing on this valve. This will prevent 4-way Valve 2 from being reversed later in the cycle unless some means is provided to shut off the

> *Cyl. 1 Advances*
> *Cyl. 2 Advances*
> *Both Cylinders*
> *   Retract Together*

Cycle Sequence

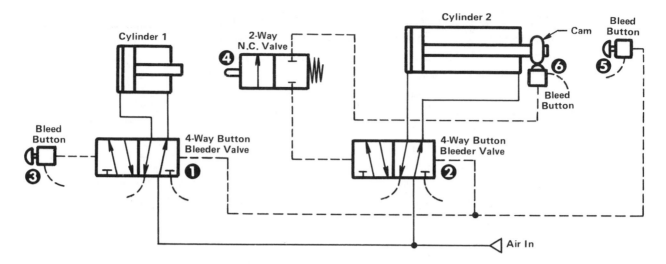

FIGURE 3-9. One Solution to a Common Circuit Problem With Button Bleeder Valves.

bleed caused by Valve 4 while Cylinder 2 is advancing. The method used here is to pipe the exhaust port of Valve 4, not directly to atmosphere, but through bleed Button 6. This button is held open by Cylinder 2 when it is at home position, but immediately closes as Cylinder 2 starts to advance. This permits Valve 2 to regain its pressure balance so it can be reversed by Button 5 when Cylinder 2 reaches the end of its forward stroke. For this circuit, Valve 4 cannot be a standard bleed button because a bleed button does not have a threaded exhaust connection. It must be a stem or cam actuated valve, of miniature size, normally closed, spring returned, and with threaded exhaust connection.

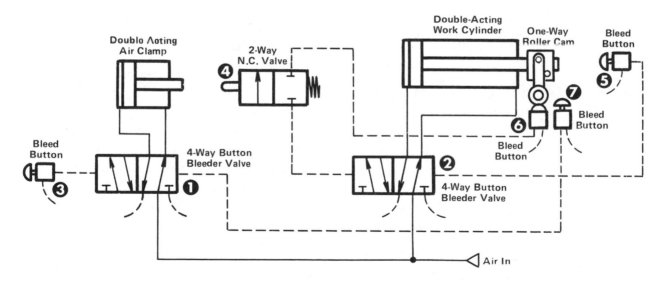

FIGURE 3-10. Clamp and Work Application With Button Bleeder Valves.

Example 3. Clamp and Work Circuit. Figure 3-10. This is one of the more difficult programs to design. The action sequence is shown in the box at the top of the next page. The air clamp is first to extend. When it actuates Valve 4 it must remain standing on this valve until the work cylinder advances then retracts to home position. The air clamp then releases and retracts.

This circuit is similar to the preceding one with the addition of bleed Button 7 to release the air clamp. The button should be placed where it will be momentarily actuated by the retracting work cylinder just before it reaches home position. The work cylinder must not actuate Button 7 while extending. The actuating mechanism, which can be mounted on any part of the mechanism moved by Cylinder 2, is a one-way roller cam hinged in a way that on the forward stroke of the work cylinder the roller will lift out of the way without actuating Button 7, but will depress it in passing on the return stroke. *Button 7 must be in its released state when Cylinder 2 reaches home position.* If possible, Button 7 should be located a few inches ahead of Button 6 to prevent false operation if the air clamp does not back off of Valve 4 quickly enough.

> Cyl. 1 Advances
> Cyl. 2 Advances
> Cyl. 2 Retracts
> Cyl. 1 Retracts

Cycle Sequence

A series bleed circuit through stem-actuated Valve 4 and bleed Button 6 is used, as it was in the previous circuit, to restore pressure balance in 4-way Valve 2 so it can be reversed later in the cycle.

Special Circuits for Button Bleeder Operation. Several special circuits can be designed as shown below if bleed lines can be kept very short, with bleed buttons within a few feet of the main valves.

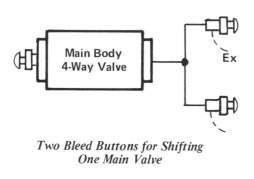

*Two Bleed Buttons for Shifting
One Main Valve*

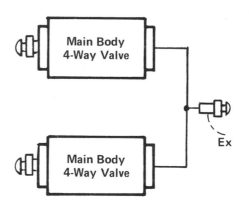

Two Main Valves With One Button

One main 4-way valve can be shifted by either of two bleed buttons connected in parallel. This circuit would permit control from two locations, or one of the buttons could be a "panic" button within easy reach of the operator.

This parallel connection of two buttons is known as an "OR" logic circuit. The main valve will shift if one button OR the other is actuated.

Two main 4-way valves can be shifted with one bleed button. A circuit example is for returning two cylinders, each controlled with a separate 4-way valve, at the end of a cycle.

For shifting more than two 1/4 or 1/2" main valves, bleed buttons should be replaced with cam or stem operated valves of larger capacity.

Notes on Button Bleeder Circuits. (1). The main valve requires a fraction of a second to recover its balance after being bled in one direction before its spool can be shifted in the opposite direction. Keep external bleed lines as short as possible to reduce this recovery time.

(2). If a cylinder or other mechanism holds a bleed button pressed down, the spool in the main valve cannot be shifted by actuating the opposite bleed button.

(3). The machine cycle should usually end with all bleed buttons released. A bleed button, while held down, will waste air at the rate of 3/4 HP when connected to a 100 PSI line.

SEQUENCE VALVES FOR PROGRAMMING CYLINDERS

We have looked at button bleeder valves as a means for programming the action of two cylinders. Sequence valves are another means for causing two (or more) cylinders to follow a regular program of operation, cycle after cycle, after the cycle has been started by an operator.

A sequence valve can be defined as a 2-way, normally closed valve. The inlet port remains closed to the outlet port until pressure at the inlet port rises sufficiently to compress the adjustable spring which holds the poppet closed. When the poppet has been opened, air can flow unrestricted through the valve to a second cylinder circuit. The valve has a characteristic which is especially valuable on some applications — that of blocking flow to the second cylinder if pressure should be lost on the first. This is called "priority action" and will be demonstrated in some of the following circuits.

Air Sequence Valve. Figure 3-11. A sequence valve and a relief valve are similar but with these important differences: The pressure in the spring chamber of both valves must be vented to atmospheric pressure. In a relief valve the internal spring chamber can be vented into the outlet port which is always at atmospheric pressure. In a sequence valve, both inlet and outlet ports may be operating at full system pressure, so the spring chamber must be vented separately, usually through a small unthreaded hole as shown in Figure 3-11.

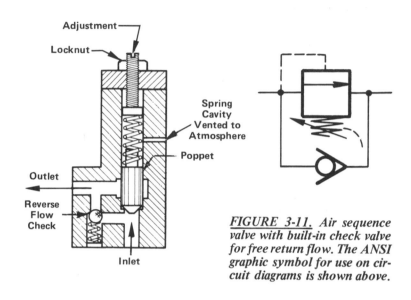

FIGURE 3-11. Air sequence valve with built-in check valve for free return flow. The ANSI graphic symbol for use on circuit diagrams is shown above.

Another difference is that sequence valves are used in circuits where air must flow in reverse around the valve later in the cycle. For convenience, a free reverse flow check valve is built into the housing of most sequence valves; reverse flow through a relief valve is never required.

Although a sequence valve provides the most effective way of programming a sequential operation of two or more cylinders, a source for purchasing air sequence valves may be hard to find, and those available may be limited to 1/4" size. In most air circuits which operate at a line pressure of 80 PSI or higher, a low range air or hydraulic relief valve, with a check valve shunted around it, can be used by slightly modifying the circuit. See more information in the box on Page 62.

Hydraulic circuits using sequence valves start on Page 161, and the difference between a sequence valve and relief valve is explained in that section. Hydraulic sequence valves of all sizes are readily available from many sources.

Priority Action of a Sequence Valve. Figure 3-13. In this basic circuit the air clamp has priority on the air flow until its piston reaches a pre-adjusted pressure level. When this level has been exceeded, even by a small amount, the sequence valve will open wide. Air can pass unrestricted to the work cylinder.

To start a cycle the operator shifts Valve 1. Air flows to the clamp and it immediately extends until it stalls against the workpiece.

During extension of the air clamp, pressure behind its piston can build up only to the level re-

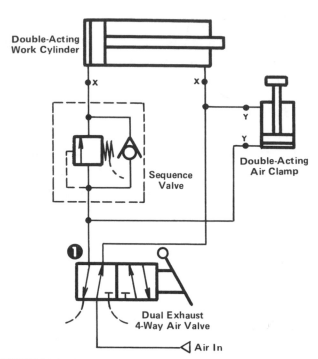

Double-Acting Work Cylinder

Sequence Valve

Double-Acting Air Clamp

❶

Dual Exhaust 4-Way Air Valve

◁ Air In

FIGURE 3-13. Basic Sequence Valve Application. A relief valve cannot be substituted unless circuit is modified as explained on Page 62.

quired to overcome friction. But when the clamp stalls against the workpiece, pressure starts to rise behind its piston. When pressure exceeds the spring adjustment which has been set on the sequence valve, the poppet opens, allowing air to flow to the work cylinder. And during the forward travel of the work cylinder the sequence valve remains wide open, but will immediately close if any condition should develop to lower the pressure behind the piston of the air clamp to less than the setting in the sequence valve. It assures that pressure behind the clamp will remain active during the work cycle, and that the work cylinder cannot move unless full clamping force remains in effect.

When directional Valve 1 is reversed, both cylinders start retraction at the same time. This circuit is limited to applications in which the air clamp can be released while the work cylinder is retracting.

Rules for Programming With Sequence Valves. Sequence valves work best in circuits where these conditions prevail:

(1). Both, or all, cylinders must be supplied through the same 4-way valve. Interaction between the cylinders caused by pressure drop through the valve, is important for good sequencing action. During the forward stroke of the air clamp, in free travel, pressure drop across the 4-way valve keeps pressure behind its piston relatively low, only sufficient to overcome friction. Then, when the air clamp stalls against the workpiece, all pressure drop across the 4-way valve disappears. The sharp rise in pressure opens the sequence valve to start the work cylinder forward.

(2). Sequence valve programming is reliable only when the first cylinder (the air clamp), is moving in free travel. If this cylinder carries a heavy load or works against high resistance throughout its stroke, the pressure rise at stall condition may not be high enough to reliably open a sequence valve.

(3). Usually, sequence valves are used for programming cylinders in one direction of travel; all cylinders are allowed to retract at the same time. These valves are more difficult to apply for sequencing in both directions of travel because all cylinders are supplied through one 4-way valve. When this valve is reversed, all cylinders are vented to atmosphere. Even though a cylinder may not actually have pressure on its piston to retract it, it may be pushed back by a reactive load. It may be possible, as demonstrated in the following circuits, to overcome this problem by adding additional components.

(4). Nearly all air circuits require speed control valves for both directions of travel on all cylinders. When pressure sensitive devices such as sequence valves or pressure switches are used, special attention must be given to placement of speed control valves so they will not interfere with sequencing action. The speed control circuit of Figure 2-29(B) on Page 44 should be used, with meter-in flow control valves installed at Points X - X in the diagram above. Read explanation near the top of Page 45. The circuit of Figure 2-27 on Page 42, in which needle valves are used in the exhaust ports of a dual exhaust control valve, should not be used in a circuit with sequence valves or pressure switches.

Sequencing in Both Directions of Travel.
Figure 3-14. On certain kinds of applications the retraction of two cylinders can also be programmed by adding a second sequence valve, 3.

In this diagram sequence Valve 2 causes the two cylinders to extend in a programmed order as described for the preceding diagram. The addition of sequence Valve 3 in the line to the air clamp will delay the retraction of the air clamp until the work cylinder has retracted and stalled.

Pilot-operated check Valve 4 retains air pressure in the clamp until the work cylinder has retracted and pressure has opened sequence Valve 3. Without it, a reactive load could push the clamp back during retraction of the work cylinder causing it to lose its grip on the work. See Page 200 for pilot-operated check valve operation.

If an air pilot-operated check valve is not available, substitute a 1/4" size hydraulic type provided the air line is well lubricated.

To add speed control, use the meter-in circuit of Figure 2-29(B) on Page 44.

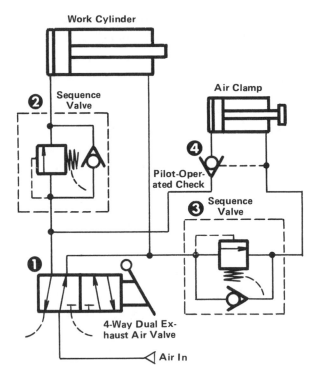

FIGURE 3-14. Sequence Valve 3 is added to program the cylinders while they are retracting. Note: To use relief valves instead of sequence valves the circuit must be modified as explained on Page 62.

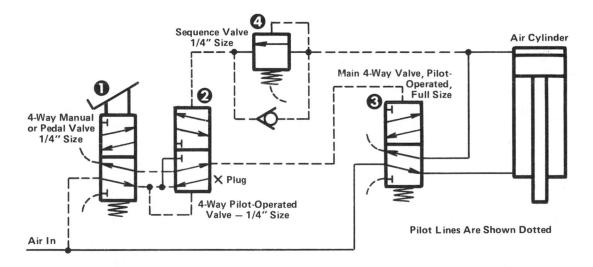

FIGURE 3-16. Manually Controlled Press Circuit for Compressed Air — Automatic Retraction. Note: A relief valve can be used instead of a sequence valve subject to limitations on Page 62.

Automatic Reversing Circuit for Air Press. Figure 3-16. In this air press circuit the cylinder automatically retracts when full air pressure appears behind its piston even though the operator may still be holding down a foot switch or pedal valve.

This is a pressure sensitive circuit using a sequence valve for automatic retraction. A small (1/4" size) pedal valve is shown for operator control, although a manual or solenoid valve could be used. Valve 3, is a pilot-operated valve sized to handle the air flow to the press cylinder. The other valving can be minimum size since it is used only for handling pilot air pressure to main Valve 3.

Sequence Valve 4 should be adjusted to open at the pressure where rated press tonnage is reached, and when it opens the cylinder will retract. The adjustment can be changed to vary press tonnage.

To operate the press, the operator stands on pedal Valve 1 and holds it down. Pilot pressure to shift the main 4-way Valve 3 comes through Valves 1 and 2. The press cylinder starts down, stalls against the work, and when full pressure has built up behind the piston, indicating full tonnage has been reached, air flows across sequence Valve 4 and shifts Valve 2. This causes the main Valve 3 to reverse, retracting the cylinder. The operator must keep the pedal valve actuated during the downstroke, and may release it either during the upstroke or after the cylinder has reached home position.

Clamp and Work Circuits for Large Air Cylinders. Figure 3-17.

Features of this 2-cylinder clamp and work circuit include the control of very large cylinders with miniature control valves, sequencing not only on the forward stroke but on retraction as well. The clamp will maintain full clamping force until the work cylinder has completely retracted.

Valve 3, the 4-way directional valve for the work cylinder, is sized for the full air flow required. All other valving is 1/4'' size for operation of the clamp cylinder and piloting of main Valve 3.

To operate the circuit the operator stands on pedal Valve 1 and keeps it actuated until both cylinders have completed their forward strokes. When he releases the pedal, the cylinders retract, with the main cylinder returning first then the clamp releasing the work and retracting. These actions can be described in greater detail:

Actuation of pedal Valve 1 causes the clamp cylinder to extend. When pressure builds up behind the clamp, it will open sequence Valve 2 and flow to the pilot of Valve 3. This valve shifts and causes the cylinder to extend.

To retract the cylinders the operator releases the pedal valve. This vents the pilot of Valve 3, causing that valve to return to its normal position and the cylinder to retract. Valve 4 remains closed, holding air pressure in the clamp, until the work cylinder has retracted. At home position of the cylinder a rise in pressure shifts Valve 4 to release the clamp.

If the clamp fails to hold during retraction of the work cylinder, the spring tension in Valve 4 can be increased, or a second sequence valve, similar to Valve 2 can be installed at Point X.

Speed control valves for the work cylinder should be the meter-in type to avoid interference with sequence valve action. If a speed control valve

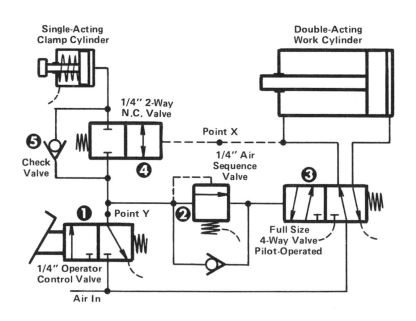

FIGURE 3-17. Clamp and Work Circuit. Single-Acting Clamp Cylinder. Automatic Sequencing in Both Directions. To substitute relief valve in place of sequence valve, see information on Page 62.

is needed for the clamp it should be placed at Point Y. It should not be placed between Valve 4 and the clamp because it would cause the sequence valve to open prematurely.

Figure 3-18. This is a variation of the preceding circuit for a double-acting clamp.

Note that Valve 4 which holds full pressure in the clamp while the work cylinder retracts, is a standard 3-position, double-piloted, closed center, 4-way valve. One side position is not used. Its pilot is vented to atmosphere through an air breather to keep out dirt.

Circuit action is as follows: The operator shifts and holds Valve 1 to start the cycle. Air flows to the clamp through check Valve 5. The clamp extends and bottoms out. Pressure rises behind its piston and opens sequence Valve 2, caus-

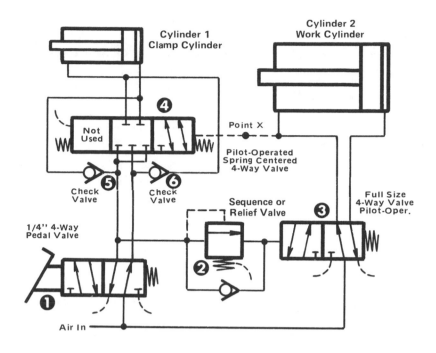

FIGURE 3-18. Clamp and Work Circuit, Double-Acting Clamp Cylinder. Automatic Sequencing in Both Directions.

ing Valve 3 to shift and start the work cylinder forward. Also, the pilot of Valve 4 vents and that valve returns to its center position. (Note: Valve 4 is held in its shifted position before the cycle starts. It must be capable of full pressure on its exhaust ports). Both cylinders are now extended and stalled. Check Valves 5 and 6 maintain full clamping pressure on Cylinder 1 while Valve 4 is in center position.

To retract the cylinders the operator releases Valve 1. This vents the pilot of Valve 3 allowing that valve to return to normal position, causing Cylinder 2 to retract. The clamp remains holding. Valve 4 should not shift while Cylinder 2 is retracting but if it does, a small sequence or relief valve (by-passed with a check valve) can be placed at Point X, and adjusted to a very low pressure. See box on Page 62 for use of a relief valve if an air sequence valve is not available. When Cylinder 2 stalls, pressure rises and shifts Valve 4, allowing Cylinder 1 to retract and the cycle is ended.

Multiple Cylinder Operation With Sequence Valves. Figure 3-19. Sequence valves can be used to program the sequential operation of any number of cylinders. Each sequence valve can be adjusted to any desired operating pressure for starting the next cylinder, and this opening pressure can be either higher or lower than that of any other sequence valve in the circuit. Sequence valves do not have to be adjusted to successively lower opening pressures. The only requirement is that inlet pressure to the system must be higher than the setting of any sequence valve.

A sequence valve does not rob system pressure from the load as a relief valve would do in the same circuit. When all sequence valves have opened (Valves 6, 7, and 8 on the diagram), full system pressure is available at the end of the line to Cylinder 4 just as it is to all other cylinders. With relief valves there would be a successive pressure loss through each valve.

In the diagram, at any given time during the sequencing operations, the pressure behind the pistons of all cylinders which have already been actuated is that of the highest pressure set on any of the sequence valves which have opened at that time. When all cylinders have extended and stalled, pressure on all cylinders reaches full system level. There is no loss of pressure through the sequence valves at this time.

Each cylinder in the string has priority on the air flow over all cylinders which follow it. The first cylinder in the string, Cylinder 1, has priority over all other cylinders.

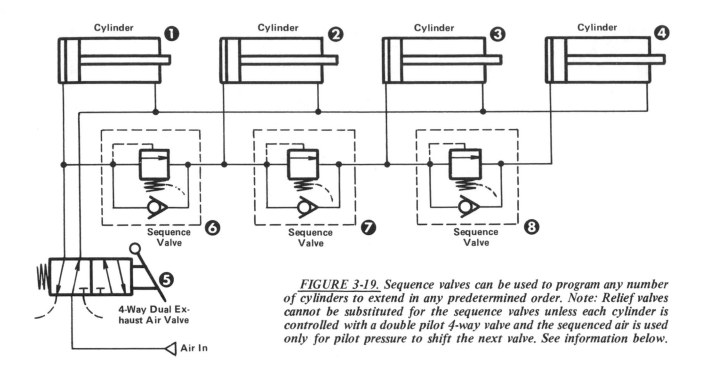

FIGURE 3-19. Sequence valves can be used to program any number of cylinders to extend in any predetermined order. Note: Relief valves cannot be substituted for the sequence valves unless each cylinder is controlled with a double pilot 4-way valve and the sequenced air is used only for pilot pressure to shift the next valve. See information below.

SUBSTITUTING A RELIEF VALVE FOR A SEQUENCE VALVE

Air sequence valves may be hard to find. A relief valve with check valve shunt will, in most circuits of 80 to 100 PSI give the same results provided the air passing through the relief valve is used to pilot a 4-way valve instead of going directly to the cylinder.

This circuit is Figure 3-17 with a relief valve in place of the sequence valve and with a check valve installed externally.

On a 100 PSI system, for example, the relief valve can be set for about 50 PSI cracking pressure. On start-up, air will not pass the relief valve until the clamp comes up to 50 PSI. Then, Valve 3 will not shift until the air passing through the relief valve to its pilot has exceeded 50 PSI by the amount needed for pilot pressure. When the machine is being set up, the relief valve can be trimmed to cause Valve 3 to shift when clamping pressure has reached 80 to 90 PSI, the pressure at which a sequence valve would be set, and Cylinder 2 can operate on the full 100 PSI system pressure.

Limitations. A relief valve must always feed into the pilot of the directional valve which controls the second cylinder, never directly to the cylinder. In Figure 3-19 above, the circuit would have to be re-arranged to use a 4-way valve, single pilot, spring return type, for each cylinder. with pressure from each relief valve feeding into the pilot of the next 4-way valve.

Pressure relief valves for air are not as readily available as they are for hydraulics. If an air relief valve of suitable size cannot be found, a low range hydraulic relief valve can be used if the air line is well lubricated.

In Figure 3-19 the cylinders extend one at a time in regular order. Cylinder 1 first, followed by Cylinders 2, 3, and 4. The order in which they extend can be changed by re-connecting the sequence valves. All cylinders start retraction at the same time when 4-way Valve 5 is reversed.

These cylinders can be programmed to retract in a regular sequence by placing additional sequence valves in the lines to their rod ports. Note: If return sequence valves are installed, and when Valve 5 is reversed all cylinders are left "floating". Even though pressure is not applied to them until the sequence valves open, their blind ends become vented to atmosphere through Valve 5, and they could be pushed back slightly if they are working against a reactive load. See Figure 3-15 for additional explanation.

For speed control of cylinders working in sequence valve circuits, a more positive sequencing action is usually obtained by connecting their speed control valves for meter-in rather than meter-out control.

SEQUENCING AIR CYLINDERS WITH SOLENOID VALVES

Types of Solenoid Directional Control Valves. Figure 3-20. A circuit designer has a choice of three valve types, each having a different action. Each has standard 4-way action and can be used for control of single or double-acting cylinders. One of these types also has a center neutral position. Each requires a different mode of electrical control, and usually the choice between them is made in favor of the one which can be designed into the simplest electrical circuit. Some of the information in this section on solenoid valves and electrical sensors may also apply to hydraulic as well as to air circuits.

Single Solenoid, 2-Position Valve. The spool shifts when the solenoid is energized. Return of the spool to its original position is by means of a spring or externally applied pilot pressure after the solenoid has been de-energized.

Current must be maintained on the solenoid to keep the spool in a shifted position, and control circuits must be designed with this in mind.

Single Solenoid Valve with Spring Return

Double Solenoid, 2-Position Valve. The spool shifts when a solenoid is energized, and will remain in the shifted position even after the solenoid becomes de-energized. To return the spool to its original position, the opposite solenoid must be momentarily energized. Most valves of this type have a detent on the spool to prevent accidental spool drift after the solenoids are de-energized. However, as a matter of extra safety we recommend that the valve be mounted with spool axis horizontal, or that current be maintained on a solenoid until time to shift the spool in the opposite direction.

Double Solenoid Valve, 2-Position, No Neutral

Double Solenoid, 3-Position Valve. Centering springs push the spool to a center neutral position after both solenoids have been de-energized. To keep the spool shifted to a side position, current must be continuously held on the appropriate solenoid. Electrical circuits must be designed with this in mind.

Double Solenoid Valve, 3-Position, Spring Centered to Neutral

FIGURE 3-20. Solenoid Valve Types

The outward appearance of a double solenoid valve may not reveal whether it has only two working positions or whether it also has a center neutral. This information can be obtained from the model code on its nameplate. Or, an end cap can be removed to determine if centering springs are present. The spool can be pushed to see if there is a spring on the opposite end. Blowing through the pressure port will also reveal if there is a center neutral.

Electrical Sensors. Figure 3-21. The limit switches, pressure switches, and timers shown here are the ones used more often for control of fluid power solenoid valves. Many other sensing and switching devices are available for circuitry. These include photocells, sonic sensors, proximity switches, motion detectors, temperature switches, electric counters, and others. For more information on the sensors mentioned here, refer to the Womack book "Electrical Control of Fluid Power".

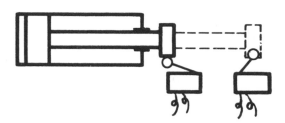

FIGURE 3-21(A). Side Actuation.

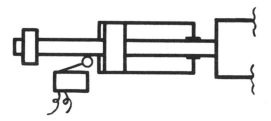

FIGURE 3-21(B). Double-End-Rod Cylinder.

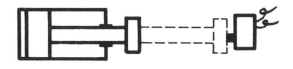

FIGURE 3-21(C). Head-On Actuation.

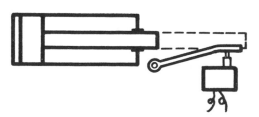

FIGURE 3-21(D). Ramp Actuation.

Side Actuation. Figure 3-21(A). Switches are mounted so they can be actuated by cross movement of a cam or cams mounted on a moving part of the machine. Switches with roller actuators are normally used. However, any side play in the mechanism or cam actuator will cause the actuation point of the switch to vary.

Actuation From a Rod Extension. Figure 3-21(B). On enclosed machines where there is no access to any moving part of the machine nor to the front end cylinder rod for mounting a limit switch, a double-end-rod cylinder can be purchased, and cams can be mounted on the rear rod extension.

Head-On Actuation. Figure 3-21(C). Useful where a greater sensitivity to small cylinder movements is needed. The limit switch should have a button or stem actuator. Of course a positive stop should be provided to protect the switch from accidental damage in case the cylinder should over-travel. Or, the switch can be mounted in a spring loaded plate which will be pushed harmlessly out of the way if the cylinder should travel too far.

Ramp Actuation. Figure 3-21 (D). Can be used with any kind of standard limit switch to keep side load off the switch stem. A gentle ramp will actuate the switch more slowly than direct actuation and prolong its life. The problem of snap-back in the switch mechanism is also eliminated.

Ramps are also useful with mechanically actuated valves which have a stem actuator.

Pressure Switches. Figure 3-21(E). They detect a rise of pressure in a fluid circuit and produce a switching signal. They are teed into a cylinder line to actuate when the system pressure has built up to their adjustment setting. They are usually provided with one set of transfer-type contacts which will give either a "make-on-rise" or a "break-on-rise" electrical switching signal. Most pressure switches used in industry have an adjustable setting.

Re-Set Timer. Figure 3-22. From the wide variety of various kinds of timers we will deal only with the "re-set" type, which is used in fluid power circuits to obtain a "dwell" at some point in the cycle.

The coil, M, is a small clock motor. When energized it starts the time cycle. At the end of the set delay period the timer contacts are actuated to give a switching signal. Most re-set timers have at least one N.C. (normally closed) and one N.O. (normally open) set of contacts. When M is de-energized, the timer re-sets itself to zero, ready for the next cycle. On fast-cycling machines, the time required for the timer to re-set should be considered. These timers are available in many timing ranges, with adjustable delay period.

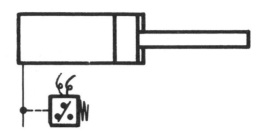

FIGURE 3-21(E). Pressure Switches.

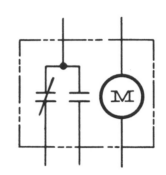

FIGURE 3-22. Electrical Re-Set Timer

Operator Controls. The more popular controls are momentary pushbuttons, maintained pushbuttons (consisting of a "START" and "STOP" button mechanically latched together), foot switches, rotary selector switches, and toggle switches having either two or three active positions.

Starting Point in the Design of Electrical Control for Fluid Power Air Circuits. A compressed air cylinder is usually operated from one end of its stroke to the other, and allowed to stall against a positive stop or against its own end cap. Air pressure is left on its piston and it is firmly held in position. Then it is run to a positive stop in the other direction. Only on unusual circuits should an air cylinder be stopped at a functional position in mid-stroke. For one thing, it is virtually impossible to stop it at a precise point. When stopped it is unstable. It is always seeking a balance between load reaction and air pressure behind its piston. If load is added or removed, the piston will automatically move to a new balance position.

In view of this characteristic of an air cylinder, circuits should be designed to use 2-position solenoid valves, either single or double solenoid types. A 3-position valve, with center neutral is seldom used.

When choosing between single and double solenoid valves, remember that electrical circuitry is often simpler with double solenoid models, where an impulse will shift the spool. Holding relays with their more complicated circuitry can sometimes be eliminated.

We suggest, then, that in the first stages of design, this type valve be considered first, to achieve circuit simplicity. Later, if the circuit is workable with single solenoid valves, they can replace the others.

On the other hand, hydraulic circuits are nearly always designed with 3-position valves with some kind of center neutral. Hydraulic circuit design will be covered in a later section.

ELECTRICAL SEQUENCING METHODS FOR AIR CYLINDERS

Sequencing is defined as the automatic starting of a second cylinder after the first one has supplied a switching signal by reaching a desired position in its travel. The switching signal is usually derived from one of the sensors previously described. The choice of sensor (limit switch, pressure switch, timer, etc.) depends on the kind of circuit action desired. There are three general methods commonly used for programming cylinders in a sequential order of operation:

(1). Position Sequencing. Figure 3-24. In this kind of operation, a switching signal is produced when the first cylinder arrives at a pre-determined position in its stroke which may or may not be its complete stroke. This signal starts the second cylinder into motion. Both cylinders eventually retract automatically to their starting positions and stop, waiting for the operator to start the next cycle.

The circuit below can be used for any sequencing program between two cylinders, and is set up for the difficult "clamp and work" sequence shown in the box on Page 56. (All limit switches may not be required for some sequences). One difficulty in a clamp and work circuit is that Cylinder 1 stands on Switch 2-LS, keeping Solenoid C on Valve 2 energized. This circuit must be broken before time for Solenoid D to be energized or Valve 2 cannot be reversed. In this circuit pressure Switch 1-PS performs the function of breaking the circuit to Solenoid C as soon as Cylinder 2 starts to advance.

A 2-position 4-way solenoid with dual exhaust ports controls each cylinder. If speed controls are required they can be placed in the valve exhaust ports or at Points X - X and Y - Y. Check Valve 4 permits pressure to actuate the pressure switch when Cylinder 2 starts to extend, and check Valve 3 vents the pressure switch at the end of the cycle. Needle Valves 5 and 6 are optional and can be adjusted to slightly delay the closing and opening of the pressure switch if this should be required.

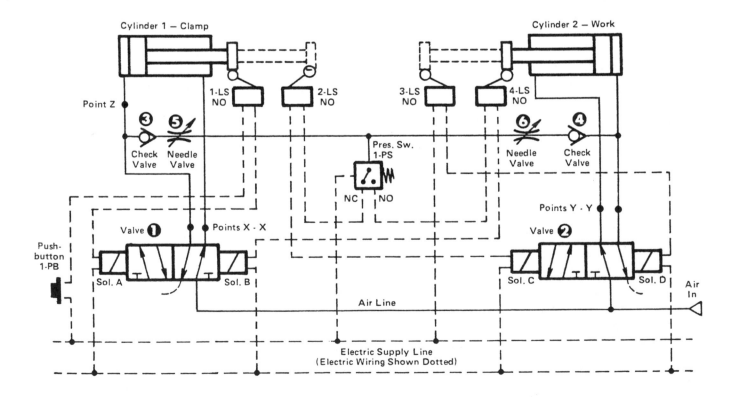

FIGURE 3-24. A general purpose position sequencing circuit – for setting up any cylinder sequence.

66

A limit switch is placed at each end of the stroke on both cylinders. In this circuit, Switch 1-LS disconnects the pushbutton as soon as Cylinder 1 has advanced a short distance; Switch 2-LS energizes Solenoid C through the contacts of pressure Switch 1-PS; Switch 3-LS actuates Solenoid D to retract the work cylinder; Switch 4-LS is held closed before the cycle starts. It opens when Cylinder 2 has advanced a short distance. When Cylinder 2 retracts, 4-LS closes and energizes Solenoid B to retract the clamp cylinder.

When Pushbutton 1-PB is momentarily pressed, Solenoid A becomes energized and Valve 1 shifts. Cylinder 1 advances and actuates 2-LS. Solenoid C on Valve 2 becomes energized and Cylinder 2 starts forward, releasing 4-LS. Cylinder 2 advances and actuates 3-LS. Solenoid D becomes energized and shifts Valve 2. The work cylinder retracts and actuates 4-LS. Solenoid B becomes energized and Valve 1 shifts. Cylinder 1 retracts and the cycle is ended.

(2). Pressure Sequencing. A second method of electrically sequencing two cylinders is to delay the start of the second cylinder until the first one has advanced far enough to build up pressure behind its piston to a pre-adjusted level. The actual distance traveled by the first cylinder is of no consequence. This method is used on presses working on parts of varying thickness such as unfinished castings. For an application of this kind it would be difficult to use a limit switch for sequencing. A pressure switch, teed into the port of the clamp, Cylinder 1, gives an electrical switching signal when clamping pressure has reached the proper level.

The circuit of Figure 3-24 can be modified to convert to pressure sequencing. Limit Switch 2-LS can be removed and the wiring to this switch can be connected to the COM and N.O. terminals of a pressure switch teed into the blind end of Cylinder 1 at Point Z. The rest of the circuit can be identical to Figure 3-24.

(3). Time Sequencing. Figure 3-25. Still a third method of sequencing two cylinders is with an electrical timer of the re-set type, the kind described on Page 65. With this method a signal to start the

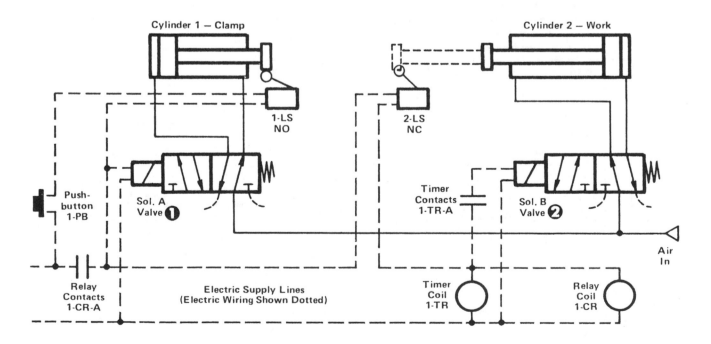

FIGURE 3-25. A time delay sequencing circuit using an electrical re-set timer.

second cylinder is obtained from a timer when a pre-set time interval has elapsed after the first cylinder has been started. However, time sequencing may waste production time because the time interval must be set longer than the estimated time required for the first cylinder to make its stroke to be certain it will have time to complete its stroke before the second cylinder starts. It may prove to be the most practical sequencing method on some applications either because

there is no practical way to use limit or pressure switches. It may be the ideal method on an application where a dwell period must be programmed into the cycle before the second cylinder starts.

Figure 3-25 is the fluid and electrical circuit for a simple time delay sequencing application with the program shown in the box. The cycle is started when an operator momentarily presses Pushbutton 1-PB. Solenoid A becomes energized and Valve 1 shifts, starting Cylinder 1 forward. Also, Relay 1-CR and Timer Motor 1-TR become energized through the NC contacts of 2-LS. A set of relay contacts, 1-CR-A, locks the relay closed and holds current on Solenoid A after the pushbutton has been released. Cylinder 1 starts forward and releases Switch 1-LS. This disconnects the pushbutton for the remainder of the cycle. Cylinder 1 advances and stalls against the work. After the time period set on the timer has elapsed, timer Contacts 1-TR-A close. This causes Solenoid B to become energized and Valve 2 to shift. Cylinder 2 advances until it actuates Switch 2-LS. This de-energizes both the relay and the timer coil, causing both solenoid valve coils to become de-energized and both cylinders to retract.

Relief Valve Sequencing With Electrical Control. Figure 3-26. Another position sequencing circuit using an air relief valve for programming Cylinder 2 to wait until pressure behind the piston of Cylinder 1 has built up to clamping pressure. Please read limitations of a relief valve as a substitute for a sequence valve in the box on Page 62.

Valve 1 is a double solenoid, 2-position, 4-way valve. Valve 2 is a single pilot, spring return 4-way

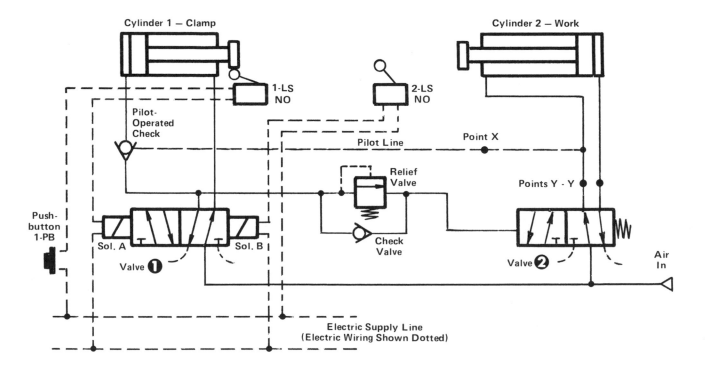

FIGURE 3-26. Using an air relief valve as a substitute for a sequence valve in programming two cylinders.

valve. Switch 1-LS is wired to disconnect the pushbutton after the cycle starts. The pilot-operated check valve retains full air pressure in the clamp while the work cylinder is retracting. At home, pressure builds up on the work cylinder and opens the P.O. check valve to release Cylinder 1. To prevent the check valve from opening while the work cylinder is retracting, it may be necessary to add a restriction in the pilot line. This can be a small, low range, relief valve (by-passed with a check valve) installed at Point **X** and discharging into the P.O. check. Set the relief valve no higher than necessary.

The cycle starts when the operator momentarily presses the pushbutton. Cylinder 1 advances. At stall, pressure builds up and goes across the relief valve to the pilot of Valve 2. Cylinder 2 advances and actuates 2-LS. This energizes Solenoid B to reverse Valve 1. Shifting of Valve 1 vents the pilot of Valve 2, allowing that valve to return to its normal position. Cylinder 2 retracts while Cylinder 1 remains clamped until the pilot-operated check valve releases its blind end allowing it to retract. For speed control of Cylinder 2, install flow control valves at Points **Y** - **Y** in a meter-in mode.

PRODUCING SHORT DURATION AIR PILOT SIGNALS

Figure 3-27. A circuit can sometimes be simplified by using a short duration air pilot signal (or impulse) to shift a pilot-operated valve. Impulses can solve the problem of a cylinder (Cylinder 1) standing on a valve (Valve 3), which provides pilot pressure to shift another valve (Valve 2) preventing that valve from being reversed later in the cycle.

Cylinder 1 is controlled with 4-way Valve 1, a manual valve with pilot pressure return. Cylinder 2 is controlled with 4-way Valve 2, a double piloted valve shifted by impulses.

The operator shifts Valve 1 to start the cycle. Cylinder 1 advances and actuates Valve 3. This connects air pressure to the pilot of Valve 2 for immediate shifting, starting Cylinder 2 forward. Pressure also builds up slowly in the VOL chamber through the needle valve, until high enough to shift Valve 5. This cuts off pilot pressure from Valve 2 and it can be reversed by the other pilot later in the cycle. Cylinder 2 advances until it actuates Valve 4. This connects pilot pressure to Valves 1 and 2 to return them to their original positions. Both cylinders retract, stall at home positions, and the cycle is ended.

The advantage of this method over impulse bleed buttons is that reliable shifting is obtained even at high cylinder speeds if the needle valve is properly adjusted.

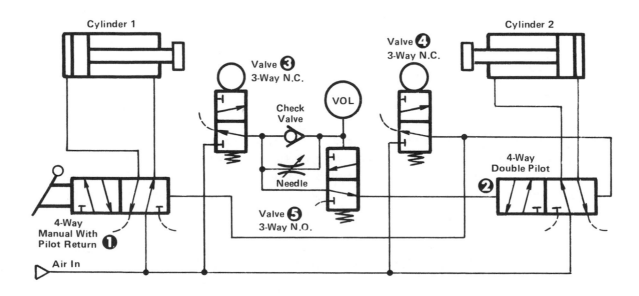

FIGURE 3-27. Producing "1-shot" impulses for shifting a 4-way valve.

SEQUENCING WITH FLOW CONTROL VALVES

Simple flow control valves, properly placed, will hold back or slow down one cylinder and give priority of flow to another. This action, if adequate for the application, is less expensive and easier to install than sequence, button bleeder, or cam valves. Good applications are for cylinders moving a latch, lever, or dog, where there is very little resistance to cylinder movement in mid-position travel and where the load has no spring-back.

Figure 3-28. Flow control Valves 2 and 3 are placed so one cylinder will be delayed on its forward stroke and the other on its return stroke.

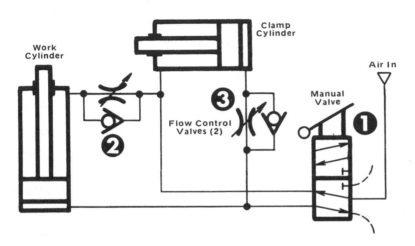

FIGURE 3-28. Simple Sequencing With Flow Control Valves.

When Valve 1 is shifted to start the cycle, air is immediately connected to the blind end of both cylinders, but flow control Valve 2 slows down the work cylinder so it moves only a short distance while the clamp moves rapidly to complete its stroke.

When Valve 1 is reversed, flow control Valve 3 delays retraction of the clamp while the work cylinder is retracting The work cylinder should be lightly loaded while retracting.

Index Table. Figure 3-29. (See diagram on next page). A momentary electrical signal to Solenoid B shifts Valve 3 to start a cycle. Air flows into the rod end of both cylinders. The small lock cylinder quickly retracts to pull the lock pin, then the index cylinder, slightly delayed by flow control Valve 4, retracts and moves the table to its next position. When limit Switch 6-LS becomes actuated, it energizes Solenoid A to reverse Valve 3. Air is then ported to the blind ends of both cylinders. Again, the small lock cylinder acts quickly to insert the lock pin while the index cylinder, slightly delayed by flow control Valve 5, extends to its starting position ready for the next cycle. Each flow control valve is adjusted for the best indexing action.

While flow control valves may give satisfactory indexing action, sequence valves can be substituted in their places in Figure 3-29 if a more positive interlock is needed between the cylinders.

Caution! Flow control valves do not give the same positive priority action obtained with sequence valves. The first cylinder to act, even though it may make its full stroke before the other one starts, will not come up to full pressure until both cylinders have completed their strokes. If the lock pin should hang, Cylinder 1 would still try to move. Lock pins should have a slight taper. Meter-in speed control may work better than meter-out control in some circuits.

SPEED CONTROLS IN SEQUENCING CIRCUITS

Special attention should be given to the method of speed control used for any cylinder operating in a sequence circuit where a sharp rise in pressure behind its piston at stall-out is used to trip a pressure

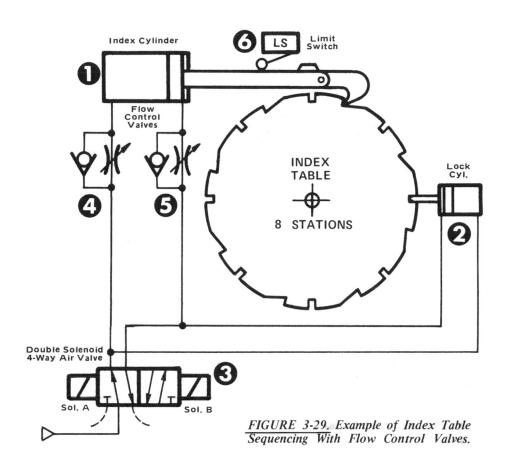

FIGURE 3-29. *Example of Index Table Sequencing With Flow Control Valves.*

switch as in the basic electrical circuit of Figure 3-25, Page 67, or to open a sequence valve as in the basic sequencing circuit of Figure 3-13, Page 58.

Although for many other applications meter out speed control is preferred because of its stability, in the applications mentioned above it is usually, but not always necessary to use meter-in speed control to obtain a pressure rise sharp enough for dependable sequencing. A rise of at least 20% as the piston stalls is recommended for stability.

SPEED CONTROL LOCATION ...

Speed Control Location With Pressure Switch. Figure 3-30. To control extension speed, a flow control valve can be placed at "**X**" to meter air into the blind end port of the cylinder. It is important that the pressure switch be connected between the flow control valve and the cylinder port.

Speed Control Location With Sequence Valve. Figure 3-31.

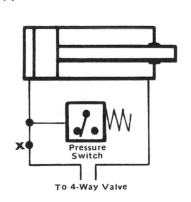

FIGURE 3-30. *Best Location for Speed Control With Pressure Switch.*

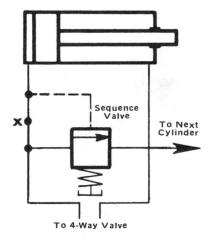

FIGURE 3-31. *Sequence Valve With External Pilot Connection.*

A meter-in speed control system should be used, with flow control valves placed at "**X**" to meter air into the blind end cylinder port. The sequence valve inlet should be ahead of the flow control so the control will not restrict air flow into other circuits when the sequence valve opens. Pilot pressure to open the sequence valve should be taken directly from the cylinder port.

If a relief valve is substituted for a sequence valve as shown in the box on Page 62, air flow through the relief valve will be very low, only enough for pilot pressure. Cylinder speed should be controlled with a standard flow control valve in the cylinder line connected for meter-in control.

4

Automatic Reciprocation

OF AIR CYLINDERS

DEFINITIONS

Cylinder Cycle. A complete cycle on a cylinder includes both the forward and return stroke back to the starting position. The forward stroke is a half cycle and the return stroke is the other half cycle.

Reciprocation. The term implies that a cylinder, when once started by an operator, will go through at least one complete cycle automatically without further action by the operator. In most cases this means that at the end of the forward stroke, there is some provision for retracting the piston automatically, and this could be done with any one of a number of devices such as a limit switch, a pressure switch, a cam valve, a sequence valve, a button bleed valve, or an electrical timer. Reciprocation may be one of two kinds:

One-Cycle Reciprocation. When the cycle is started by an operator the cylinder goes through one cycle and stops at its original position. The next cycle must be initiated by an operator.

Continuous Reciprocation. When once started, the cylinder continues to reciprocate, cycle after cycle, indefinitely until stopped by an operator.

RECIPROCATION WITH BUTTON BLEEDER VALVES

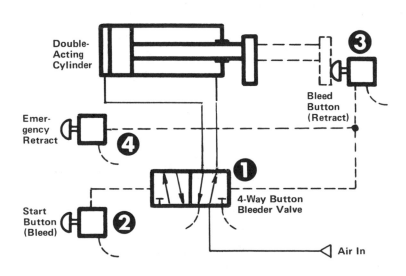

FIGURE 4-1. *Example of One-Cycle Reciprocation.*

One-Cycle Reciprocation. Figure 4-1. The student may wish to review the action of button bleeder valves starting on Page 53. Additional information will also be found in Volume 1, Chapter 4.

Both bleed buttons should be unscrewed from the body of the main valve. One of them, Button 2, can be mounted as a "START" pushbutton for the operator. The other, Button 5, can be mounted where it will be "bumped" by a moving part of the machine at the end of the forward stroke of the cylinder. Buttons should be connected to the main valve end caps with 1/4" I.D. hose or tubing.

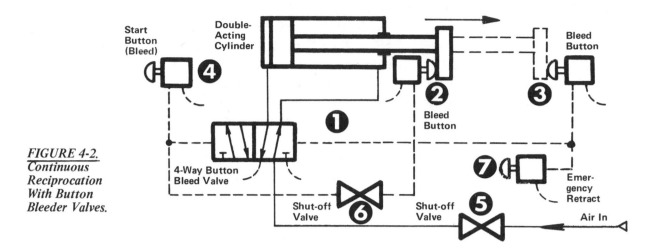

FIGURE 4-2.
Continuous
Reciprocation
With Button
Bleeder Valves.

Figure 4-2. Continuous Reciprocation. Bleed Buttons 2 and 3 are placed at the stroke reversal points. If placed for head-on actuation, positive stops must be provided to prevent damage to the buttons if the cylinder should overtravel.

This circuit is suitable for either continuous or one-cycle reciprocation. For continuous cycling, Valve 6 must be left open. Air power is admitted by opening line Valve 5. In some circuits the cylinder may not immediately start if it is not standing on a button, because master Valve 1 spool may have jumped to a center stall position when air was suddenly admitted from the line. Refer to Volume 1, Chapter 4 for discussion of this problem. Pressing Button 4 will immediately start the cylinder.

For one-cycle reciprocation, Valve 6 must be closed. Valve 5 is then opened to admit inlet air. When the operator presses start Button 4, the cylinder makes a forward and reverse stroke and stops.

CYLINDER RECIPROCATION BY ELECTRICAL MEANS

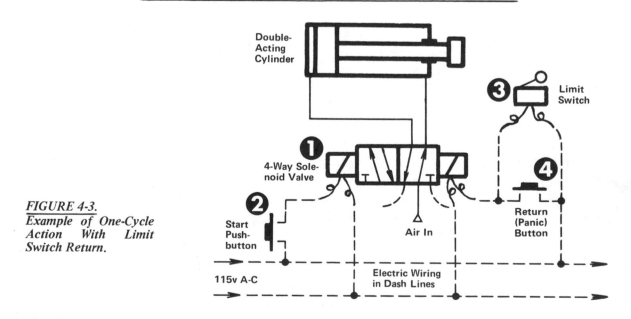

FIGURE 4-3.
Example of One-Cycle
Action With Limit
Switch Return.

One-Cycle Reciprocation, Limit Switch Return. Figure 4-3. This circuit uses a pushbutton for starting the cylinder and a limit switch for return, with a double solenoid, 2-position, 4-way valve.

When Pushbutton 2 is momentarily pressed, Solenoid A becomes energized and Valve 1 shifts. This causes the cylinder to start its advance. At the end of the forward stroke a cam on the cylinder or mounted elsewhere on the machine actuates limit Switch 3. This action energizes Solenoid B and Valve 1 returns to its original position, causing the cylinder to retract. The cylinder stops at its home position and remains there under full air line pressure. This ends the cycle.

Valve 1 is a double solenoid valve with no center neutral and no return springs. When momentarily actuated, as by a momentary pressure on a pushbutton, its spool will shift and will remain in shifted position even though the pushbutton is released.

The piston of the cylinder, when it has once started forward, cannot be stopped in mid-stroke by the operator (except by cutting off its air supply). We recommend the addition of a second pushbutton, 4, as a safety feature, to permit the operator to instantly reverse the cylinder during its forward movement if it should become necessary. Pushbutton 4 is wired in parallel with limit Switch 3, and has the same action in the circuit as if the limit Switch 3 had been actuated. It is sometimes furnished with a red button, and is usually referred to as a "panic button".

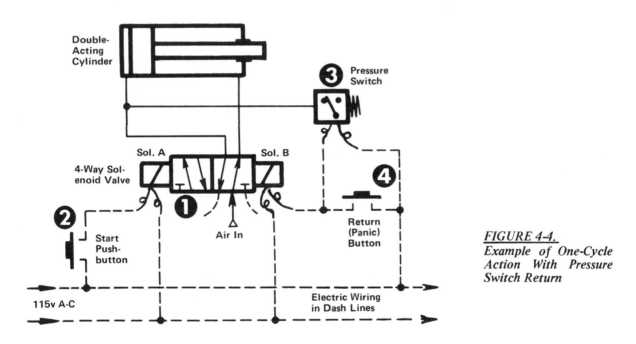

FIGURE 4-4.
Example of One-Cycle Action With Pressure Switch Return

One-Cycle Reciprocation, Pressure Switch Return. Figure 4-4. This circuit has a pushbutton for starting the cylinder advance and a pressure switch for starting its retraction. The pressure switch may be preferred to a limit switch used in the preceding circuit on those applications where the requirement is for the cylinder to reach a certain force level rather than a certain length of stroke before it retracts. The directional valve is a double solenoid model with no center neutral.

Circuit action is similar to that described in the preceding circuit. Care must be used in adding speed controls to this pressure sensitive circuit. Refer to information on Pages 44 and 45.

Safety. Please note that an air cylinder which is in motion cannot be instantly stopped by cutting off its air supply. It will continue to travel until the air pressure behind its piston expands and reaches a balance with the load against the piston rod. It may travel an inch or two depending on its speed and the nature of the load. A panic pushbutton, No. 4 in the figure, should be included on all electrically controlled air circuits, and will cause the cylinder instantly to reverse. Cylinders controlled with manually operated 4-way valves will instantly reverse if the valve is shifted to its opposite side position.

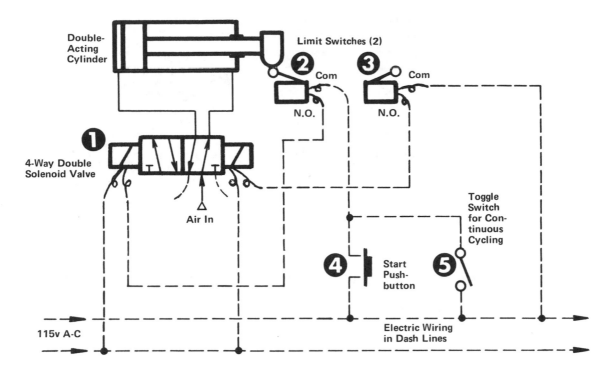

FIGURE 4-5. This circuit may be used for either continuous or one-cycle reciprocation.

Continuous Reciprocation With Limit Switches. Figure 4-5. This circuit is arranged so the operator can get one-cycle or continuous reciprocation, depending on whether Switch 5 is open or closed.

For continuous reciprocation, Switch 5 must be closed. When the air supply is first turned on, the cylinder will cycle back and forth between stroke limits set by the position of Switches 2 and 3. Several features of this circuit are of particular interest:

(A). Limit Switches 2 and 3 may be positioned to reverse the cylinder before it reaches stall at the end of its stroke, forward and reverse.

(B). If the electric control current is cut off while the cylinder is traveling in either direction it will not immediately stop; it will continue in the same direction until reaching stall in that direction. In doing this it may override the limit switch if the switch happens to be located at a point ahead of stall position.

(C). Figure 4-6. If a cylinder can override a limit switch, its cam must keep the switch actuated during the override travel. If not, the circuit will not reactivate itself when Switch 5 is again closed.

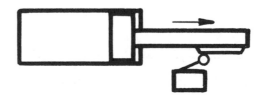

FIGURE 4-6. Cam must be long enough to keep limit switch actuated if cylinder overrides it.

For one-cycle operation, Switch 5 is left open. When the cylinder is started with Pushbutton 4, it will make a forward and reverse stroke, travel to the reverse stall position and stop.

Reciprocation With Pressure Switches. Pressure switches may be substituted for limit Switches 2, 3, or both in Figure 4-5. Reversal of a cylinder with a pressure switch is described in Figure 4-4. Action may be operator's choice of one-cycle or continuous by opening or closing Switch 5.

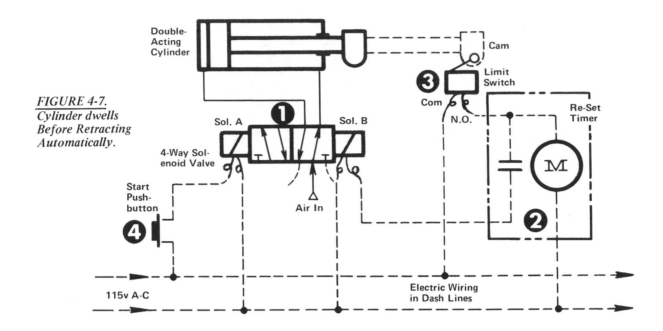

FIGURE 4-7.
Cylinder dwells
Before Retracting
Automatically.

Reciprocation With Dwell. Figure 4-7. The cylinder, after extending, dwells at the end of its forward stroke before retracting automatically. The action of a re-set timer is covered on Page 65.

Pushbutton 4 starts the cycle. Valve 1 shifts and the cylinder starts forward. Speed control during the forward stroke is discussed later. When the cam on the piston rod (or on a moving member of the machine) actuates limit Switch 3, an electrical circuit energizes timer motor, M, and starts the dwell timing. At the end of the dwell period, the timer contacts close. This causes the right solenoid of Valve 1 to become energized through Switch 3. The valve spool shifts back to its original position (provided the operator has released Pushbutton 4). The cylinder retracts to home position and stalls with full pressure against its own end cap. This ends the cycle.

Because cylinder reversal does not take place immediately when Switch 3 is actuated, a positive stop must be provided for the cylinder to stall against during the dwell period.

Note: When designing circuits using a re-set timer, be sure to allow enough time for the timer to re-set back to zero before initiating another cycle. Usually the timer will re-set rapidly, in less time than it takes for the cylinder to make its return stroke. However, on very short strokes and/or very rapid cycling, this point should be considered.

Speed Control in Reciprocation Circuits. Speed control valves should normally be installed for both directions of movement in most air circuits. Various speed control methods have been shown on Pages 42 through 45 for air cylinders. As a rule, meter-out methods of Figures 2-27 and 2-28 will give greater stability. However, in Figures 3-16 through 3-19 where a pressure rise behind the piston at stall is used as a signal to reverse the cylinder, meter-in speed control methods are preferred. Suggested placement of these controls for pressure switch and sequence valve applications is shown in Figures 3-30 and 3-31. Flow control valves for air are described on Page 178.

Mixed Sensors. We should point out in the reciprocation circuits shown, whether electrical or other, identical sensors need not be used at both ends of the stroke. Reversal methods can be mixed. For example, a sequence valve could be used on the forward stroke to sense load build-up, and a cam valve or limit switch at the other end of the travel.

RECIPROCATION WITH SEQUENCE VALVES

If necessary, review the construction and action of air sequence valves starting on Page 57. Also, review the limitations on placing speed control valves in sequence valve circuits on Pages 44 and 45.

On preceding pages we have shown sequence valves used for delaying the start of a second operation (such as the advance of a work cylinder) until the first operation (such as the advance of a clamp cylinder) has been completed, and until pressure in the clamp cylinder has risen to an adjusted minimum level. On this page we are showing sequence valves for automatically reversing the same cylinder when pressure behind its piston has risen to the selected minimum level. A sequence valve is a useful alternative in a non-electrical circuit to pressure switches, limit switches, and solenoid valves.

One disadvantage to sequence valve operation is that opening pressure on the sequence valve must be set no higher than about 90% of the supply pressure to prevent the valve from prematurely opening. The cylinder is, therefore, robbed of about 10% of the supply pressure while it is in motion. When it stalls, pressure behind its piston will reach supply pressure level and open the sequence valve.

One-Cycle Reciprocation. Figure 4-8. When Valve 1 is shifted, the cylinder will start its advance and continue until it stalls either at the end of its stroke or when it encounters an overload. Pressure behind the piston will build up to full system pressure, pass across the sequence valve and return Valve 1 to its original position. The cylinder will automatically retract, stall at home position, and the cycle is ended.

Valve 1 is a manually operated air valve with pilot pressure return. It is sized to match the cylinder.

Valve 2 is an adjustable air sequence valve set to open at about 90% of the air inlet pressure. If an air sequence valve is not available, a relief valve, by-passed with a check valve, can be substituted. Refer to Page 62 for more information.

Usually flow control valves will be required for cylinder speed control. They can be placed at Points **X** in a meter-in mode. Refer to Page 44, Figure 2-29B.

Continuous Reciprocation. Figure 4-9. This cylinder is arranged to automatically reciprocate between two stall positions in its stroke which can be the complete stroke or at midpoints where load resistance builds up sufficient pressure to open a sequence valve.

Valve 1 is a 4-way, 2-position, double piloted valve. Valves 2 and 3 are adjustable air sequence valves, or they can be air relief valves as described on Page 62.

To start the cylinder, inlet air Valve 4 must be opened. The cylinder will reciprocate automatically between stall points until the air is cut off by closing

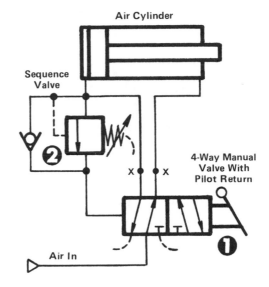

FIGURE 4-8. One-Cycle Reciprocation Circuit.

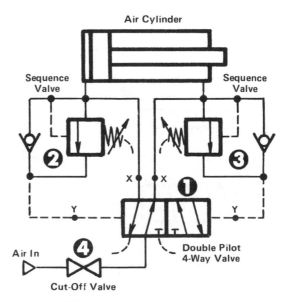

FIGURE 4-9. Continuous Reciprocation.

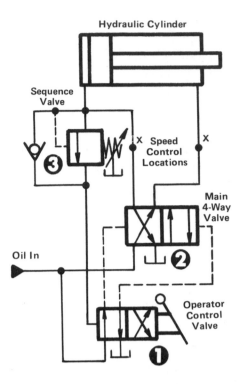

FIGURE 4-10. Hydraulic Reciprocation Circuit.

Valve 4. The cylinder will stop at a random point.

To cause the cylinder to complete its forward or return stroke before stopping, a cut-off valve can be placed at either or both Points **Y**.

As in the preceding figure, flow control valves for cylinder speed can be placed in a meter-in mode at Points **X**.

Caution! These circuits cannot be used on hydraulics. They work on air because of its compressibility, but on hydraulics, the spool of Valve 1 would stall at center position and would not complete its travel.

Hydraulic Reciprocation Sequencing. In the two preceding circuits, if used on hydraulics, the sequence valve could not cause the cylinder to retract. In Figure 4-8, for example, at the completion of the forward stroke, the sequence valve would open and admit hydraulic pilot oil into the pilot chamber of Valve 1. But when the spool of Valve 1 reached its mid travel point, hydraulic pressure would be cut off and the spool could not complete its travel.

Hydraulic Sequencing. Figure 4-10. For successful hydraulic operation, an intermediate pilot-operated valve must be interposed between operator control valve, 1, and the cylinder. Valve 2 is selected to match cylinder size. Valve 1 can be a miniature 4-way valve, manually operated, with pilot return.

The spool in main 4-way valve, 2, cannot stall in mid travel because, even if pilot pressure from the sequence valve should barely crack open the passage through Valve 1, Valve 2 spool could make its full shift.

RECIPROCATION WITH CAM VALVES

Cam valves are another non-electrical method of automatically retracting an air cylinder when the end of its stroke has been reached, or to cause it to continuously reciprocate between two cam valves placed at desired points of reversal.

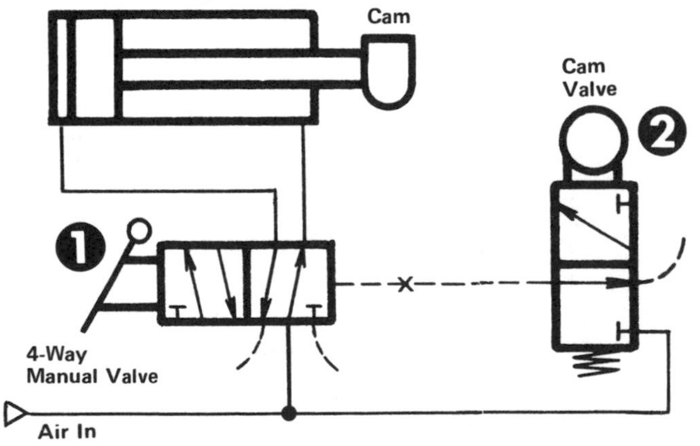

One-Cycle Reciprocation. Figure 4-11. The operator shifts manual Valve 1 to start the cylinder forward. The handle of this valve remains in its actuated position.

When the cam on the cylinder reaches and actuates the cam valve, line air pressure is directed into the pilot of Valve 1. Shifting of this valve causes the cylinder to retract and the valve handle to return to normal position.

Speed control can be handled as in the preceding sequence valve circuits. Continuous reciprocation can be produced by adding another cam valve at home position.

FIGURE 4-11. One-Cycle Reciprocation With a Cam Valve.

5
Miscellaneous Air Circuits

SYNCHRONIZING AIR CYLINDERS

Two (or more) air cylinders cannot be made to move precisely in step with each other by any easy means. In Chapter 2 it was pointed out that speed of an air cylinder depends mainly on how much overpowered the cylinder is with respect to the load resistance. Therefore, even if air flow to two cylinders could be divided evenly between them, they would still travel at different speeds according to how heavily each was loaded. However, fairly accurate synchronization can be obtained through the use of air power combined with oil metering. Several circuits are presented in Chapter 12 which will give good results.

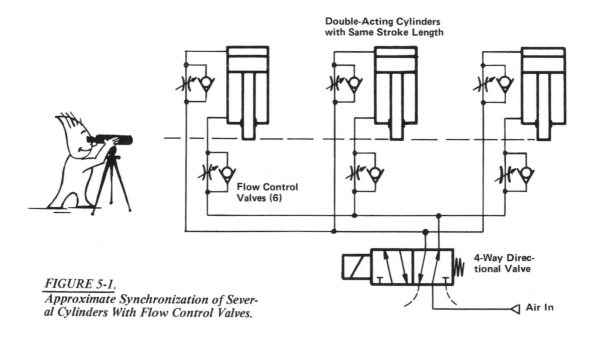

Double-Acting Cylinders with Same Stroke Length

Flow Control Valves (6)

4-Way Directional Valve

Air In

FIGURE 5-1.
Approximate Synchronization of Several Cylinders With Flow Control Valves.

Flow Control Synchronization. Figure 5-1. Where only approximate synchronization is required, a pair of flow control valves can be installed at each cylinder. Then, with the system in operation, the flow control valves can be experimentally adjusted until the cylinders travel pretty well together. As long as the loading on each remains the same they will stay reasonably close together provided they are allowed to bottom out at each end of their stroke. If the load changes on any one of the cylinders, the system must be re-adjusted.

Mechanical Yoking of Air Cylinders. Figure 5-2. The best way to make two air cylinders on the same machine travel together is to have them mechanically linked, as for example through a platen with sturdy guides. Other means may be devised for mechanically joining them together. Keep the following points in mind when designing mechanisms powered with air cylinders:

(1). Use one larger cylinder, mounted on the load centerline in preference to several smaller cylinders distributed over the platen and connected in parallel. This may mean that the platen will have to be mechanically reinforced to properly distribute the thrust, but this is better than trying to keep several small cylinders from cocking the platen, binding the guides,

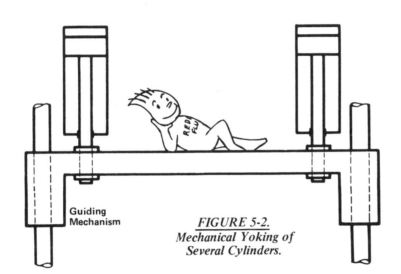

Guiding Mechanism

FIGURE 5-2.
Mechanical Yoking of
Several Cylinders.

(2). If multiple small cylinders must be used, mount them as close together as practical, and at equal distances from the load centerline.

(3). Make sure that all work is centered in the press as closely as practical.

DECELERATION OF AIR CYLINDERS

As the piston of an air cylinder reaches the end of its stroke, it will impact against the end cap. Most air cylinders are built to take a moderate amount of this, but in the case of unusually fast moving cylinders or those carrying a high momentum load, some means should be installed for reducing speed near the end of their stroke. A high momentum load is defined as one having heavy weight (and mass) but very little sliding friction. Repeated impact can cause damage of various kinds: fracture of the cylinder cap, piston, or tie rods; stretching of the tie rods causing barrel end seals to blow out; stripping of the threads on the piston rod where attached to the piston (or to the load); possible external damage to the load.

Impact damage can be minimized by installing external positive stops to keep the piston from hitting the end caps, but this does not altogether preclude damage to rod threads and external load. We recommend positive stops on every installation, where practical, whether the load be light or heavy. But in addition, a means of deceleration should also be included where extremely fast movement and/or massive loads are involved.

Cushioned Cylinder. Figure 5-3. Most cylinders can be purchased with a cushion on either or both ends. The cushions are manufactured as part of the cylinder and cannot be added in the field. They usually increase the overall length of the cylinder.

For hydraulic cylinders, cushioned ends are very effective, but on air cylinders they are of doubtful value while adding considerably to the cost. They become active when the piston is about one inch from the end of its stroke, and a 1-inch column of air is usually not very effective for absorbing the energy of high momentum loads. However, they may be of some value on high speed retraction of an unloaded cylinder, or for operation of low mass loads consisting largely of frictional resistance.

For maximum effectiveness, meter-out rather than meter-in speed control should be used so the air in the cushion pocket will be pre-compressed to some degree at the time cushioning starts.

A cross sectional view of a cush-ioned cylinder and a more complete description of the cushioning princi-ple will be found in Volume 1, Chapter 2.

Cam Valve Deceleration. Figure 5-4. A cam Valve 2 may be placed a short distance before the end of the stroke to start deceleration. Positioning of this valve is experimental, to find the point giving a suitable reduction of impact without unduly impairing the efficiency of the circuit.

In operation, while the cylinder is traveling forward, exhaust air from its rod end flows freely through cam

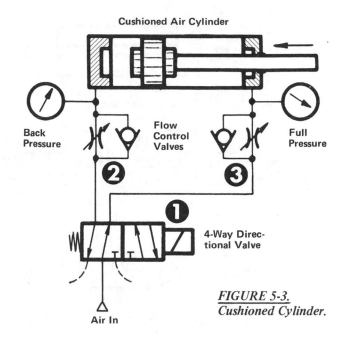

FIGURE 5-3.
Cushioned Cylinder.

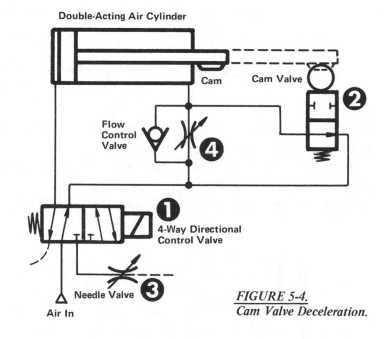

FIGURE 5-4.
Cam Valve Deceleration.

Valve 2. Steady travel speed is controlled by needle Valve 3 in the 4-way valve exhaust port. When cam Valve 2 has been actuated, exhaust air must now pass through flow control Valve 4, where it can be restricted to provide the degree of deceleration required.

If necessary to decelerate the cylinder in the return direction, another set of Valves 2, 3, and 4 must be installed in a similar circuit to cushion the cylinder at the end of its retraction stroke.

To adjust this circuit for deceleration, first close Valve 4 completely and adjust Valve 3 for the desired mid-stroke travel speed. Then, experimentally adjust the position of the cam valve and its flow setting for the amount of deceleration desired.

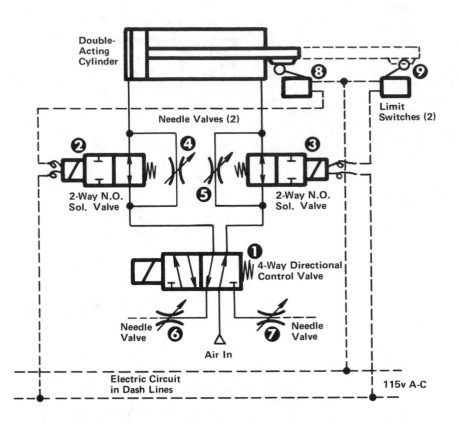

FIGURE 5-5. *Example of deceleration of an air cylinder at the end of its stroke when it is operated with solenoid valves and limit switches.*

Electrical Deceleration. Figure 5-5. On systems which use solenoid 4-way valves it may be easier to use electrical deceleration. This circuit provides for deceleration on both ends of the cylinder stroke. Limit Switches 8 and 9 should be placed experimentally at positions sufficiently ahead of the ends of the stroke to allow the cylinder to decelerate to a safe speed before impacting against the end caps.

Electrical wiring is shown in dash lines; compressed air lines are shown solid. Needle Valves 6 and 7 are placed in the 4-way valve exhaust ports to regulate the steady speed during mid-stroke travel. Directional Valve 1 may be either a single or a double solenoid model.

Limit Switches 8 and 9 are wired to deceleration solenoid Valves 2 and 3. These are 2-way normally open valves. They remain open for minimum flow restriction until energized through one of the limit switches. When they close, needle Valves 4 and 5 control deceleration by restricting exhaust air.

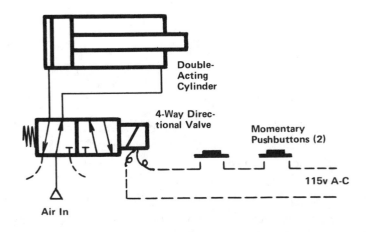

FIGURE 5-6. *Two-Hand Electrical Control.*

TWO-HAND CONTROL CIRCUITS

Circuits shown under this heading are not offered as all-purpose safety circuits. However, they do require the simultaneous actuation of two pushbutton switches or two air valves before a cylinder can start its cycle. They do offer a certain measure of safety on those machines manned by two operators, where each operator has a starting switch or valve located out of reach of the other operator. One operator is then prevented from accidentally starting the machine when the other operator is not ready.

For genuine safety circuits requiring one operator to use both hands to start a machine, please refer to the Womack book "Practical Fluid Power Control" under the heading of safety circuits.

Electrical Two-Hand Control. Figure 5-6. On circuits using solenoid 4-way valves, a pair of N.O. (normally open) pushbuttons may be wired in series so both will have to be pressed at the same time in order to energize the solenoid valve. If either button is released during forward travel, the cylinder will immediately retract. For additional safety, use recessed-type pushbuttons and mount them vertically so they cannot be weighted down with a heavy object.

Non-Electrical Two-Hand Control. Figure 5-7. The main Valve 1 can be sized as large as necessary to handle the air flow. It is a pilot-operated, spring return type.

Control Valves 2 and 3 may be of small size, 1/8 or 1/4'' since they handle only pilot air into the main valve. They are connected in series as shown by the dash lines.

Both pilot valves must be actuated in order to shift the main valve. If either is released during forward travel, this will vent pilot pressure from the main valve and it will reverse itself, retracting the cylinder.

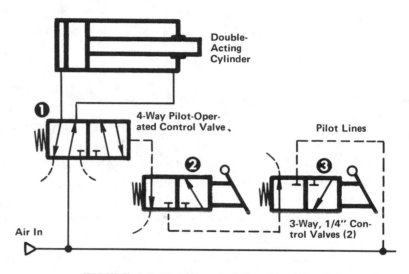

FIGURE 5-7. *Non-Electrical Two-Hand Control.*

MULTIPLE INDEXING WITH AIR CYLINDERS

Although an air cylinder cannot be accurately stopped in mid-stroke by cutting off its air supply, individual cylinders may be stacked end-to-end for accurate indexing to pre-determined positions.

Figure 5-8. In this example, two cylinders with 1'' and 2'' strokes are stacked to get four fixed positions spaced 1 inch apart. Position 1 is with both cylinders retracted. Positions 2 and 3 are reached with one extended, the other retracted, Position 4 with both extended.

Electrical Control for Above Circuit. Figure 5-9. Each cylinder must have its own 4-way valve, as it will have to be extended or retracted indepen-

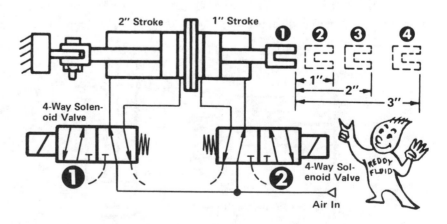

FIGURE 5-8. *Two Air Cylinders for Multiple Indexing.*

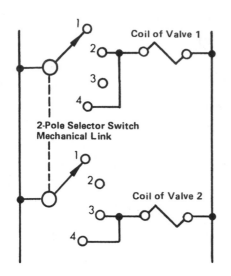

FIG. 5-9. Electrical Circuit for Multiple Indexing.

dently of the other one.

Two separate switches, such as toggle-type, could be used. It is more convenient, however, to have a single control for the operator, and this could be a rotary selector switch having four positions and wired as shown in Figure 5-9. It must also have two poles, or decks, one to switch each solenoid valve.

With the switch in Position 1, neither solenoid is energized so both cylinders are retracted and the assembly has retracted to its shortest position, Position 1 on the fluid diagram of Figure 5-8.

To index the cylinder assembly to the 1-inch extension, Position 2, Valve 2, only, should be energized by rotating the selector switch to No. 2 position. For the 2-inch extension, Valve 1, only, is energized in Position 3 of the switch. For the 3-inch extension, both valves are energized in Position 4 of the selector switch.

Alternate Methods of Electrical Control. Electrical circuitry, as such, is not within the scope of this book. A general rule, though, is that any switching circuit must use a switch with at least as many poles as the number of solenoid valves in the stack. And, a separate solenoid valve must be used for each cylinder. Other devices than shown here can be used such as momentary or maintained pushbuttons or lever switches, or the solenoid coils can be tied in with the programmed electrical circuit of another machine. A more detailed treatment of the subject of multiple indexing will be found in the Womack book "Electrical Control of Fluid Power". Circuits are shown for the use of holding relays with single solenoid valves and for the use of double solenoid, 2-position valves which will shift with a momentary signal from a pushbutton without the use of holding relays.

OPERATION FROM TWO OR MORE LOCATIONS

Sometimes an air cylinder must be operated from more than one control point. Examples would be a door powered by an air cylinder in which control must be exercised from either side of the door, or a large machine such as a press where the operator must be able to control its action from either of two points, or where either of two operators must be able to start or reverse the cylinder.

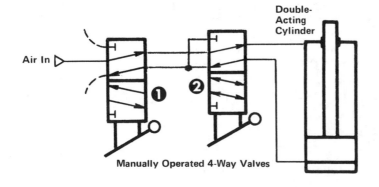

FIGURE 5-10. Manual Operation From Two Locations.

Figure 5-10. Where requirements are simple and the cylinder is small, manual Valves 1 and 2 may be installed at the two desired operating locations and piped as shown.

The operator must be able to view the cylinder, as he cannot tell which position it may be in by the position of the valve handle.

Spring return valves are not suitable for this circuit. The spool must stay in the last position to which it was actuated.

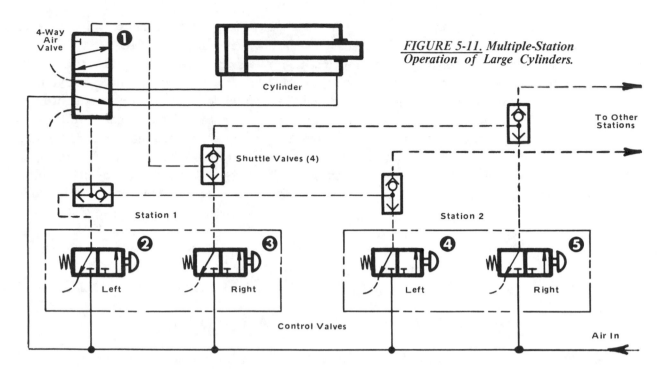

FIGURE 5-11. Multiple-Station Operation of Large Cylinders.

Figure 5-11. For larger cylinders, and where the cylinder is out of view of the operator, this circuit or a variation of it, is suitable for either manual or solenoid control.

Shuttle valves are used to isolate control stations. See their description in Volume 1, Chapter 3. Two control stations are shown but others may be added and isolated with additional shuttle valves. Control Valves 2, 3, 4, and 5 are miniature 3-way valves with stem or lever actuators and spring return. Valve 1 handles the main air flow and is sized for the cylinder being used.

Electrical Control. The circuit above can be directly adapted to electrical operation by substituting miniature 3-way valves, operated with momentary pushbuttons, in place of Valves 2, 3, 4, and 5. Or, a double solenoid, 2-position valve may replace Valve 1. It can be operated with momentary push-buttons which will replace the shuttle valves and Valves 2, 3, 4, and 5.

CYLINDER DWELL

Figure 5-12. When this cylinder reaches cam Valve 2 on its forward stroke, it dwells before retracting.

Directional Valve 1 may be a manual, solenoid, or pedal type having only two working positions with pilot pressure (instead of spring) return of the spool.

All delay or dwell circuits for compressed air use use the principle of slowly metering air into a chamber until the pressure rises high

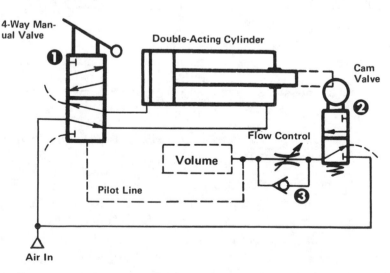

FIGURE 5-12. Dwell at the End of Stroke.

85

enough to produce an action like operating a pressure switch or shifting a pilot-operated valve. Length of delay is governed by metering rate and chamber volume into which the air is metered.

The operator starts the cylinder forward by shifting, then releasing, Valve 1. This valve stays in its shifted position because at this time there is no pilot pressure to return its spool. When the cylinder reaches and actuates cam Valve 2, it stalls against a positive stop while air is being metered through flow control Valve 3 into the pilot of Valve 1. When air pressure becomes high enough to shift Valve 1, the cylinder retracts. The auxiliary "volume" chamber shown in dotted lines may be added to get better repeatability on long delays by allowing a faster metered flow, giving a less critical adjustment of the flow control valve. It may be omitted for very short delays. An air line lubricator of 1/2 pint capacity (when empty), makes a convenient volume chamber. The check is an integral part of flow control Valve 3, and allows air to escape quickly from the pilot of Valve 1 when that valve is actuated.

Air metering for dwell timing is reasonably accurate for short delays of a few seconds. Where longer delays are needed, or where high accuracy is needed even on short delays, an electrical system is preferred. A circuit using a re-set timer for dwell delay is shown on Page 69.

LUBRICATION OF SHORT STROKE AIR CYLINDERS

Short stroke or small bore air cylinders operating through long connecting lines from a 3-way or 4-way valve may not receive oil from the system lubricator. If the internal volume of the connecting lines is large compared to cylinder volume, the cylinder may not receive sufficient lubrication or may receive none at all. As the cylinder strokes back and forth, air trapped in the cylinder flows out into the line, then returns to the cylinder. New air containing oil mist may never reach the cylinder.

One good solution is shown in Figure 2-32 on Page 46. Quick exhaust valves at the cylinder ports will vent used air to atmosphere, allowing fresh lubricated air to enter the cylinder on the next cycle. Quick exhaust valves of 1/4" size have more than ample capacity for small air cylinders. They can be used with single-acting or double-acting cylinder models.

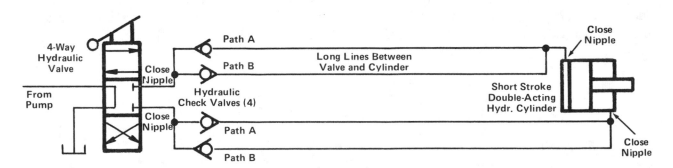

FIGURE 5-13. Double lines conduct new fluid (air or hydraulic) to cylinder on every cycle.

Dual Lines. Figure 5-13. Another solution is shown in this figure and can be used with air cylinders to introduce new lubricant or can be used with small hydraulic cylinders to discharge hot, used oil, to allow cool and filtered oil to enter the cylinder on the next cycle. This will prevent overheating in the cylinder and connecting lines.

Double lines are used to both ends of the cylinder, isolated with check valves. Oil flowing toward the cylinder will follow Paths A, and oil returning from the cylinder will follow Paths B. Since the check valves prevent the oil from reversing, a new charge of oil (or air) enters the loop on each cycle.

6

INTRODUCTION TO

Hydraulic Circuitry

SIMPLIFICATION OF DIAGRAMS

Most of the graphic diagrams in this book are simplified, for ease of understanding, and illustrate a simple idea or principle. Only the components which relate directly to the basic idea are included. Other components, even though they may be important, are omitted to keep from complicating the circuit and obscuring the presentation of the main thought.

We believe that ideas are communicated more clearly through very simple sketches. If the student is able to understand each individual basic idea, he then should be able to combine several of them into a composite circuit that will exactly meet his needs. We have purposely refrained from presenting highly complicated diagrams of an entire machine that would be laborious to follow and of little interest to the student. Every hydraulically-powered machine is different, and requires a different circuit.

In addition to the circuits in this chapter which are aimed specifically toward hydraulic fluid power, some of the air circuits in previous chapters may be adaptable to hydraulic operation. Special circuits using a combination of both air and oil are covered in Chapter 12.

This book, Volume 2, concentrates on circuitry for cylinders, and on the valving most commonly used for controlling them. We have reserved our study of rotary devices, air and hydraulic motors, hydraulic transmissions, and rotary actuators, for Volume 3.

Block Diagram. Figure 6-1. Most hydraulic fluid power systems may be divided into three general divisions: Section ❶, the initial conversion of mechanical energy into an equivalent amount of fluid power; Section ❷, the regulation of this fluid power with various kinds of valving; and finally, Section ❸, the re-conversion into mechanical energy by means of a cylinder or other device.

Conversion of Mechanical Energy Into Fluid Power. Figure 6-1, Section 1. Power into a fluid system may come from an electric motor or from an engine as mechanical power applied to the pump shaft, to then be converted into an equivalent amount of fluid power. Positive displacement pumps of the gear, vane, gerotor, and piston types are used almost entirely in industrial fluid power systems in preference to impeller types. Positive displacement pumps are described in Volume 1, Chapter 5. In contrast to impeller pumps, they can operate at high pressures. At higher pressures, power can be transmitted through smaller pipes, and all components — pumps, valves, cylinders — can be relatively small in size and weight; the horsepower per pound ratio is higher, and the efficiency greater.

Non-positive displacement pumps of the impeller, centrifugal, or turbine type may sometimes be used for light duty chores while positive displacement pumps handle the high power transmission. Light-duty jobs may include coolant pumping, supercharging the inlet of high pressure pumps, and moving or transferring oil from one vessel to another at very low pressure.

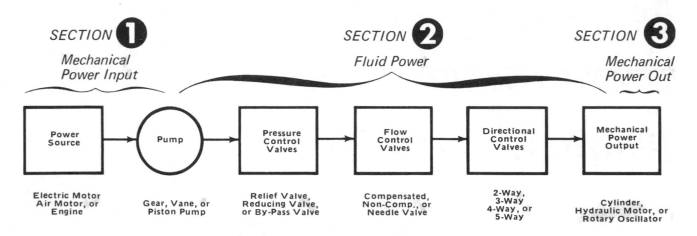

FIGURE 6-1. Block Diagram of a Typical Hydraulic Fluid Power System.

Regulation of Fluid Power, Figure 6-1, Section 2. Fluid power is present in a system only when the fluid (air, oil, water) is flowing and is also at a pressure above atmospheric. The amount of fluid power flowing in a pipe is directly proportional to its PSI gauge pressure and also to its rate of flow (GPM). Various valves are used to regulate the maximum pressure and to regulate the rate of flow. Pressure control valves include relief, pressure reducing, and by-pass types. Needle or flow control valves regulate the rate of flow. Simple valves of these types were covered in Volume 1; they will be covered in greater detail in this book with example circuits.

In systems using cylinder output, and sometimes in hydraulic motor circuits, fluid flow must be reversed to reverse the direction of motion of the cylinder or motor. Several types of directional control valves, 2-way, 3-way, or 4-way may be used. These valves have been given preliminary treatment in Volume 1. The present book will cover them in more detail.

Conversion of Fluid Power Into Mechanical Power, Figure 6-1, Section 3. Finally, since fluid power is not often used in its fluid state (except in spray nozzles), it must be converted into an equivalent amount of mechanical power. This book is concerned with conversion by means of cylinders. Volume 3 is concerned with conversion into rotary power by means of other actuators.

HYDRAULIC SYSTEMS DEFINED

Since there are several basic kinds of hydraulic systems, anyone intending to design a hydraulic system must choose the type system which he thinks will do the best job on his particular application. The basic systems shown on the following pages cover the majority of those used in both mobile and industrial hydraulics. After deciding on the most appropriate basic system, he can then tailor it to his particular use by adding accessories or additional components. Various modifications of the basic system will be shown in succeeding chapters.

Jacking System. Figure 6-2(A) and (B). The simplest hydraulic system consists of a hand pump and ram. Its prime purpose is not to transmit *power*, but to multiply and transmit *force*. If the multiplied force is to be used at the same location as the hand pump, a self-contained system, Part (A), is usually employed. Hand pump, ram, reservoir, and valving are all contained within a common housing. However, if the hand pump must be located remotely from the ram, each item must be separately purchased and plumbed together as in the diagram, Part (B).

FIGURE 6-2(A). Hydraulic Jack.

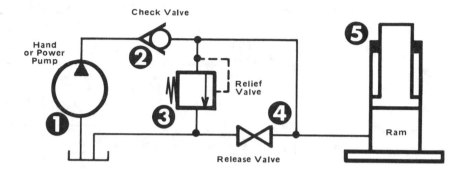

FIGURE 6-2(B). Simple Jacking System.

A simple jacking system may be the best choice if high force at very low speed is all that is needed. Even with hand operation a very high force can be produced on the ram, up to 100 tons and more. Because of their slow operating speed, jacking systems are not ordinarily suitable for high production work. They are more practical for occasional jobs, intermittent, specialty, or custom work where speed is unimportant.

Small rotary power pumps operated from an electric motor, or air-driven pressure intensifiers, may be used to produce high pressures. They may be substituted for the hand pump on hydraulic shop presses to reduce operator fatigue or to speed up the work.

The operation of a jacking system is very simple. Pumped fluid passes from the pump through the check valve and into the ram. The ram will raise if the release valve is closed. When the operator stops pumping the check valve will prevent backflow through the pump. The ram retracts, when the release valve has been opened, either by gravity, by springs, or by mechanical force.

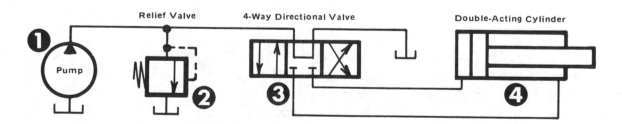

FIGURE 6-3. Typical Open Loop Hydraulic System.

Standard Open Loop System. Figure 6-3. An open loop system may be defined as one in which return oil is discharged into a reservoir at atmospheric pressure, to be picked up later by the pump and re-circulated. This is in contrast to a closed loop system in which return oil, instead of returning directly to the reservoir, is re-circulated under low pressure back to the pump inlet.

A circuit designer usually considers the open loop arrangement of Figure 6-3 first on a new system. But, if it will not meet application requirements he may consider one of the open loop variations next described, or he may consider a closed loop arrangement.

A fixed displacement gear, vane, or piston pump, 1, operates a cylinder through 4-way directional Valve 3. A means for unloading the pump (removing the pressure load from it) is recommended for periods when the cylinder is not moving. In this circuit, directional Valve 3 is a tandem center type, but other methods of unloading the pump could be used. To the basic circuit suitable relief, flow control, and other valving, plus filtering, can be added as needed by the application.

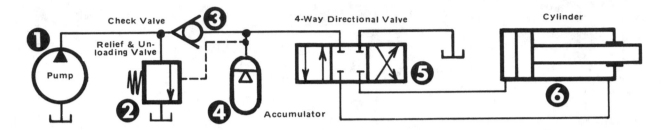

FIGURE 6-4. Typical Pressure Manifold Open Loop Hydraulic System.

Pressure Manifold Variation of the Open Loop. Figure 6-4. This, too, is an open loop circuit because the oil discharged from the cylinder is returned, at atmospheric pressure, to the reservoir and retained until again picked up by the pump and re-circulated.

However, in this variation, when the 4-way valve is in neutral, the pressure line is blocked. Oil is pumped into the accumulator and stored under pressure. When this pressure reaches the desired level, the unloading valve opens to allow pump oil to circulate back to reservoir at near zero pressure. When the next cycle of the machine is started, stored oil flows out to join the steady flow from the pump to provide a greater cylinder speed than possible with pump oil alone.

For additional description of accumulators, their circuits and applications, and how to calculate accumulator capacity needed, refer to Volume 1, Chapter 6.

This variation of the open loop system may give better results than the common open loop of Figure 6-3 on some applications for one of these reasons:

(1). Where more than one cylinder branch circuit must operate from the same pump. A greater volume of usable oil is available because the pump can produce stored oil during periods in the cycle when none of the branch circuits is working.

(2). Where a very high flow rate, for rapid advance of the cylinder is needed, but only for periods of short duration. A relatively small pump, sized so it will run loaded about 75% of the total elapsed time, will serve in place of a larger and more costly pump sized to produce the entire flow of oil for short periods and allowed to idle for long periods. A significant saving can also be realized on electric motor size since the larger pump would have to be driven by a motor sized for peak horsepower even though it might be working only a small percentage of the elapsed time.

(3). On long-holding applications where high pressure must be maintained on a workpiece for a long time under stationary conditions. Instead of the pump discharging across a relief valve during the holding periods, it can charge the accumulator then be unloaded through the unloading valve while active pressure from the accumulator is maintained on the workpiece. Valve spool leakage or cylinder seal leakage during the holding period can be made up from the accumulator while the pump remains unloaded. However, there is a slow drop in holding pressure as oil is metered out of the accumulator.

All accumulators require periodic testing and replenishment of the pre-charge gas pressure. Circuits using them are usually limited to "in plant" use where a trained maintenance crew can give them proper attention. Their use may not be desirable on mobile or other applications where no regular maintenance program is followed. Consider this point before designing them into the system.

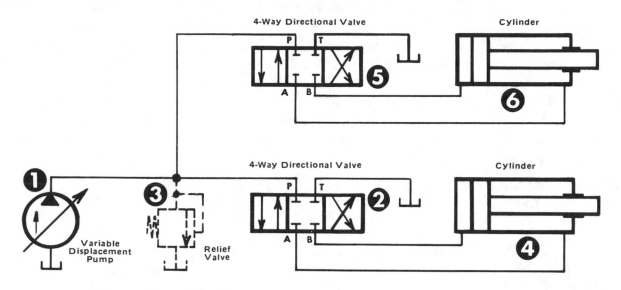

FIGURE 6-5. Typical Open Loop System Using Pressure Compensated Pump.

Pressure Compensated Pump Variation of the Open Loop. Figure 6-5. A variable displacement pump with pressure compensator is used with closed center 4-way valve(s). When the 4-way valves are centered, blocking the pump flow, the compensator automatically reduces pump displacement (GPM) to near zero. During normal operation, the compensator allows the pump to operate at full flow until pressure reflected from the load reaches the pre-set adjustment of pressure on the compensator. At this point a pilot signal picked up from the pump outlet port causes pump displacement (flow) to be reduced as much as necessary to keep outlet pressure from rising any higher. When the output circuit reaches deadhead (stall) condition, the compensator reduces pump displacement to near zero, with only enough oil being pumped to make up for normal circuit leakage. Please refer to Page 203 for further description of the pressure compensator principle as applied to piston pumps.

The cut-off or "firing" pressure of the compensator is usually adjusted with a knob or screw, and is set for the maximum pressure required on the load, or no higher than the maximum rating of the pump, or in any case no higher than can be produced by the available horsepower from the driving source without overload. Some compensators are built with an additional feature which makes them responsive not only to maximum pressure in the system but also to maximum flow or to a combination of pressure and flow. These compensators limit the HP demand on the power source to prevent overloading if the combination of flow and pressure would create a HP demand greater than the pump rating or greater than the input HP available.

Directional Valves 2 and 5 must have their pressure inlet port closed in neutral. Full closed center porting is used in most circuits. With both valves centered, the pump operates at the pressure setting of its compensator and produces a small flow, just enough to make up leakages across the 4-way valve spool, the cylinder piston, and the pump itself. In this condition, the power consumption and heat generation are very low. Relief Valve 3 can sometimes be omitted if the pump has a fast-acting compensator and if there is a sufficient volume of oil trapped in the line between pump and 4-way valve to absorb (by oil compression) the transient pressure spikes generated during the time required for the compensator to act. In most cases a relief valve should be used, although it can be very small in size, just large enough to discharge the very small volume of the pressure transient.

The unique characteristic of this system is its ability to maintain full pressure against a stalled load while consuming very little input power. Ideal applications are cable tensioning, bonding and curing presses, and many other applications where a constant high pressure must be maintained against a load for a long time, at stall. These are applications where a fixed displacement pump such as a gear pump would have to continually discharge across a relief valve, soon overheating the oil. This system should also be considered on systems where several branch circuits operate from one pump and where maximum pressure may be required on more than one branch at the same time. However, pumps without a case drain line, such as variable displacement vane pumps, could overheat if held at stall too long because the slippage oil would re-circulate back to the inlet many times and could eventually reach a dangerously high temperature.

During periods in the cycle when the cylinder is in motion, the action of this system is quite similar to any other closed loop system, and the valving circuits described in later chapters are applicable.

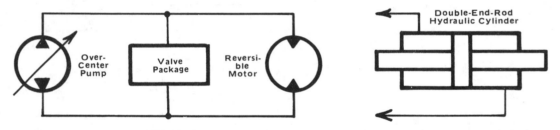

FIGURE 6-6. Typical Closed Loop Hydraulic System.

Closed Loop Hydraulic System. Figure 6-6. In a closed loop system the return oil, instead of being drained to reservoir, is ported directly to the pump inlet at low pressure. Closed loop circuits are used extensively in hydraulic transmissions for driving hydraulic motors, although with suitable circuitry cylinders can also be operated. Both of the loop lines connecting pump to hydraulic motor are under pressure at all times, one line at high pressure, the return line at low pressure. Features of closed loop operation are high efficiency, close control of both positive and negative loads, and infinite control of motor speed in both forward and reverse.

Piston-type pumps and motors are almost a necessity. Gear and vane-type units have too much internal slippage for successful closed loop operation. Most closed loop systems use a variable displacement, over-center pump driving a fixed displacement hydraulic motor.

A closed loop system is more costly to build than an open loop system, primarily because of the higher cost of piston-type pumps and motors. But the extra cost is well justified if the important advantages of closed loop operation are needed. Closed loops are especially advantageous on vehicular wheel drive, winches, cable reels and tensioners, where infinite speed variation, forward and reverse, complete control of overrun by the load, and ability to maintain long-holding force without overheating are necessary. These systems operate at high efficiency, and since they usually operate in the 3000 to 5000 PSI pressure range, their size and weight are small in relation to the horsepower they handle.

For further information on closed loop hydraulic transmissions, refer to Volume 3. They will not be covered further in this book.

GRAPHIC SYMBOLS FOR HYDRAULIC CIRCUITS

Symbols for hydraulic circuits are identical, where applicable, to most of those used for air circuits except that open triangles ($\triangle$) are used on air circuits to indicate a gaseous medium, especially on pumps, motors, and piloting medium, and solid triangles ($\blacktriangle$) are used on hydraulic circuits to indicate a liquid medium. A complete list of ANSI (American National Standards Institute) symbols may be obtained from the National Fluid Power Association, 3333 N. Mayfair Rd., Milwaukee, Wisconsin 53222. Write for quotation on Publication Y32.10-1967, Graphical Symbols.

If correctly drawn, graphic symbols are self-explanatory without further labeling. However, in this book they are usually captioned to help the student learn their meaning. In this chapter, only the symbols for pumps and motors are shown. Symbols for other components will be given as each type of component is covered. In most cases, the official ANSI symbols are used although occasionally a pictorial symbol may be used for a special reason.

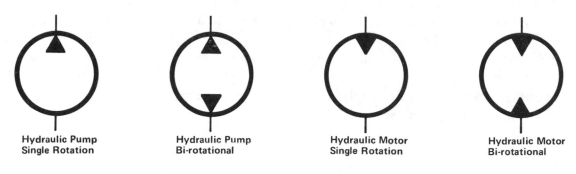

FIGURE 6-7. *Graphic Symbols for Fixed Displacement Hydraulic Pumps and Motors.*

Fixed Displacement Pumps and Motors. Figure 6-7. Solid triangles are used inside the circular envelope to indicate these are hydraulic units rather than air compressors or air motors. The triangle also indicates direction of fluid flow, pointing outward for pumps, inward for motors. Bi-rotational units are capable of shaft rotation in either direction with no mechanical modification of the unit. This capability is indicated by the use of two triangles. The type of pump, gear, vane, gerotor, piston, etc., is not indicated by the symbol, but if important, may be indicated by a caption alongside the symbol.

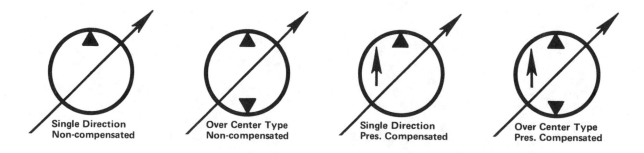

FIGURE 6-8. *Graphic Symbols for Variable Displacement Hydraulic Pumps.*

Variable Displacement Pumps and Motors. Figure 6-8. A slash arrow through the usual pump symbol indicates that some means is provided to change pump displacement to change output flow. It may be a semi-permanent adjustment by means of a handwheel which can be set while the pump is stopped, to limit the HP demand which the pump places on the driving motor or engine. Or, it may be a small lever for use by the operator to change displacement while the pump is operating, for controlling speed of the system actuator, hydraulic motor or cylinder.

Variable displacement hydraulic motors are indicated by a similar symbol, with the triangle at the inlet port with tip pointing inward. Hydraulic motors are never built for over-center operation.

A pressure compensator on a variable displacement pump is indicated with a small vertical arrow inside the symbol envelope. A pressure compensator can also be used on a variable displacement hydraulic motor, and when so used it is for the purpose of increasing the motor torque when required by the load. Normally, the compensator allows the motor to operate at minimum displacement for maximum speed. But if load demand should increase to the point where system pressure cannot supply sufficient torque, the compensator will automatically increase motor displacement to provide more torque at a reduced maximum speed.

Two triangles in a pump or motor symbol indicate that the unit can accept flow (if it is a motor), or produce flow (if it is a pump), in either direction through its ports. In some pumps this means that the shaft rotation can be reversed to produce flow in the opposite direction with no mechanical modification being required inside the pump. On other pumps, with variable displacement, the shaft rotation can remain the same but the pumping elements can be moved across a center neutral position to change the timing of inlet and outlet ports. Most hydraulic motors are built for reversible rotation and are shown with two triangles.

7
Directional Control
OF HYDRAULIC CYLINDERS

SINGLE-ACTING CYLINDERS

Single-acting cylinders (or rams) produce power in one direction, usually while extending, then retract either by reaction force (or weight) of the load or by spring force. There are several ways in which they can be controlled.

Two-Way Valve Control. Figure 7-1. When using a hand pump, any kind of 2-way valve — gate, plug, globe, or even a spool-type — may be used as a release valve. To raise the cylinder, the release valve must be closed and the hand pump operated until the cylinder reaches the desired extension. While holding a load suspended, the check valve prevents oil leaking back through the hand pump. To lower the cylinder, the release valve must be opened. Load weight will retract the cylinder, causing discharge oil to drain back to tank.

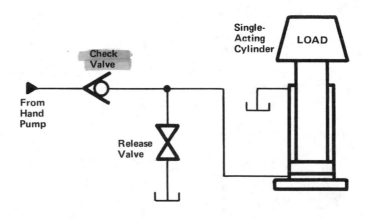

FIGURE 7-1. *Single-Acting Cylinder With 2-Way Valve.*

Three-Way Valve Control. Figure 7-2. When using a power driven pump, a 3-way spool or poppet type valve may be used. To raise the cylinder, the valve must be shifted to connect pump oil into the cylinder. The valve shown has no neutral position, so the cylinder can only be stopped in mid-stroke by stopping the pump. The check valve will prevent backflow when the pump is stopped. To lower the cylinder, the 3-way valve must be shifted to the position shown.

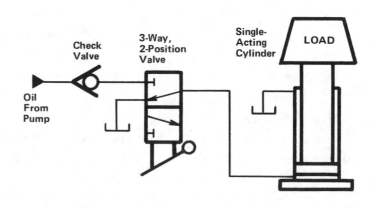

FIGURE 7-2. *Single-Acting Cylinder With 3-Way Valve.*

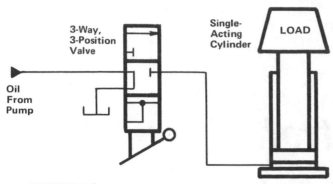

FIGURE 7-3.
Single-Acting Cylinder With 3-Way, 3-Position Valve.

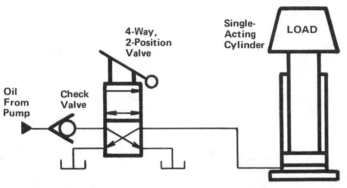

FIGURE 7-4.
Single-Acting Cylinder With 4-Way, 2-Position Valve.

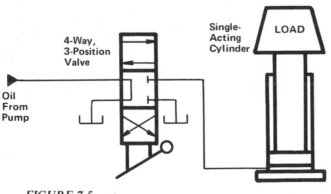

FIGURE 7-5.
Single-Acting Cylinder With 4-Way, 3-Position Valve.

3-Way, 3-Position Valve. Figure 7-3.

A center neutral position is required on the valve to provide a holding position for the cylinder in mid-stroke without stopping the pump.

In this figure the top position of the valve is the raise position for the cylinder; the center position is the hold position in which the hydraulic pump is unloaded to tank; the bottom position on the valve is for lowering the cylinder. The pump as well as the cylinder, is drained to tank in this position.

4-Way, 2-Position Valve. Figure 7-4.

There are two independent flow passageways through a 4-way valve. In the valve position shown, one passage unloads the pump to tank and the other passage drains the cylinder while it is retracting. When the valve is shifted, pump oil flows through one passage to raise the cylinder; the other passage is non-functional at this time.

Since there is no neutral position on this valve, the cylinder cannot be stopped in mid-stroke except by stopping the pump. The check valve prevents backflow from the cylinder while the pump is stopped.

We recommend, if practical, two tank return lines be used rather than combining both ports on the valve into a common tank line. If combined, back pressure from pump flow might generate enough back pressure to cause the cylinder to retract too slowly.

4-Way, 3-Position Valve. Figure 7-5. If the cylinder must be stopped and held at some point in mid-stroke, a neutral position must be provided on the control valve. A tandem center 4-way valve is the easiest way to provide this holding position as well as to unload a power pump. In this figure the top block on the valve is raising position for the cylinder; center block is holding position for the cylinder with pump unload function; the lower block is the lowering position for the cylinder.

In all circuits shown here, flow control valves can be added in the cylinder lines to regulate both extension and retraction speeds.

DIRECTIONAL CONTROL OF DOUBLE-ACTING HYDRAULIC CYLINDERS

Four-Way, 2-Position Valve Control. Figure 7-6. Most hydraulic cylinders are controlled, for direction of travel, with 4-way valves. A 4-way spool-type valve has two grooves on the spool and can handle two independent flows at the same time — flow going toward the cylinder and flow returning from the cylinder. The 4-way valve in this figure has two working positions but does not have a center neutral. Therefore it is limited in its control because it cannot stop the cylinder in mid-stroke, nor can it unload the hydraulic pump. However, it may be suitable on unusual applications, especially those where the cylinder, when started, may operate for long periods, cycling back and forth, without stopping.

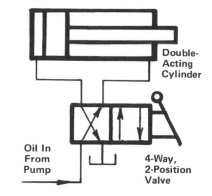

FIGURE 7-6. Double-Acting Cylinder With a 4-Way, 2-Position Valve.

Four-Way, 3-Position Valve Control. Most hydraulic circuits use a 3-position, 4-way valve with a center neutral position on the spool. Double solenoid valves of this type have built-in centering springs to bring the spool to neutral from either side of center after both solenoid coils have been de-energized. The spools in manually operated valves are usually spring centered to a neutral position, or have a detent in each working position. The spool centers shown in the following diagrams are those most often used in hydraulic valves for controlling double-acting cylinders.

Tandem Center Spool. Figure 7-7. Cylinder ports become blocked in center neutral position of the spool, holding the cylinder in an "oil locked" condition. The pressure port is open to the tank port, unloading the pump automatically without special attention from the operator. Provides a simple and economical way to unload the pump with a minimum number of components and circuitry. This type spool works well on low volume, low pressure systems. Standard spools of this type usually have closed center crossover (see Figure 7-12). This may cause a pressure spike to appear in the pump line as the spool is shifted, and may cause noise in the system and shock to the components. When specified, tandem center valves are available with open crossover porting. This reduces the shock of spool shift but may allow the load to drop if used on vertically mounted cylinders, especially on manually operated valves when the spool is shifted slowly.

On high power systems a better pump unloading system should be used — one which does not create the massive shock which a tandem center spool may sometimes generate. A good unloading system for electrically operated (solenoid) valves is shown on Page 115. To minimize shifting shock on manually operated, multiple branch systems, bank or stack valves should be used.

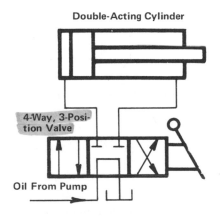

FIGURE 7-7. Tandem Center Spool.

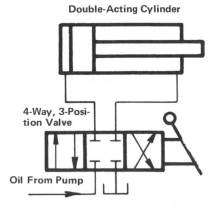

FIGURE 7-8. Closed Center Spool.

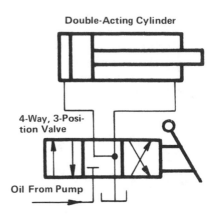

FIGURE 7-9. Float Center Spool.

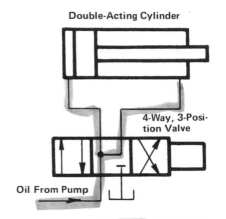

FIGURE 7-10. Regenerative Center Spool.

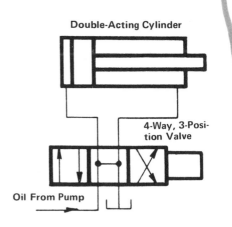

FIGURE 7-11. Full Open Center Spool.

Closed Center Spool. Figure 7-8. All ports become blocked from each other and from tank when the spool is centered. Since the valve spool does not provide unloading for the pump, some other means for unloading must be used. The unloading system shown on Page 115 is often used.

Closed center spools are necessary when operating two or more branch circuits from one pump, where more than one branch must operate at the same time. They are often used with pressure compensated pumps as in Figure 6-5, or in open loop accumulator circuits as in Figure 6-4.

Closed crossover porting (Figure 7-12) is standard on most closed center valve spools.

Float Center Spool. Figure 7-9. In center position inlet pressure becomes blocked, but the two cylinder ports become connected to each other and drained to tank. This relieves "fluid lock" on both sides of the cylinder piston, allowing it to be positioned manually or mechanically. On some machines the hydraulic cylinder is attached to a moving member of the machine. It may furnish power for part of the cycle then may "float" with the machine movement for the remainder of the cycle. Open center crossover porting is preferred with float center spools.

Another application for float center action is on the spool of a miniature solenoid valve mounted "piggy-back" on top of a large valve to control shifting of its spool. When both solenoids are de-energized, the solenoid valve spool returns to its center "float" position. This permits the main spool to seek its spring centered position without possible interference from the solenoid spool because of "pressure lock".

Regenerative Spool. Figure 7-10. Both cylinder ports become connected to pressure when the spool is centered. This causes a single-end-rod cylinder to advance very rapidly because of an unbalance of fluid force caused by difference in areas on opposite sides of the piston. The application of this spool to regenerative circuitry starts on Page 124.

Regenerative spools have been used in miniature solenoid pilot valves mounted "piggy-back" on large main valves which have pressure centered spools.

Open Center Spool. Figure 7-11. All ports are open to each other and to tank when the spool is centered. It is used more often with hydraulic motors than with cylinders, and is sometimes called a "motor spool"

An open center spool may be used to unload a hydraulic pump while leaving the cylinder ports vented. However, both cylinder ports become exposed to back pressure created by pump flow to tank.

Single-end-rod cylinders operated with an open center spool may drift forward when the spool is in center position because of back pressure creating an unbalance of forces on opposite sides of the piston. See Figure 7-14(B) for additional information.

If an open center spool is used in one section of a bank valve to operate a single-end-rod cylinder, it should be placed nearest the tank port, as the last section downstream from the pump. If used further upstream, back pressure which appears when a downstream spool is shifted may cause the handle of the open center spool section to shift by itself.

Crossover Porting. Figure 7-12. A valve spool, in passing from a working position to center neutral, must travel through a porting pattern between these positions where momentarily all ports are either blocked (closed crossover porting) or are open to each other (open crossover porting). Solenoid valves pass through this porting pattern very quickly without stopping, but on manually operated valves the spool could linger in this position if the operator shifts the spool slowly. On some types of spools the crossover porting (open or closed) could be important. A typical spool is shown below:

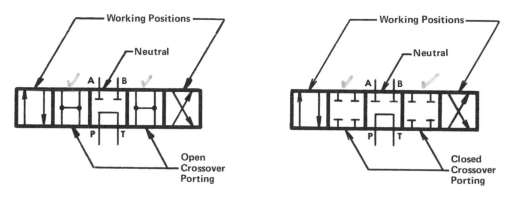

FIGURE 7-12. *Crossover Porting Patterns for a Tandem Center 4-Way Valve Spool.*

Special Spools. Figure 7-13. On certain applications one cylinder port must become vented to tank in neutral position of the 4-way valve. For example, when using a pilot-operated check valve, the cylinder port which furnishes pilot pressure for the check must be vented so the check valve can close. On hydraulically powered winches having a pressure released brake, the cylinder port which supplies release pressure to the brake must be vented in neutral so the brake can set. One of the spools shown in this figure should be used for these applications.

(a). Pressure Port Closed.   *(b). Pressure Port Open to Tank.*

FIGURE 7-13. *Other Less Popular Spools Available From Some Manufacturers.*

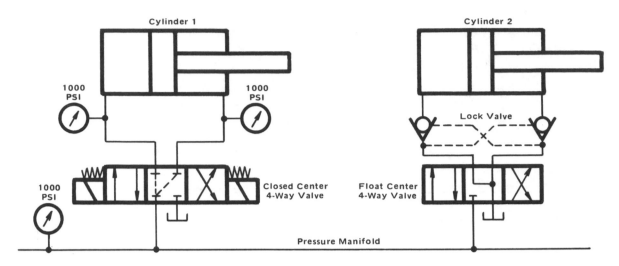

FIGURE 7-14(A). Cylinder Drift Problem With a Closed Center, Pressure Maintained, System.

CYLINDER DRIFT PROBLEMS

Closed Center 4-Way Valve. Figure 7-14(A). This is a system which may include two or more branch circuits operating in parallel from one pump. Full pressure remains on the inlet ports of all 4-way valves when in neutral position. A variable displacement pump, pressure compensated, is normally used on these systems. Leakage across a spool valve, shown in dotted lines on Valve 1, may cause full pressure to appear on both cylinder ports. Due to a difference in area on opposite sides of the piston, the cylinder may drift outward unless the static load against the piston rod is greater than the force produced for drifting. See example calculation in the figure below.

On the original design the solution to this kind of drift may be to use a lock valve as shown for Cylinder 2. This is two pilot-operated check valves in one housing, cross piloted. A float center instead of a closed center 4-way valve must be used to allow free action of the pilot-operated check valves.

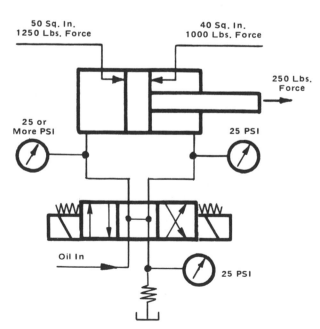

FIGURE 7-14(B). Drift With Open Center Spool.

Open Center 4-Way Valve. Figure 7-14(B). A high back pressure in the tank return line will cause a drifting force to be produced across the piston of the cylinder because of unequal areas on opposite sides of the piston.

For example, assume an 8″ bore cylinder with a 3½″ diameter piston rod, and with a 25 PSI back pressure in the tank return line. Full piston area is 50 square inches and the net area is 40 square inches. Residual force producing forward drift is 25 PSI x 50 sq. in. = 1250 lbs. Resisting force on the rod side of the piston is 25 PSI x 40 sq. in. = 1000 lbs. Net force causing drift is 1250 – 1000 = 250 lbs. The piston will drift forward unless the static load on the system is greater than 250 lbs.

Therefore, it is very important, when using a full open center 4-way valve, to keep back pressure in the tank return line to a minimum.

100

VALVE ACTUATORS AND THEIR GRAPHIC SYMBOLS

Common Actuators for Directional Control. The majority of in-plant industrial hydraulic systems use solenoid controlled directional valves. The intricate nature of most control systems required to operate complicated machines almost precludes the use of many actuators used on compressed air systems. Manually controlled 4-way valves are sometimes used on low power, simple hydraulic systems. But the use of other actuators such as cam, foot, and pilot, has almost been discontinued, and these valves are now available mostly in miniature size for occasional use on non-electrical control circuits.

On mobile systems, working outdoors, manually controlled 4-way valves of the bank or stack type are by far the more popular type because of the precise control which can be exercised over cylinder motion by modulating flow with handle position. Each handle of a bank valve can be used not only for direction but also for speed control. Solenoid valves, on the other hand, are seldom used on mobile equipment because they lack the fine control of manual valves.

Solenoid Actuators. Figure 7-15(A). A solenoid is shown by a rectangle with a slash line through it. Double solenoid valves have a symbol on both ends of the block.

The symbol in this figure is used for direct-acting solenoid valves and is modified slightly as shown later, for pilot-operated solenoid valves.

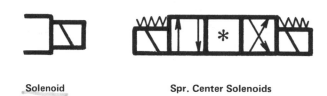

Solenoid   Spr. Center Solenoids

FIGURE 7-15(A). Symbols for Solenoid Actuators.

Most solenoid valves for hydraulics have three valving positions with neutral being in center position. Internal springs move the spool into neutral when both solenoids are de-energized. Pressure centering the spool to neutral has been largely discontinued by the industry but is still used on some brands of high capacity valves operating at high pressure.

Manual Actuators. Figure 7-15(B). Manual actuators usually consist of a hand lever, although in some applications the valve may be operated remotely by means of a cable attached directly to the valve spool.

Most manually actuated hydraulic valves have two working positions plus a center neutral. Neutral porting may be one of the spool types previously shown.

Lever   Stem

Spring Centered

Actuators Seldom Used for Hydraulics . . .

FIGURE 7-15(B). Symbols for Manual Actuators.

Pilot Actuators. Figure 7-16. Either the simplified or complete pilot actuator symbol may be used. Although the simplified symbol, consisting only of a dash line drawn to the actuator block, is usually preferred, it is sometimes necessary to use the complete symbol to indicate whether the pilot source is liquid (oil, water, etc.) or compressed air. A solid traingle inside the block indicates hydraulic operation while an open triangle indicates air piloting.

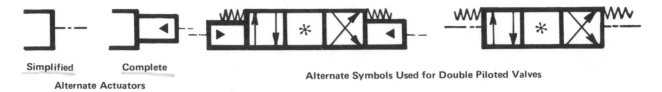

Simplified   Complete

Alternate Actuators   Alternate Symbols Used for Double Piloted Valves

FIGURE 7-16. Graphic Symbols for Pilot Actuators for 4-Way Valves.

Cam Actuators. Cam valves are difficult to locate near the point of actuation, they are inconvenient to install and connect, too large and heavy for some applications, and may require too much physical force to actuate. Since they are limited to two-position operation, they are difficult to use in many control circuits.

Pedal and Treadle Valves. They are bulky, heavy, and hard to install and plumb. Even in the small sizes the force to actuate them is fatiguing to an operator, especially since the pedal or treadle sits relatively high off the floor. They are seldom available any longer in any but 1/4" size for hydraulics, and are most often used to pilot a larger valve which carries the main flow.

Palm Button Valves. Never popular for hydraulics, they have become almost unavailable. Too great an effort is required to operate them even in the small 1/4" size.

The above valves have largely been replaced by solenoid valves controlled with limit, hand, and foot switches. Components for electrical circuitry are smaller and lighter, easier to install and connect. A much greater separation is possible between operator and machine than is practical with fluid piping, and a much wider range of control functions is possible.

Logic Actuators. Figure 7-17. Although not within the scope of this book to go into miniature logic circuitry, a few of the logic symbols are shown since the reader may encounter them on schematic diagrams.

Miniature valves built for logic functions may have double actuators on the same end of a valve; this is rare on larger valves built for power handling.

To illustrate, a pilot-operated 2-way valve may be shown with two separate pilot actuators on the same end of the valve.

Note: Solid triangles are used when the piloting medium is hydraulic and an open triangle is used if the piloting medium is compressed air.

(A). "OR" Logic. The 2-way valve body is normally closed (N.C.). If the two pilots are stacked side-by-side, this indicates that either one OR the other, by itself, will move the spool and provide a flow path through the valve.

(B). "NOR" Logic. This is the opposite of OR. The valve body is 2-way normally open (N.O.). If neither pilot NOR the other is actuated, there is a flow path through the valve. If either is actuated, the spool shifts and cuts off the flow path.

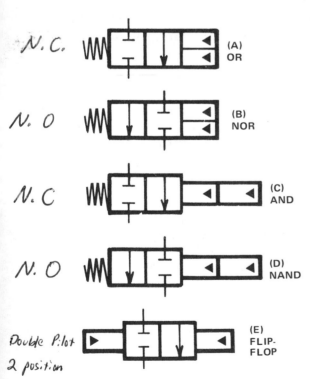

N.C. — (A) OR

N.O — (B) NOR

N.C — (C) AND

N.O — (D) NAND

Double Pilot 2 position — (E) FLIP-FLOP

FIGURE 7-17.
Logic Actuators for
Fluid Power Circuitry.

(C). "AND" Logic. If the two pilot operators are stacked end-to-end on a 2-way, N.C. valve body, this indicates that neither one, by itself, will shift the valve. It requires one pilot AND the other pilot, energized together, to shift the valve.

(D). "NAND" Logic. This is opposite to AND. On a 2-way, N.O. valve, a flow path is open through the valve unless both pilots are actuated at the same time. This shifts the spool and cuts off flow through the 2-way valve body.

(E). "FLIP-FLOP" Logic. This is equivalent to a double-piloted, 2-position, no-spring standard valve. A momentary signal on one pilot will shift the spool and it will stay shifted.

Button Bleeder Valves. Very popular for air circuitry, but impractical for hydraulics because their operation depends on the compressibility of the fluid.

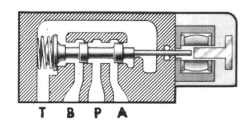

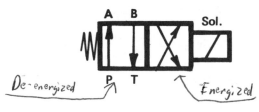

FIGURE 7-18.
Cross Section and Graphic Symbol for
Direct-Acting Single Solenoid 4-Way Valve.

DIRECT-ACTING SOLENOID VALVES

Figure 7-18. A direct-acting solenoid valve is one where the solenoid armature is directly linked to the spool and provides the necessary push or pull "muscle" for shifting it.

In this single solenoid valve the spool is driven in one direction by the solenoid force and in the other by spring action after the solenoid is de-energized. Porting through the valve when de-energized and energized is shown by the left and right blocks respectively of the graphic symbol. This is the same as for any standard 4-way valve.

The usual construction of spool valves naturally results in having an exhaust discharge path at each end of the spool. On air valves these dual exhausts are often brought out individually, but on hydraulic valves, to simplify plumbing, they are usually combined internally and brought out as a single connection, T, for tank.

Electrical Requirement. The single solenoid valve operates on a "maintained" electrical signal to stay in its shifted position. Breaking the signal allows it to return to its "normal" position.

Maximum Size. Direct-acting valves of modern design are usually, but not always, limited to 1/4" up to 1/2" in size. The high current and heavy impact of larger solenoids to operate larger size valves creates many operational problems both electrically and mechanically.

Double Solenoid Valve, Two Positions. Figure 7-19. The valve spool will shift and remain in shifted position on a momentary electrical impulse. An impulse on the opposite solenoid will return it to its original position. There are only two active valving positions, as the spool cannot be stopped in a center neutral position. Additional valving must be used if a cylinder is to be stopped in mid travel.

If current is maintained on both solenoids at the same time, the coils will "fight" each other, and the one which is unable to close the armature gap will draw excessive current and burn out in a short time. To prevent this, electrical interlocking can be provided by extra contacts on relays or pushbuttons.

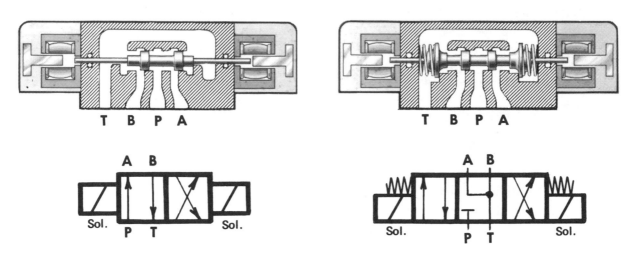

FIGURE 7-19.
Cross Section and Graphic Symbol for Double Solenoid, Two-Position, 4-Way Solenoid Valve.

FIGURE 7-20.
Cross Section and Graphic Symbol for Double Solenoid, Three-Position, 4-Way Solenoid Valve.

Although the valve spool will shift on a momentary electrical signal, greater reliability is obtained by holding the current on one coil or the other throughout the entire cycle. An occasional failure may occur if operation is based entirely on impulse signals. See Page 63. All solenoid valves should be mounted with spool axis horizontal to minimize the possibility of self-drift under gravity or vibration.

Double Solenoid Valve, Three Positions. Figure 7-20. The valve spool will shift from its center neutral position when a coil is energized. Current must be maintained on the coil to keep the spool shifted. When both coils are de-energized, centering springs move the spool to its center neutral.

On A-C solenoid coils, extreme care should be used to be sure both coils are not energized at the same time. Even a slight overlap of double energization on every cycle, may over a period of time cause overheating and possible burn-out of a coil. Too frequent cycling will also cause overheating of A-C coils on air-gap type solenoids. Oil immersed or wet armature type solenoids should be used on applications where the solenoid is cycled continuously on less than 15-second intervals.

Three active valving positions make possible a third cylinder function when the valve spool is in center or neutral position. This function could be cylinder stop, float, or regenerative, as needed. See Pages 98 and 99 for description of center positions on 3-position valve spools.

Dry Armature Solenoid Coils. The cross section views of Figures 7-18, 7-19, and 7-20 illustrate the principle of dry armature solenoids in which an encapsulated solenoid coil operates in a dry environment protected by a solenoid housing. The push pin connecting solenoid armature to the valve spool goes through a dynamic seal, such as an O-ring, into the hydraulic drain chamber at the end of the spool. Eventually the seal may leak, allowing hydraulic oil to leak out into the solenoid chamber. Oil seeping from around the solenoid cover usually means that the pin seal must be replaced.

Wet Armature Solenoid Coils. Most valve manufacturers are now using a wet armature construction in which the moving armature is contained in a pressure-tight tube. The inner end of the armature tube is open to the hydraulic fluid, and the armature operates submerged in the fluid. This eliminates the pin seal with its leakage problems. The solenoid coil itself slips over the outside of the pressure-tight tube. The only moving seals in the armature circuit are the manual override seals. Since these are used infrequently, they are unlikely ever to develop a leak. If they should leak, the leakage goes outside the valve rather than into the coil cavity.

Advantages claimed for the wet armature construction are improved shifting reliability by eliminating the friction of the push pin seals, cushioned impact of the armature, reduced noise, longer life, and elimination of leakage into the solenoid coils.

SOLENOID-CONTROLLED, PILOT-OPERATED VALVES

General Description and Advantages. While direct-acting solenoid valves are usually preferred in 1/4" and smaller sizes, on larger valves ordinary solenoid coils do not have sufficient power to reliably operate the spools on a direct push or pull. In the early days of solenoid valve operation the problem was solved by building larger solenoid structures with more power. But the large size and weight, the tremendous impact on closure, and the high inrush current created many operational problems. The valves were short lived and subject to frequent breakdowns.

To solve these problems, valve manufacturers developed solenoid valves which were controlled by solenoid action but which used power available from the inlet fluid line for shifting the spools. These valves are called "solenoid controlled, pilot-operated valves". The reader is referred to Volume 1 for a schematic layout of such a valve, with a detailed description of how it works.

Four-Way "Piggy-Back" Operator. Figure 7-21.
A solenoid controlled, pilot-operated valve is built in two stages. The first stage includes the main body and spool, sized to handle the main flow. The second stage includes a miniature direct-acting single or double solenoid valve usually mounted "piggy-back" on top of the first stage. Motive power to shift the main spool is taken internally from the main pressure inlet, manifolded to the second stage through O-ring seals, then manifolded back again to the ends of the main spool. Drain oil from the solenoid valve is likewise manifolded down to the main body and into the tank port.

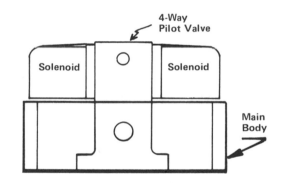

FIGURE 7-21. Example of Two-Stage Piggy-Back Valve With Double Solenoid Operator.

When using a double solenoid "piggy-back" second stage valve, if both solenoid coils are energized at the same time, the armature of the last solenoid to be energized cannot seat and its coil will burn out.

Solenoid controlled, pilot-operated valves have several important advantages over direct operation of large valves with large solenoids:

(1). Because they use miniature solenoids they are quieter in operation. The small coils do not have the high level of electrical hum and the high intensity impact of large solenoid structures.

(2). The shifting speed of the main spool can be reduced with sandwich flow control valves manifolded between the "piggy-back" solenoid valve and the main valve. This reduces shifting shocks. No attempt should ever be made to reduce shifting speed of the spool in a direct-acting solenoid valve. This will cause extra heating in the coil with possible burn-out if cycled often.

(3). The inrush current of an A-C solenoid (as used on most hydraulic 4-way valves), may be from 5 to 10 times its steady-state current. By using miniature solenoids, only as operators, the inrush current can be held to a level low enough so it can be handled reliably with standard limit switches with long switch life, and without the addition of extra relays or contactors to handle the high inrush current.

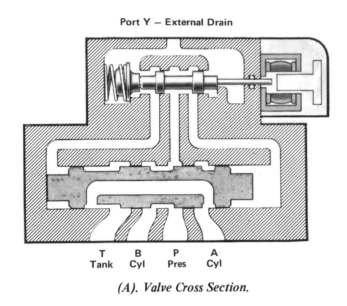

Port Y — External Drain

T — Tank   B — Cyl   P — Pres   A — Cyl

(A). Valve Cross Section.

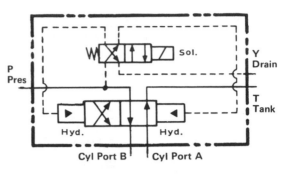

P — Pres   Sol.   Y — Drain   T — Tank

Hyd.   Hyd.

Cyl Port B   Cyl Port A

(B). Complete Graphic Symbol.

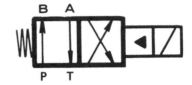

B   A

P   T

(C). Simplified Graphic Symbol.

<u>*FIGURE 7-23.*</u> *Single Solenoid Controlled, Pilot-Operated, Two Position, Spring Returned Valve.*

Single Solenoid Valve. Figure 7-23. Part (A) of this illustration shows the miniature solenoid operator mounted "piggy-back" on top of the main body. The direct-acting solenoid operator in this case includes a single solenoid driving a 4-way pilot spool which is spring returned.

The main oil flow is handled by the large spool in the main body. Although the pilot valve spool is spring returned, the main spool is powered in both directions by hydraulic pressure obtained through the "piggy-back" pilot valve.

As the main spool shifts, the discharge oil from its end cavity is routed back through the piggy-back spool and is discharged to the reservoir through a drain line connected to the pilot drain. Back pressure in this drain line should be kept as low as possible. Any back pressure in the drain line or in the pilot spool cavity causes a mechanical load on the solenoid coil because of unequal spool areas; the pilot spool has a larger area on its left end. The additional load on the solenoid, on some applications, can cause heat to build in the coil windings resulting in possible burn-out. On existing equipment, if frequent coil burn-out is experienced with no apparent cause, the pilot drain back pressure should be investigated. High back pressure causes excessive wear on the seal into the solenoid compartment, with possible oil leakage into the solenoid compartment. External draining of a solenoid valve is always preferred when practical.

If the drain line is internally connected to the main tank port the valve is said to be internally drained. This method is used if it is impractical to run a separate drain line to reservoir. Some manufacturers permit internal draining if the tank line back pressure will not exceed a certain amount. Follow manufacturers instructions.

Part (B) of Figure 7-23 shows the complete graphic symbol for a single solenoid valve. Side-by-side comparison with the cross section, Part (A), will show the meaning of each part of the symbol. On circuit drawings the complete symbol may be used if necessary to exactly define the envelope contents. However, for most schematic drawings the simplified symbol of Part (C) is used because of its simplicity and because no further details of the valve need to be defined on the drawing.

The simplified symbol of Figure 7-18 can be used but a note should be added that the valve is pilot-operated, if this information could be important.

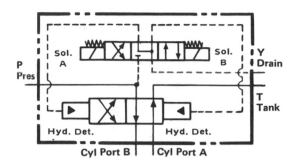

(B). Complete Graphic Symbol.

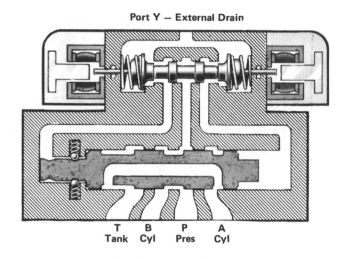

T   B   P   A
Tank   Cyl   Pres   Cyl

(A). Valve Cross Section.

(C). Simplified Graphic Symbol.

FIGURE 7-24. *Double Solenoid Controlled, Pilot-Operated, Two-Position, No-Spring Valve.*

Double Solenoid, Two-Position Valve. Figure 7-24. The valving action in the main body is the same as its counterpart, the direct-acting version of Figure 7-19. A momentary impulse of electric current to one solenoid causes the pilot ("piggy-back") valve spool to shift. This directs hydraulic oil, taken from the main inlet port, into the main spool end cavity to shift the spool. At the end of the current impulse, the pilot spool goes back to center but the main spool remains in its shifted position. It is retained in the shifted position, against accidental drift, by a detent mechanism.

Even though the main spool has only two working positions (no center neutral), a 3-position, spring centered, piggy-back solenoid valve is ordinarily used. This spool moves into a side position and remains there until the solenoid becomes de-energized, then centers itself. There are several advantages to using a 3-position pilot valve rather than one having only two positions: Reliability of shifting is increased because the centering springs help the solenoids when their gap is widest and they have the least power. Shifting of the pilot spool is faster because it is already half-way in its travel when the electric signal is received. It also relieves oil pressure lock between main spool and pilot spool, minimizing the possibility of spool drift under certain operating conditions. To relieve this pressure lock the pilot spool must have a float center neutral.

For safety, modern 2-position hydraulic valves include a detent, either ball and spring or a friction detent. Without detents it is possible for the spool to drift from mechanical vibration, from excessive flow or flow surges, or from gravity if the spool is not mounted with its axis vertical. Overloading a spool-type valve with excessive flow causes a self-shifting effect on the spool.

If the pilot spool has equal areas on both ends, as in this figure, the effect of back pressure in the pilot drain is not as likely to cause coil burn-out. Still, back pressure does load the end seals for increased spool friction, and should be avoided as much as possible. Refer to discussion of this point on single solenoid valve description.

The complete graphic symbol of Part (B), Figure 7-24, is seldom used. The simplified symbol of Part (C) is usually acceptable. Or, the symbol of Figure 7-19 is often used.

We recommend that all spool-type hydraulic valves be mounted with spool axis horizontal, whether the spool is detented, spring offset, or spring centered. This may become important on large valves where the spool may be quite massive.

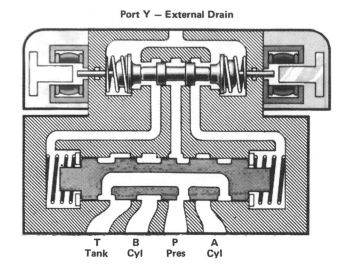

Port Y — External Drain

T — Tank   B — Cyl   P — Pres   A — Cyl

(A). Valve Cross Section.

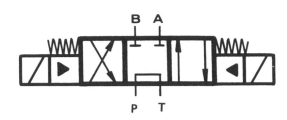

(B). Complete Graphic Symbol.

(C). Simplified Graphic Symbol.

FIGURE 7-25. Double Solenoid Controlled, Pilot-Operated, Three Position, Spring Centered Valve.

Double Solenoid, Three-Position Valve With Spring Centered Spool. Figure 7-25. The valving action in the main body is the same as in the direct-acting solenoid valve of Figure 7-20. When both solenoids are de-energized, the pilot spool is centered. It is a float center spool, to allow centering springs on the main spool to bring it to center without a pressure lock between pilot and main spools. When a solenoid is energized, the pilot spool shifts and directs hydraulic pilot oil obtained from the main valve inlet to the end of the main spool. Since both solenoids are attached to opposite ends of a common spool, energizing both solenoids at the same time will overheat the solenoid whose armature is not able to seat, and the coil will eventually burn out. Electrical circuitry should be designed so this cannot happen, perhaps using the back contacts of switches or relays so when a solenoid is energized it breaks the circuit to the other solenoid, thus positively eliminating the danger. On applications where the solenoids are cycling frequently, an overlap of even a few milliseconds on each cycle, where both solenoids are energized, can cause a slow build-up of heat over a period of time, and eventually burn out the coil.

To keep the spool shifted to a side position, current must be maintained on one of the solenoids. When both are de-energized, the pilot spool goes to float center, allowing springs on the main spool to center it also.

A wide variety of spool types is available, the more popular ones being shown in Figures 7-8 to 7-12. The tandem center spool is illustrated above.

A point worthy of note: Spool-type hydraulic valves, like air valves, usually have a fluid exhaust at both ends of the spool. To simplify external plumbing, these dual exhausts are normally joined internally in hydraulic valves and brought out on one tank port, T. In the typical construction shown here, the spool center provides a connecting passage to join the exhausts. Other manufacturers choose, instead, to use a cored hole through the body casting to join the two exhausts.

The complete graphic symbol of Part (B), Figure 7-25, is seldom used. Normally the simplified symbol of Part (C) or the symbol of Figure 7-20 is acceptable for circuit drawings.

Double Solenoid, Three-Position Valve With Pressure Centered Main Spool. Instead of centering springs, hydraulic pressure taken from the valve inlet is used to bring the spool to a center neutral position when both solenoids are de-energized. General information on spring centered valves applies to these types.

Pressure centering is a more positive, a more powerful way of forcing the main spool back to center when the solenoids are de-energized. It is more costly and more complicated and so is usually limited to use on applications where maximum reliability is essential, where high volume flows are to be handled, and particularly to high pressure operation above 3000 PSI, although it can be used with equally fine results at lower pressures.

Spool-type valves, operated at high pressure, are subject to spool "freeze". This is apt to occur if the spool remains shifted for a long time with high pressure on it. When, finally, the solenoids are de-energized, the spool sticks tightly to the side of its bore and centering springs are unable to move it to center. It is impractical to build the springs with sufficient strength to overcome spool freeze. Pressure centering provides a more powerful way to break the spool away to center neutral.

MOUNTING AND CONNECTING SOLENOID VALVES

Valve sizes 1-inch and smaller are usually mounted either on a subplate or on a manifold constructed by the user. Connections come out through the base of the valve and enter the subplate through O-ring seals. Circuit connections are made through NPTF dryseal or SAE straight thread ports on the bottom or sides of the subplate.

Valves larger than 1-inch usually have SAE 4-bolt pads on the side of the valve body for the four main ports plus NPTF or SAE straight thread ports for the auxiliary connections.

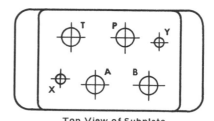

Top View of Subplate

FIGURE 7-26. Typical Subplate.

Subplate. Figure 7-26. Subplates for pilot-operated solenoid valves have four main portholes and two smaller auxiliary portholes. for external pilot and drain connections when and if required. The main ports are P (pressure inlet), T (tank return), A and B (cylinder connections). Auxiliary ports include Port "X" for connection of external pilot pressure if required, and Port "Y" for connection of a drain line to tank if required. If either of these ports is not used it should remain plugged.

Direct-acting solenoid valves mount on a similar subplate, usually smaller in size, having four main portholes as above, plus one external drain port.

PILOT SOURCE FOR PILOT-OPERATED SOLENOID VALVES

Pilot-operated solenoid valves must have a source of pilot pressure for shifting the main spool. This pressure is under the control of the miniature solenoid valve manifolded on top of the main body. On some circuits the pilot pressure can be obtained inside the valve itself from the inlet port. Pilot oil is taken up through an O-ring seal into the solenoid structure. When pilot pressure is obtained in this way the valve is said to be "internally piloted". Or, pilot pressure can be obtained from a source outside the valve, such as a separate pump. External pilot pressure can be connected to Port X on the subplate, and the valve is said to be "externally piloted".

In either case, for the main spool to shift, the solenoid must be energized, *and in addition*, pilot pressure of at least 75 to 100 PSI must be present to furnish shifting power.

Know

Lack of Pilot Pressure. Figure 7-27. Tandem and open center spools, and other types in which the inlet is open to tank in center position for pump unloading, must either be externally piloted or must have a restrictor placed in the pump line for retaining at least 75 PSI for pilot pressure when the spool is in center position.

In new designs, if the valve will not shift out of neutral when a solenoid is energized, the most likely cause is insufficient pilot pressure. Check this by temporarily connecting a 500 PSI pressure gauge to the pump line or at the valve inlet port.

Caution! Do not energize either solenoid while this gauge is connected; if the spool should shift, the gauge may be damaged.

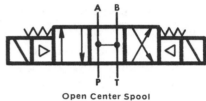

FIGURE 7-27. *Open Flow Spools.*

Once the spool has shifted into a side position, reactive pressure from the load will always be sufficient to keep the spool in its shifted position. Lack of pilot pressure is a problem only when attempting to shift the spool out of center. In this position the pump will be in an unloaded condition and there will be little or no pressure available for piloting. Use the manual override on the end of a solenoid coil. If the main valve spool will not shift in response to the override, this usually indicates lack of sufficient pilot pressure.

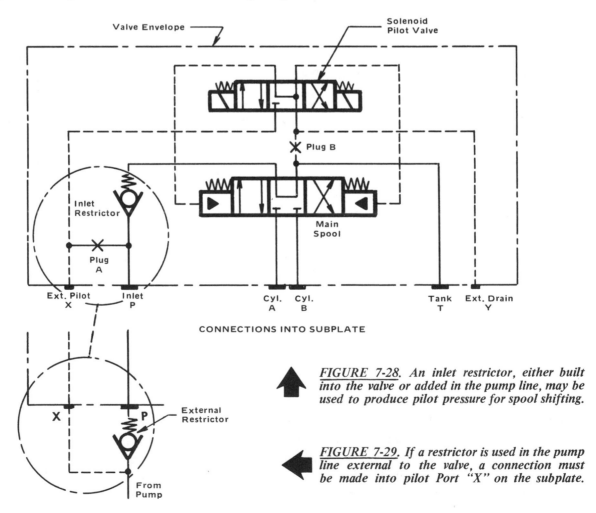

FIGURE 7-28. *An inlet restrictor, either built into the valve or added in the pump line, may be used to produce pilot pressure for spool shifting.*

FIGURE 7-29. *If a restrictor is used in the pump line external to the valve, a connection must be made into pilot Port "X" on the subplate.*

110

Internal Piloting. Figures 7-28 and 7-29. Valves which have any kind of a spool which unloads the pump line when the spool is centered, such as those in Figure 7-27, and in sizes of 1" or smaller, can be factory furnished with a restrictor built into the inlet port to retain at least 75 PSI in the pump line to be used as internal pilot pressure to move the spool out of center position. As shown in Figure 7-28, Plug "A" must be removed to provide an opening for the pilot oil to reach the solenoid structure. Also, the external pilot port, "X" must be plugged. On valves larger than 1" in size, since there is usually not sufficient space inside the valve body to add a restrictor, the user must add an external low pressure relief valve or a needle valve, adjusted for 75 PSI, in series with the valve inlet port as shown in Figure 7-29. Then the pressure upstream of the restrictor will serve as pilot pressure. It should be connected through a 3/8" steel tubing line into pilot port, "X" on the subplate, and Plug "A" must be installed inside the valve.

On most hydraulic systems, especially those of smaller size, it is more convenient and economical to use internal piloting. However, on larger systems where pump volume is quite high, the HP loss from an inlet restriction of 75 or more PSI may be excessive. A separate pump for pilot pressure may be a better alternative. Valves which have the pressure port blocked in center position never need an inlet restrictor. If one is present it should be removed.

External Piloting. An inlet restrictor in the pump line should never be used on any valve which is to be externally piloted, and if present should be removed. External pilot pressure must be connected to Port "X" on the subplate and Plug "A" must be added inside the valve body.

Tank Line Restriction. A restrictor can be added in the tank return line to develop pilot pressure, but this is not as desirable as in the pressure inlet. If used, the solenoid valve *cannot* be drained internally. An external drain line to tank from Port "Y" is *always* required. See bottom of this page.

Plug "A" Installation and Removal. Figure 7-30. Valve construction varies with different manufacturers, but Plug "A" can usually be installed and removed without disassembling or dismounting the valve. Common practice is to remove an access plug on the outside of the valve body. This exposes a passage for installing or removing Plug "A" with an Allen wrench. Whenever internal piloting is used, Plug "A" must be removed to open the passage. The solid plug can be removed and an orifice plug installed to reduce shifting speed of the main spool. When using internal piloting, Port "Y" is a convenient place to measure pilot pressure. Otherwise, Port "Y" should be plugged.

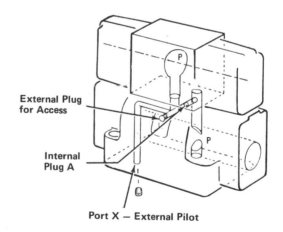

FIGURE 7-30. Phantom view of a pilot-operated solenoid valve showing internal passages for external/internal piloting changes.

CHOICE OF SOLENOID DRAINING METHOD

Internal and External Draining. Figure 7-31. Draining refers to the method by which the drain oil from the solenoid pilot valve is returned to tank. On some circuits the drain oil can simply be ported through a passage on the inside of the valve to the main tank port (internal draining). On other circuits a separate drain line must be run directly to tank from Port "Y" on the subplate. Whether or not an external drain is needed depends mainly on the level of back pressure which may appear on the main tank port of the valve. If there is a restrictor in the tank line or if the tank port is to be connected to another valve instead of directly to tank, as in a tandem circuit involving several valves in series, then the external drain line must be provided from Port Y to tank.

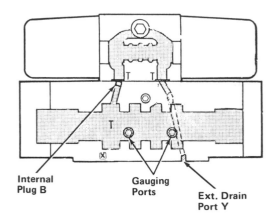

Internal Draining of Solenoid Valve. Figure 7-31. The drain cavities on both ends of the pilot spool are brought through the manifold surface and into the main valve body. One of these passages leads into the tank groove of the main spool and is used for internal draining. This passage must be opened by removing Plug "B". The solenoid valve will have to be unbolted and removed from the main body for access to this plug. Port "Y" on the subplate is left plugged.

Internal
Plug B

Gauging
Ports

Ext. Drain
Port Y

FIGURE 7-31. Passageways which must be opened or closed for changes in the method of draining the solenoid pilot valve.

External Draining of Solenoid Valve. Figure 7-31. The other passage into the main body leads to Port "Y" on the subplate. For external draining, install Plug "B" to block the internal drain and connect an external drain line to Port "Y".

If a pilot-operated solenoid valve will not shift out of center position when the appropriate solenoid is energized, this could either be a symptom of insufficient pilot pressure or could be related to a problem in the electrical control circuit.

On 4-way valves which have a closed pressure port in neutral, insufficient pilot pressure is rarely the problem, although if there is any doubt, it can be measured with a gauge. On these valves the problem is more likely to be electrical.

On 4-way valves which have the pressure port open to tank in neutral, and which are internally piloted, there must be at least 75 PSI on the pressure port with the spool in center position. This can be determined with a pressure gauge. If the valve is externally piloted, the source of pilot pressure must be checked to be sure it is adequate.

To determine whether the problem is hydraulic or electrical, we recommend these tests be made in the order given:

(1). First, try shifting the valve by using the manual overrides. These are located on the end of each solenoid structure. They shift the pilot valve without the solenoids being energized. If the machine starts working, this indicates sufficient pilot pressure, and the problem is electrical.

(2). By inserting a small rod into the manual override and holding it loosely while actuating the "Start" button, it is easy to tell if the pilot spool is shifting in response to the Start button.

(3). Check out the electrical circuit by measuring voltage across the solenoid coil when the Start button is pressed. Small pilot lamps permanently connected across each solenoid coil are a valuable aid in making a quick check of the electrical circuit.

(4). Check the voltage rating stamped on each solenoid coil to be sure it matches the control circuit voltage of the machine. Check each coil to be sure it is not burned out.

(5). If the machine does not start working in response to the manual overrides, and if the electrical circuit is OK, test the level of pilot pressure with a low range pressure gauge as previously described.

SEVERAL BRANCH CIRCUITS OPERATING FROM ONE PUMP

Some of the information under this heading is a summary of information given elsewhere. Where applicable, the reader is referred to other parts of this volume or to another volume in the "Industrial Fluid Power" series of textbooks.

Although these circuits are shown for solenoid operation, most of them can also be used with individual manual valves. Multiple branch operation with sectional valves will be covered later.

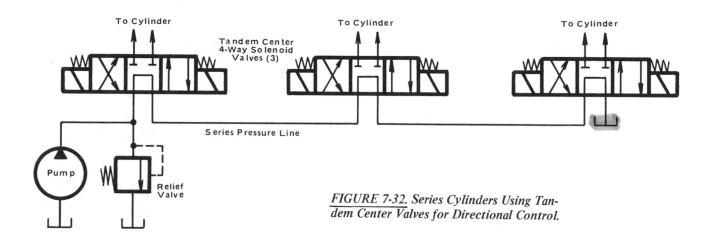

FIGURE 7-32. *Series Cylinders Using Tandem Center Valves for Directional Control.*

Series Valves. Figure 7-32. The 4-way valves for all branch circuits are connected in series with each other. The circuit uses tandem center 4-way valves, and is intended for fixed displacement pumps. Oil from the pump loops from valve to valve and finally is discharged to tank at the last valve in the string. When all valves are centered, the pump can unload through the valve centers.

Any one of these branch circuits can operate in either side position, but only one branch at a time can be operated with full pump pressure available to the branch. If more than one of the valves is shifted at the same time, pump pressure will divide between the two branches in direct proportion to the load resistance against each branch, with no branch having priority over any other branch. Full pressure is not available to either branch.

Additional information will be found in Volume 1, Chapter 4.

Series-Parallel Valves. Figure 7-33. This is a variation of the straight series circuit shown in the preceding diagram. It is also intended for fixed displacement pumps. When used on applications for which it is suitable, it gives slightly higher efficiency and will allow the simultaneous operation of several branch circuits at full pressure, under certain conditions. Valve 1 is always a tandem center valve for pump unloading. Valves 2 and 3 may be either two position or three position valves either single or double solenoid. They do not have to unload the pump. All valves are connected in parallel on the pump line, so all may have access to full system pressure at the same time.

The restriction to the use of this circuit is that Valve 1 must always be shifted to a side position before any of the other valves have pump pressure available to them. It is the ideal circuit for clamp and work applications. Valve 1 operates the clamp cylinder. It is always the first valve to be shifted and is the last one to be released at the end of a cycle. While it is in either side position, the other valves can operate in any sequence, together or separately, and in either side position with full pump pressure available to both valves.

113

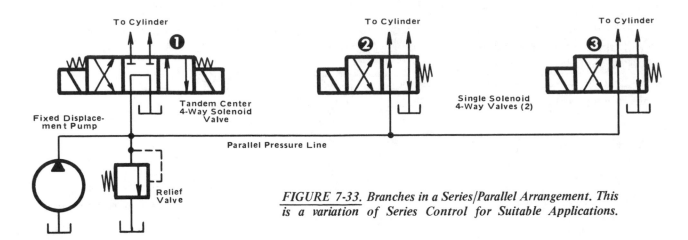

FIGURE 7-33. Branches in a Series/Parallel Arrangement. This is a variation of Series Control for Suitable Applications.

This circuit offers several advantages over a full parallel circuit. Valves in Branches 2 and 3, being single solenoid, are less expensive than double solenoid models. It allows the use of less expensive pumps, such as gear pumps, in place of the more expensive variable displacement models.

Parallel Valves, Variable Displacement Pump. Figure 7-34. Four-way valves in all branches are connected in parallel to the pump line. This permits several branches to be operated simultaneously with full pump pressure available to all branches in which the 4-way valve is shifted to a side working position. The pump flow will normally divide between the branches in inverse proportion to the load resistance in each, the more heavily loaded branches receiving the least flow. If a more equitable division of oil between the branches is required, flow control valves may be placed either in the pressure inlet to each 4-way valve or in the lines connecting valve to cylinder in each branch.

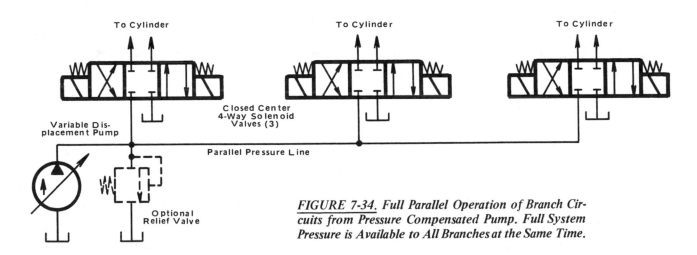

FIGURE 7-34. Full Parallel Operation of Branch Circuits from Pressure Compensated Pump. Full System Pressure is Available to All Branches at the Same Time.

In this circuit a variable displacement, pressure compensated pump is used. Instead of unloading through the valve spool centers, this pump will unload itself by means of its pressure compensator when all 4-way valve spools are centered. It will maintain full system pressure but will pump only enough flow to make up for slippage losses in the pump itself and in the 4-way valves.

Full closed center valves are shown for all branches. However, other spools as shown on Page 98 may be used, but any spool used in this circuit must have a closed pressure port when centered.

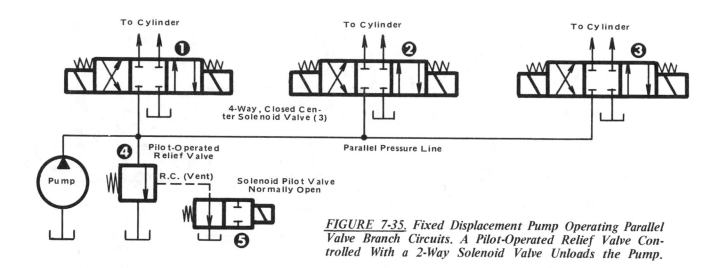

FIGURE 7-35. Fixed Displacement Pump Operating Parallel Valve Branch Circuits. A Pilot-Operated Relief Valve Controlled With a 2-Way Solenoid Valve Unloads the Pump.

Parallel Valves, Fixed Displacement Pump. Figure 7-35. This circuit resembles the preceding circuit for full parallel operation of several branch circuits, but also includes an electrical unloading system for the pump, allowing the use of a gear, vane, or other fixed displacement pump.

The pump relief, Valve 4, is a pilot-operated type relief valve with remote control (R.C.) venting port for dumping the pump by remote control. This relief valve and circuits for its use are fully described starting on Page 142 of this volume.

The R.C. port of the relief valve can be vented to tank by de-energizing the solenoid of Valve 5, or it can be blocked by energizing Valve 5 solenoid. When the R.C. port is vented, the relief valve cannot build up pump pressure. The pump can discharge to tank through the main ports of the relief valve at a low back pressure, about 75 PSI. When the R.C. port is blocked (by energizing the solenoid of Valve 5), the relief valve can resume normal operation and will allow pump pressure to build up to the pressure setting of the relief valve.

When designing an electrical control circuit, Valve 5 solenoid remains de-energized between cycles to keep the pump unloaded, but must be energized when any of the 4-way valve solenoids is energized. When energized, it causes relief Valve 4 to return to its normal function of maximum pressure protection. Electrical circuits for this type of unloading control are shown in Chapter 5 of the Womack textbook "Electrical Control of Fluid Power". See book listings inside rear cover.

The circuit above shows a standard pilot-operated relief valve controlled with a separate miniature 2-way, normally open, solenoid valve located a short distance away. For circuit compactness and to save plumbing, pilot-operated relief valves are available with solenoid operators mounted on them.

Spool-Type Flow Dividing Valves. Figure 7-36. Branch circuits may be isolated from one another by using a flow divider valve (or valves) to split the pump flow equally (or in some ratio) between two separate branch circuits. Load or speed changes in one branch will have no effect on the operation of the other branch. In this figure, assuming flow divider Valve 4 has a 1:1 split ratio, half of the pump flow will be available only to 4-way Valve 1 serving Branch 1. The other half will be split again by divider Valve 5 with half going to Branch 2, the other half to Branch 3. When all 4-way valves are in center position, the pump will be unloaded, and will be working only against the normal flow loss through the dividers which may be in the range of 100 to 200 PSI.

Spool-type flow dividers would seem to be an ideal answer to the problem of using one pump to supply flow and pressure to several isolated branch circuits. But if the application is not suited to

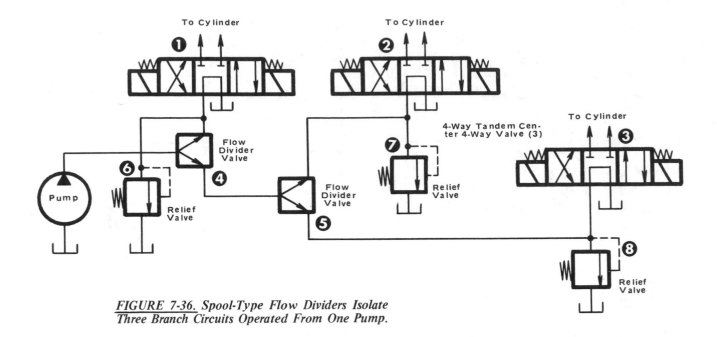

FIGURE 7-36. *Spool-Type Flow Dividers Isolate Three Branch Circuits Operated From One Pump.*

them, they can produce very high power losses and can quickly overheat the system. To illustrate, suppose Valve 1 is shifted to a working position while Valves 2 and 3 are left in center position. The pump is now required to produce the pressure required in Branch 1 at the full pump flow. However, only one-half its flow is producing useful output in Branch 1, while the other half is being converted into heat since it is not producing useful output in Branches 2 and 3. The heat is produced in the flow dividers as their spools move to a throttling position to produce an artificial load equal to the load in Branch 1. Great care must be exercised to make sure the particular application is the kind that will use the maximum amount of pump output and generate the least amount of heat. Spool-type flow dividers are covered in great detail in Volume 3, and the reason for their unsuitability in certain applications is explained.

Relief valves must be used to protect each output leg of a flow divider. Valves 7 and 8 protect both legs of divider Valve 5, and in turn also protect the bottom leg of divider Valve 4. The upper leg of divider Valve 4 is protected with relief Valve 6. When these relief valves are used, a relief valve on the pump line is not necessary.

Rotary-Type Flow Dividers. Figure 7-37. A rotary flow divider is like two gear pumps (or motors) with shafts coupled together so one cannot rotate without the other being forced to rotate at the same speed. Their inlets are joined and receive the pump flow. Their outlets deliver equal flow, assuming both have equal internal displacement.

Because of possible pressure intensification in the circuit we recommend using only one rotary flow divider per pump. In this diagram, assuming a 1:1 flow split in the divider, half of the pump flow goes to 4-way Valve 1. The other half goes to Valves 2 and 3 which happen, in this instance, to be in a tandem circuit. This represents only one possibility in the application of these units. Also because of pressure intensification we recommend keeping the flow split ratio at 1:1.

Rotary dividers act differently than spool-type dividers. When the 4-way valve in one branch, Valve 1 for example, is centered, the oil flow assigned to that branch goes to tank at low power loss, but the pressure which is represented by that power is transferred through internal shaft torque to the other branch, diverting the full pump power output to Branches 2 and 3. Therefore, rotary dividers will fit a

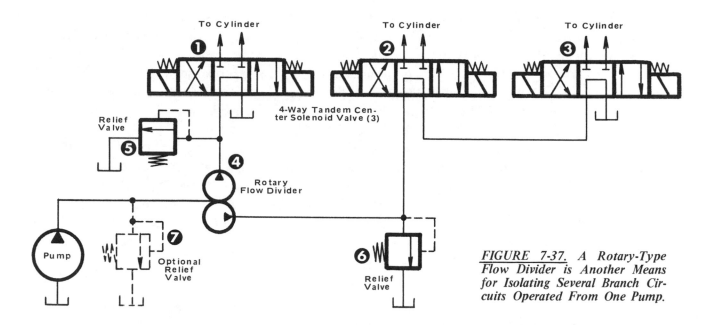

FIGURE 7-37. A Rotary-Type Flow Divider is Another Means for Isolating Several Branch Circuits Operated From One Pump.

wider range of applications with more efficient operation than will spool-type dividers. For a more complete treatment of both spool-type and rotary dividers see Volume 3 textbook.

Usually a relief valve is recommended on the pump line and in each branch. The pump relief Valve 7 will not give complete circuit protection because of pressure intensification in one branch circuit when the other branch is unloaded.

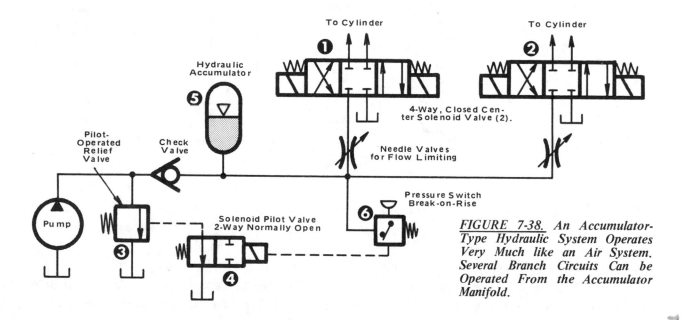

FIGURE 7-38. An Accumulator-Type Hydraulic System Operates Very Much like an Air System. Several Branch Circuits Can be Operated From the Accumulator Manifold.

Accumulator Systems. Figure 7-38. This is another way of operating several branch circuits in parallel so they all can operate at full system pressure simultaneously. Accumulator systems are also used on many applications where the running time of the pump under full pressure is very short compared to its idling time between cycles. This permits the use of smaller pumps and electric motors, which results in a more economical design. Accumulator systems are covered in Chapter 6 of Volume 1.

If desired, the pump can be a variable displacement type with a pressure compensator to limit pressure. In this case, the solenoid valve, 4, and the pressure switch, 6, can be eliminated. However, in most systems, fixed displacement pumps are used and there must be a means for unloading them when the accumulators become fully charged. An electrical unloading system is used in this diagram. When the accumulators come up to full charge, usually 3000 PSI, the "break-on-rise" pressure switch opens. This de-energizes solenoid Valve 4, which vents the main relief Valve, 3, and allows the pump (still running) to unload its flow to tank through the main ports of the relief valve.

On most parallel branch circuits, flow control valves may be required in each branch to not only limit the speed of the cylinder but to achieve a satisfactory division of flow to the two branches. They can be installed in the inlets of the 4-way valves as shown, or in the lines connecting to the cylinders.

Instead of the electrical unloading system shown in the diagram, hydraulic unloading could be used in which a pilot-operated accumulator unloading valve, installed across the pump line, snaps open when the accumulators reach full charge, then snaps closed when sufficient oil has been used out of the accumulator to cause system pressure to drop to about 80% of full charge pressure.

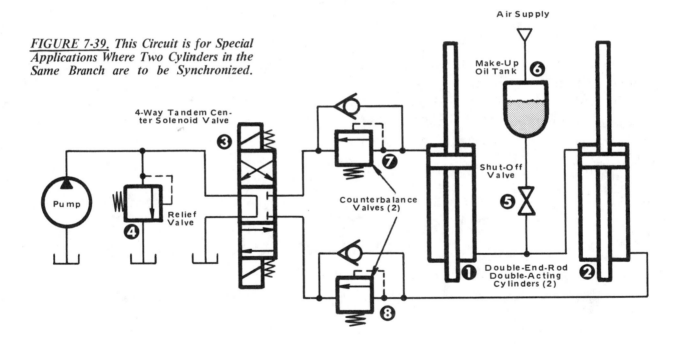

FIGURE 7-39. This Circuit is for Special Applications Where Two Cylinders in the Same Branch are to be Synchronized.

Series Cylinders. Figure 7-39. This is a means of operating a pair of equal bore, equal stroke, double-end-rod cylinders from one pump and keeping them moving in synchronization throughout their stroke. See Chapter 12 of this book for more details of the application.

As Cylinder 2 moves forward, oil discharged from in front of the advancing piston is ported to the back end of Cylinder 1. This forces Cylinder 1 to move at the exact speed of Cylinder 2 regardless of the amount of load carried by each. On retraction, the oil forced from underneath the piston of Cylinder 1 forces Cylinder 2 to move at the same speed.

Since the cylinders are in series they must divide the system pressure between them in proportion to the load carried by each. The total force produced by both cylinders is the same as could be produced by one of them working alone on the same pressure.

Counterbalance Valves 7 and 8 must be installed as shown, Valve 7 to prevent Cylinder 1 from overrunning the oil supply while extending, and Valve 8 to prevent Cylinder 2 from overrunning the oil supply while the cylinders are retracting.

SECTIONAL VALVES FOR DIRECTIONAL CONTROL

A sectional (or bank) valve is a composite assembly of two or more valve sections bolted together. They are intended for operation of several branch circuits from one hydraulic pump. Although some manufacturers offer monoblock valves in which several spools are contained in a single casting, the sectional construction seems to be more popular because of its versatility. A complete valve bank can be assembled from individual sections having different spool types or different features. Manufacturers and distributors can give better delivery to customers, and can make up a bank valve to exact requirements.

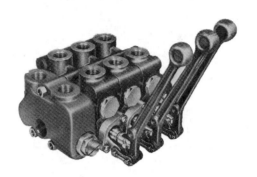

FIGURE 7-40
Typical 4-way Sectional Valve.

A sectional valve includes the front section with inlet port and built-in pressure relief valve. Then, any number of 4-way or 3-way sections with choice of center neutral porting can be added. Finally, the rear section completes the assembly with outlet port and sometimes with a power beyond port. Individual valve sections can include various optional features such as a load holding check valve in the inlet, relief valves in one or both cylinder ports, a pilot-operated relief valve in one or both cylinder ports, and sometimes other optional features.

Sectional valves find their best application with manual lever rather than solenoid actuators because the spool can be modulated to split the pump flow between two branch circuits in any proportion. With solenoid operation the spool shifts quickly from neutral to full flow in a side position and the flow cannot be split except on those servo-type electrically operated valves with torque motor actuator. Solenoid operated sectional valves are used primarily where remote control is necessary or on very large valves which it may be impractical to operate manually.

Sectional valves are offered with 2 to 12 (or even more) sections. On open center types, in neutral, the pump flow is ported directly through all spool centers and to tank at low back pressure. An operator can shift two spools at the same time, and by modulating with the spool nearest the valve inlet, can split the flow in any proportion between the two branch circuits.

Sectional valves are used primarily on mobile applications where an operator is continually monitoring the operation of all cylinders. Solenoid valves would be awkward for this type of operation. On the other hand, solenoid valves work out much better for industrial machinery where the machine repeats the same cycle automatically every time.

Closed center sectional valves have blocked center flow when all handles are in neutral. They cannot be used for unloading a hydraulic pump. For this reason they are used principally with variable displacement, pressure compensated pumps.

Sectional valves have important advantages over several individual valves plumbed together with pipe nipples. The assembly is more compact, taking less space on the machine. But most important, valving functions are possible which would be impossible with individual valves. Additional cored passages can be provided so the valve can be open center when all handles are in neutral and yet two or more branches can be operated simultaneously with full pressure on each.

Manufacturers offer many variations in construction and action. Each spool can have a different action. Some spools may be 3-way and others can be 4-way with variations in neutral porting. Some may have a fourth position with float or regenerative action. Some spools may be spring centered while others may be detented, or detented in only the fourth position. While manufacturers do have some standard catalog models, most of the market is for custom assemblies with number and type of spools and optional features to suit a specific application.

In the diagrams to follow, all spools are shown in neutral position. To aid in tracing oil flow when the spool is shifted to a side position, some students may find it helpful to make a movable overlay using a piece of tissue or tracing paper. Lay the paper over one valve spool and trace the block outline and all porting inside the blocks. Move this paper up and down, aligning it in turn with the external connections. This will reveal the flow paths for the oil when the valve is shifted to each position.

Most sectional valves include a relief valve in the front end cover to serve as the system relief valve. This saves the cost of an external relief valve and its plumbing.

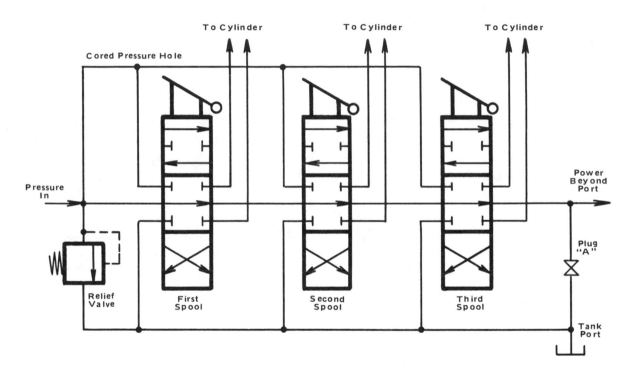

FIGURE 7-41. Schematic Diagram of a 3-Section, Parallel-Type Sectional Valve With Power Beyond.

Parallel Ported Sectional Valves. Figure 7-41. In this type valve the incoming oil from the pump is connected through cored holes to the inlet of all spools in a parallel connection. Thus, full pump pressure is available, in a parallel arrangement, to all spools which are shifted into a side position. When two branch circuits are in parallel, the oil naturally flows more freely into the branch with the least load. To cause cylinders in both branches to move, the operator may find it necessary to throttle in the lightly loaded branch to force more of the oil into the heavily loaded branch.

This figure shows open center porting in which pump oil flows freely through all spools to tank when they are centered. To change to closed center porting, the power beyond port should be plugged. With Plug "A" also in place, center flow through the valve is closed.

In a parallel type sectional valve, pump oil has equal access to all spools which are shifted. The first spool, for example, does not have flow priority over downstream spools.

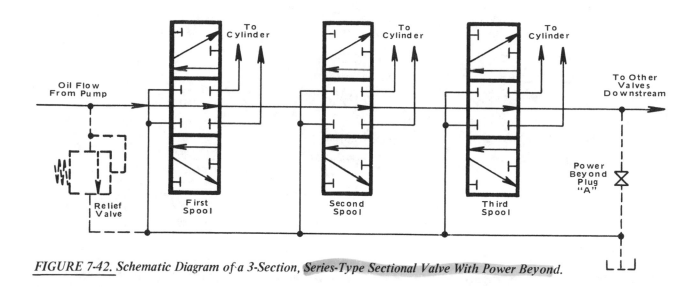

FIGURE 7-42. *Schematic Diagram of a 3-Section, Series-Type Sectional Valve With Power Beyond.*

Series-Type Sectional Valves. Figure 7-42. When all handles are centered, pump oil can flow through the center of all spools. This center flow can be diverted freely to tank to make an open center valve, or can be blocked to make a closed center valve. Open center operation is preferred when using a fixed displacement pump, such as a gear pump, and closed center operation is preferred when using a variable displacement pump with pressure compensator.

If any spool should be shifted to a side position, the flow of inlet oil is diverted into that branch circuit, and is cut off from all spools downstream. One advantage of a series type sectional valve over several individual valves connected together with pipe nipples is that return oil from all branch circuits drains into the valve case and goes out the tank port at a minimum pressure loss without having to go through all downstream valve centers before going to tank.

In a series type sectional valve, if any two spools are shifted at the same time to a full side position, the spool nearest the inlet will receive the entire flow, and all other spools downstream will receive none. With this priority characteristic in mind, it is good design to connect the more important branch circuits to spools nearer the inlet. If oil priority is not important, then the branch circuits should be assigned to valve sections which have some kind of physical orientation with the cylinder (or hydraulic motor) operated by each; that is, cylinders on the front of the machine should operate with handles on the front of the valve bank, etc. This makes it easier for an operator to learn and remember the operation of a complicated machine.

If two branch circuits must be operated at the same time, the operator must shift the downstream spool to its full side position, then modulate the flow with the handle nearest the valve inlet. This will divide the flow and control the proportion going into each branch. Full pressure is available to both branches as long as the upstream handle is in a partially shifted position.

Tandem Center Sectional Valves. There is a third kind of sectional valve which is called a tandem center type. The action is the same as if several individual valves were externally connected together with pipe nipples. It has series action but return oil from branch circuits does not go into the valve case and directly to tank; it goes out the tank port, and this port is connected to the inlet of the next section. This creates a higher pressure loss than in a series valve. In other respects it acts very much like a series type sectional valve. Flow can be split by modulating with the handle nearest the valve inlet, and full pressure is available to two branches while the first section is in a partially shifted position.

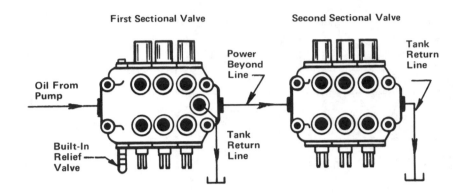

FIGURE 7-43.

When two sectional valves are connected in series, only the first section is required to have the power beyond feature.

Power Beyond Feature. Figure 7-43. This is an optional feature available for both parallel and series type sectional valves. All sectional valves must have this feature when either a single valve or another sectional valve is to be added downstream to operate from the same pump flow. Or, when the discharge from the outlet port of the sectional valve is to be used for any purpose which would create high back pressure. One manufacturer calls this feature "High Pressure Carryover".

Power beyond requires two outlets on the valve, usually on the rear cover. One outlet is the regular tank port and must always be connected directly to tank. The other outlet is the power beyond port which carries the center flow through the valve. This is the outlet which must be connected to the inlet of the next valve downstream. Plug "A" (see Figures 7-41 and 7-42) must be in place (or already installed) inside the valve to keep center flow from coming out the tank port.

On applications where the valve has a power beyond outlet and power beyond operation is not desired, Plug "A" must be removed and the power beyond outlet should be plugged.

Most sectional valves are of the open center type where the center flow goes directly to tank when all spools are centered. To make a closed center sectional valve, the valve must have a power beyond port which can be plugged. Also, Plug "A" should be installed. The main tank port should never be plugged because the internal case cavity is not rated to withstand high pressure.

Figure 7-43 shows two valve assemblies connected to operate from one pump. The first assembly must have a power beyond port. The second assembly does not need one. The relief valve built into the inlet of the first assembly serves as relief protection for both sections.

Caution! Serious damage can be caused by connecting two valve assemblies in series unless the first assembly has a power beyond port. The first assembly would operate normally but the damage would be caused when a spool in the second assembly was shifted. This would cause load pressure to back up against the tank port of the first assembly and against the outlet side of the pressure relief valve in the first assembly. The relief valve would become useless for pump protection.

If both assemblies have a built-in relief valve, and if a spool in each assembly is shifted at the same time, pump pressure could rise to the sum of both relief valve settings, possibly doubling the pressure in the pump line and possibly destroying the pump.

Back pressure from the second assembly, backing up on the tank port, could possibly burst the casting in the first assembly, or if the area on opposite ends of the spools is not the same, the spools could shift by themselves. Also, back pressure might blow out the spool seals in the first assembly or make the spools very difficult to shift.

Limitations on Power Beyond Operation. Refer to Figure 7-43. When two sectional valves are joined, although each one may be designed for parallel operation of all spools within its own assembly, the parallel operation does not carry through from the first to the second assembly. That is, the spools in the second assembly are in parallel with each other but not in parallel with spools in the first assembly. For example, if a spool in the first assembly and a spool in the second assembly are shifted at the same time they will not operate in parallel; they will be placed in series with one another and will act as described for a series type sectional valve. The spool in the first assembly will have priority on the oil flow and would have to be modulated if flow were to be split off to the second assembly. Therefore, when designing a machine with more than one sectional valve on the same pump flow, all spools which must operate in parallel must be in the same assembly.

However, when two series or tandem type sectional valves are connected together, the action is the same as if all circuits were in one assembly.

Adding a Second Sectional Valve. Figure 7-44. If necessary to add one or more additional circuits to an existing sectional valve installation, the preferred method is to purchase the additional sectional valve with power beyond feature and install it upstream from the existing sectional valve. The relief valve in the new section will serve for both sections. The old relief valve can be blocked or screwed down to maximum pressure. Note that spools in one section will not operate in parallel with spools in the other section.

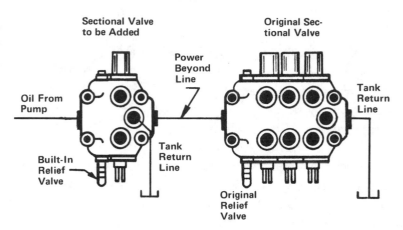

FIGURE 7-44. *If additional valving must be added to an existing system, it should be added upstream, and the added valve must have the power beyond feature.*

Spools Types for Sectional Valves. The spool types described here are the more popular ones. Some of them may not be available on some brands. On all listed spools the inlet is open to downstream when the spool is centered. On some brands additional features may be available such as load holding checks, pilot-operated checks in one or both cylinder ports, cylinder port relief valves, etc.

(A). Standard 4-Way. For double-acting cylinders. Both cylinder ports are closed in center.

(B). Standard 3-Way. For single-acting cylinders. The cylinder port is closed in center position.

(C). Motor*. All ports connected to tank including pressure port. For hydraulic motors only. Free running in center position. See footnote warning on cylinder operation with this spool.

(D). Float. For connecting both cylinder ports together and to tank while keeping them isolated from center flow through the valve. For either cylinder or motor operation.

(E). Regenerative*. Both cylinder ports connected together and to inlet. Normally used only in a 4th (not neutral) valve position for rapid advance of single-end-rod cylinders.

Warning! Do not use these spools for operating a single-end rod (unbalanced piston area) cylinder except as the last section downstream with no appreciable back pressure beyond it. Exposure of both cylinder ports to pressure at the same time might cause drifting in a single-end-rod cylinder.

REGENERATIVE CIRCUITS FOR DIRECTIONAL CONTROL

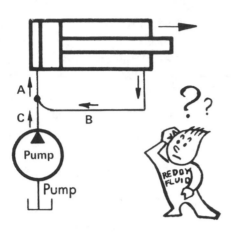

FIGURE 7-45. *Basic Principle of Fluid Power Regeneration.*

Regenerative Principle. Figure 7-45. Regeneration in any device is the principle of taking a part or all of its output and feeding it back to the input in order to further increase the output. In fluid power, regeneration is obtained by taking discharge oil from the rod end of a cylinder and feeding it back to the blind end, to join the pump oil, to increase cylinder travel speed. In its simplest form it consists of a direct connection of rod port back to blind end port. This causes the cylinder to travel forward at a faster speed rate than with pump oil alone. Suitable means of retracting the cylinder will be discussed later

Rod oil can enter the blind end against pump pressure because of pressure intensification that takes place in the rod end of any single-end-rod cylinder. Pressure intensification is explained in Volume 1, Chapter 2.

The purpose of regeneration is to increase travel speed of a cylinder in one direction without increasing pump size. Normally it is used only on hydraulic circuits; seldom, if ever, on air circuits. It is applicable only on single-end-rod, double-acting cylinders; that is, those which have a substantial difference in area on opposite sides of their piston. Generally the most practical results are obtained with 2:1 ratio cylinders; those having a piston area approximately twice the rod area, although it can be used with other piston/rod ratios. It can work only on the extension, never the retraction stroke.

Regenerative circuitry is most suitable for two general applications: (1), those needing a rapid approach to the work at low tonnage. Valving can be added to take the cylinder out of regeneration at some point in its stroke and into full tonnage at a lower speed. This gives an efficient 2-speed action without flow controls and with only a single-speed pump. (2), those applications needing approximately equal tonnage and speed in both directions. (A 2:1 ratio cylinder must always be used for this).

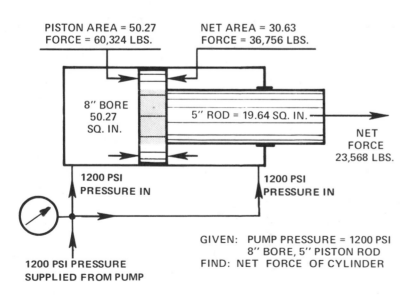

FIGURE 7-46
Force Diagram for a Regenerative Cylinder Calculation.

Regenerative Circuit Calculations. Calculations are made a little differently than for standard circuits, but if an orderly procedure is followed, solving for force, travel speed, and oil flow volume, always in that order, calculations are easily made.

Calculations of Force. Figure 7-46. Force is sacrificed in order to gain a greater travel speed. It is apparent in this figure that the entire piston area does not produce force. The annular area around the rod, having full PSI on it, cancels an equal area on the blind side. The final result is that force is produced only on an area equivalent to the rod area. This force-producing area is shown in this figure.

124

FORCE OF REGENERATIVE CYLINDER

The force developed in regeneration is equal to the rod area times the PSI gauge pressure.

See regenerative cylinder design chart on Page 248

Rule-of-thumb

Solution of Sample Problem. Figure 7-46. Cylinder force while in regeneration may be found with the rule-of-thumb given above or as follows: Piston force in the forward direction = area x gauge pressure, or 50.27 (the area of an 8″ diameter piston) x 1200 PSI = 60,324 pounds. Counterforce (in the opposite direction) = annulus (net area) x gauge pressure, or 30.63 x 1200 PSI = 36,756 pounds. The annulus area was found by subtracting rod area of 19.64 from full piston area of 50.27. The net forward force while in regeneration is, then: 60,324 − 36,756 = 23,568 pounds. The same answer could be obtained with the rule-of-thumb: 19.64 x 1200 PSI = 23, 568 pounds.

Calculation of Speed. Figure 7-47. Regenerative oil from the rod end an equal volume behind the piston; this is shown as a shaded area. Therefore, the pump needs only to fill the center space of the unshaded area. Cylinder bore has nothing to do with its speed while in regeneration, and for this reason is not given with the problem. After the cylinder is taken out of regeneration, its speed is determined by its bore and the rate of pump flow.

The hydraulic speed table on Page 255 can be used to determine regenerative speed by using rod diameter as piston diameter. See also Page 248.

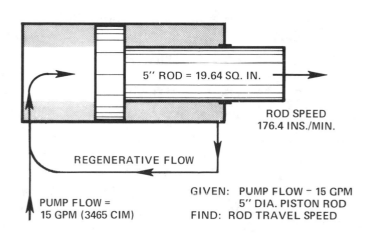

5″ ROD = 19.64 SQ. IN.

ROD SPEED
176.4 INS./MIN.

REGENERATIVE FLOW

PUMP FLOW =
15 GPM (3465 CIM)

GIVEN: PUMP FLOW – 15 GPM
5″ DIA. PISTON ROD
FIND: ROD TRAVEL SPEED

FIGURE 7-47. Regenerative Speed Calculation.

Solution of Sample Problem. Figure 7-48. Use rule-of-thumb: *Speed = Pump Flow ÷ Piston Area.* Pump flow must be in CIM (cubic inches per minute). As a reminder, 1 GPM = 231 CIM. Piston area is effectively the rod area while the cylinder is in regeneration as explained in the paragraphs above. Travel speed will be in inches per minute. Solution: Speed = 3465 ÷ 19.64 = 176.4 inches per minute.

Rule-of thumb

SPEED OF REGENERATIVE CYLINDER

(Inches per Minute)

Speed = Pump Volume (CIM) ÷ Rod Area
All data must be in minutes and inches.

See regenerative cylinder design chart on Page 248 of Design Data Section

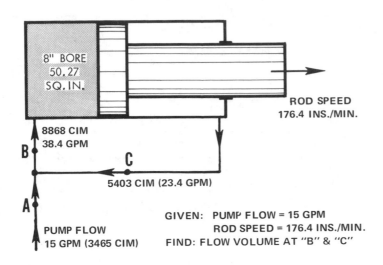

FIGURE 7-48. Regenerative Flow Volumes.

GIVEN: PUMP FLOW = 15 GPM
ROD SPEED = 176.4 INS./MIN.
FIND: FLOW VOLUME AT "B" & "C"

Flow Volumes. Figure 7-48. To properly size the valving needed in parts of the regenerative circuit, GPM flow in blind end and rod end should be calculated. The same cylinder example is used here as for Figures 7-46 and 7-47.

First, the piston and rod travel speed should be calculated as explained in Figure 7-39. This was 176.4 inches per minute. Obviously, the GPM at Point B, whatever it may be, certainly must be sufficient to produce 176.4 inches per minute displacement when working against full piston area. GPM to produce this speed against an 8" diameter piston can be calculated as in the example below or taken from the speed table on Page 255. Next, GPM at Point C is calculated by subtracting pump GPM (at Point A) from flow into the cylinder at Point B. C (GPM) = B (GPM) – A (GPM).

Solution to Sample Problem of Figure 7-48. Travel speed of this cylinder with an 8" bore and 5" rod has been determined to be 176.4 inches per minute from the preceding problems. To find flow at Point B when the cylinder is supplied from a 15 GPM pump, multiply piston area times length of travel in one minute's time: 50.27 x 176.4 = 8868 cubic inches per minute. Divide by 231 to convert to GPM: 8868 ÷ 231 = 38.4 GPM. To find flow at Point C, we know this must of necessity be the difference between pump GPM and actual GPM entering the cylinder at Point B: 38.4 GPM – 15 GPM = 23.4 GPM.

Rule-of-thumb

TO FIND FLOW VOLUMES IN REGENERATIVE CYLINDERS

1. Find piston travel speed, inches per minute. (Page 255).
2. Solve for GPM entering blind end, Point B, Figure 7-48.
3. Find regenerative flow, Point C, by subtracting pump GPM.
Note: See Volume 1 for method of calculating cylinder speed.
See also regenerative cylinder design chart on Page 248.

Estimating Oil Flow in Regenerative Cylinders. The chart below shows oil flow on exact 2:1 or 4:1 ratio cylinders. Catalog cylinders are seldom exact ratios (the one in Figures 7-46, 7-47, and 7-48 is not), but the chart may be useful for estimating flows or as a rough check against calculations.

FOR QUICK ESTIMATION OF FLOW VOLUMES

See regenerative cylinder design chart on Page 248.

Rule-of-thumb

Circuit	2:1 Ratio Cyls.	4:1 Ratio Cyls.
Regenerative Flow, Point C	*1 x Pump GPM*	*3 x Pump GPM*
Total Flow at Point B	*2 x Pump GPM*	*4 x Pump GPM*
Flow out Blind End, Retraction	*2 x Pump GPM*	*1.33 x Pump GPM*

GENERAL NOTES ON REGENERATIVE CIRCUITS

(1). When in regeneration, the cylinder force produced will be that of pump PSI times rod area. The remainder of the piston area is neutralized by an equal and opposing area on the rod side exposed to the same pump pressure.

(2). If and when the full cylinder force is needed, the circuit must be valved out of regeneration by disconnecting the rod end of the cylinder from pump pressure and re-connecting it to tank. Some of the following circuits show how this may be done.

(3). Regeneration is ordinarily used only with large rod cylinders, approximately 2:1 ratio between blind and rod end areas. With smaller rods, cylinder thrust (force) becomes extremely small, the travel speed very fast, and the large oil flows generated require very large valving. The advantage of regeneration is to be able to use a smaller pump, and with small-rod cylinders the return speed may be unreasonably slow in comparison to the regenerative forward speed.

(4). Because of multiplication of flow volumes discharged from the blind end of 2:1 ratio cylinders, valving in this circuit should be sized with this in mind. For a review of this subject refer to Volume 1.

(5). Pressure intensification takes place in the rod end of any cylinder with unequal areas on opposite sides of its piston if oil flow is restricted or blocked. The larger the rod the greater the intensification. Pressure rating of the chosen cylinder and its associated valving should be sufficiently high to avoid damage by intensification. Review this topic in Volume 1. For safety, an auxiliary relief valve is sometimes installed in the rod end, discharging to tank.

(6). The regenerative portion of the cycle is usually for moving a machine member rapidly into proximity with the work, and the force required from the cylinder may be relatively small. Therefore, if this is the case, it is permissible, and usually good practice to allow oil velocity to be high during this part of the cycle. The pressure loss, though high, should not affect the speed, and the expense saved by using smaller valves and plumbing may be considerable. We suggest permitting oil velocities up to 40 or 50 feet per second. Volume 1 gives recommended flow velocities for various parts of a hydraulic system, with pipe size recommendations.

(7). Because large rod cylinders are used in regenerative systems, the oil level in the reservoir will rise and fall each cycle more than if small-rod cylinders were used. Reservoir capacity should be such that the pump will not cavitate as the cylinder approaches full extension.

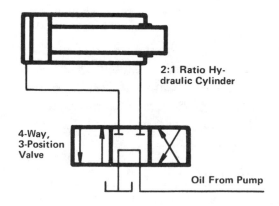

Part (A). Conventional Circuit.

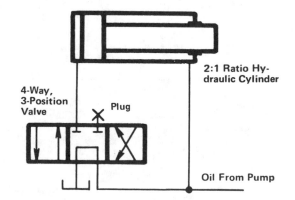

Part (B). Conversion to Regenerative Circuit.

FIGURE 7-49. Standard circuit (on the left) can be converted to regenerative operation (on the right).

Sample Regenerative Circuits. Figure 7-49. These two circuits show the basic difference between a conventional circuit and a regenerative circuit. Each circuit has three cylinder functions – forward, reverse, and neutral stop. They both retract at the same speed when operated from the same oil supply, Part (A), standard circuit travels forward at low speed, high force, while Part (B) travels forward at higher speed, lower force. On some applications it may be possible to make a field conversion to a regenerative circuit by re-connecting the cylinder rod end from the 4-way valve to the pressure line, plugging the unused port on the valve. This type of conversion, obviously to increase forward speed, should only be undertaken on cylinders of 2:1 ratio between piston and rod areas, and only on systems where the prime mover can supply the additional horsepower which may be needed for the faster speed. (The same forward force at twice the speed would require twice the input horsepower).

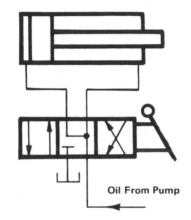

Regenerative Valve Spool. Figure 7-50. A valve with three active positions will produce three cylinder functions; in this case two forward speeds plus reverse. The right side position of the valve will give full force, normal speed forward; the left side position gives retraction; center position gives high forward speed at reduced force. There is no stop position on the valve; the cylinder piston must be stalled against one end cap or against an external stop. Nor is there any valve position for unloading a hydraulic pump.

This is a "regenerative" valve spool. It cannot carry full flow in center position because spool grooves are only halfway open to the cylinder ports. Check manufacturers flow rating before designing a valve with this spool into a circuit.

Solenoid valves are normally used with this spool type. The system pump must be unloaded by means other than through the valve spool. See Page 115 for relief valve unloading.

FIGURE 7-50.
Regenerative Spool on a 4-Way Valve.

Four-Function Regenerative Circuit. Figure 7-51. With the combination of Valves 1 and 2, four active valving conditions are obtained and four functions are possible with the cylinder. Valve 1 gives directional control, mid-position stop, and pump unloading in neutral. Valve 2, when energized, adds regeneration for rapid advance, forward only.

The circuit may be put into or taken out of regeneration either by locating a limit switch at the desired point in the cylinder travel, or by using a pressure switch to sense build-up of load pressure behind the piston. Sample electrical circuits are shown on the next page.

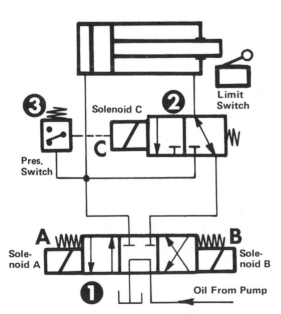

FIGURE 7-51.
Four-Position Regenerative Circuit.

Circuit Condition	Energize Solenoids:
Neutral (Cylinder Stopped)	*None*
Rapid Advance (Regeneration)	*B and C*
Full Tonnage Advance (Slow Speed)	*B Only*
Return of Cylinder	*A Only*

128

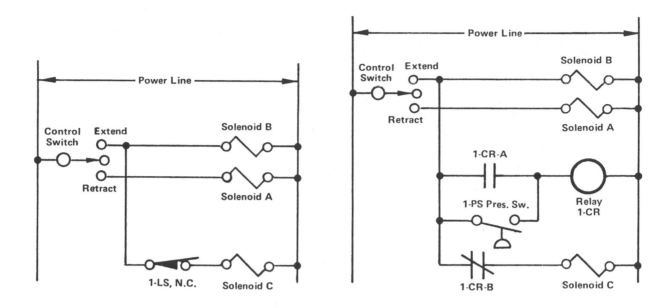

(A). With Limit Switch Control.   *(B). With Pressure Switch Control.*

FIGURE 7-52. Sample Electrical Diagrams for Controlling Regeneration.

Sample Electrical Diagrams. Figure 7-52. Two basic circuits are shown for switching the fluid circuit of Figure 7-51 into and out of regeneration. For simplicity, a 3-position rotary or toggle switch is used for operator control. In the Womack book "Electrical Control of Fluid Power" more sophisticated circuits are shown with pushbutton control, and using holding relays to keep valve solenoids energized.

Limit Switch Control. Figure 7-52, Part (A). One method of control is to locate a limit switch at the stroke position where regeneration should be stopped and full force developed by the cylinder. The N.C. (normally closed) set of contacts on limit Switch 1-LS is used. When the rotary switch is thrown to "Advance", Solenoid Coils B and C become energized. The cylinder regenerates forward until 1-LS is actuated. This de-energizes Solenoid C, releasing Valve 2. The circuit goes out of regeneration.

When the rotary switch is thrown to "Retract", solenoid Coil A, only, is energized and the cylinder retracts. When the cylinder reaches home position, the rotary switch should be thrown to center off.

Pressure Switch Control. Figure 7-52, Part (B). There is a special problem when a pressure switch instead of a limit switch is used to take the circuit out of regeneration. When load pressure reaches the take-out level, the pressure switch is actuated, de-energizing Solenoid C and releasing Valve 2. Load pressure immediately drops to approximately one-half its original level, thus allowing the pressure switch to cause Solenoid C to be energized and the circuit goes back into regeneration. This causes a flutter until load pressure rises high enough to prevent it. To correct this condition, a holding relay, 1-CR, must be used to lock out the pressure switch for the remainder of the cycle as soon as it trips.

Circuit action is as follows: When the rotary switch is thrown to "Advance", solenoid Coils B and C become energized. The cylinder advances in regeneration. The N.O. contacts on 1-PS pressure switch close when the load pressure setting has been reached. This energizes Relay 1-CR. Contacts 1-CR-A, normally open, will close and lock the relay closed electrically, by-passing the pressure switch. When the rotary switch is thrown to "Retract", only Solenoid A of Valve 1 becomes energized. This causes the cylinder to retract. At home position the operator should place the rotary switch in center off.

For more complete information on electrical diagrams for this and other applications, refer to the Womack book "Electrical Control of Fluid Power". See inside rear cover of the present book.

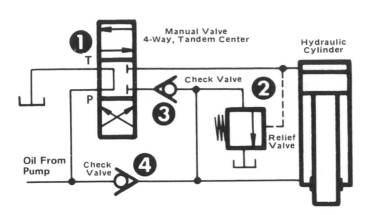

FIGURE 7-53. Non-electrical regenerative circuit using a by-pass valve to automatically take it to a slower speed, at full force, when a pre-set tonnage has been reached.

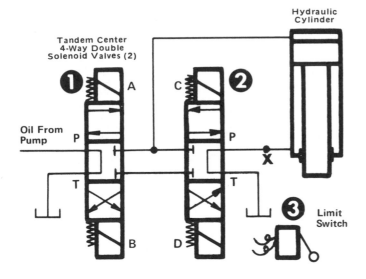

FIGURE 7-54. Electrical regenerative circuit using a limit switch to take the circuit out of regeneration at a pre-determined stroke position.

Cylinder Condition	Solenoid Condition
Cylinder Stopped	All De-energized
Regenerative Advance	A and C Energized
Normal Advance	A Only, Energized
Retract	B and D Energized

Non-Electrical Regenerative Circuit. Figure 7-53. This circuit automatically goes out of regeneration and into slow speed at high force at any point in its stroke where build-up of load pressure becomes too high to continue in the high regenerative speed. Switching from one condition to the other is by means of by-pass Valve 2.

In operation, when directional Valve 1 is shifted to its crossed arrows position to start the downstroke, oil flows to the cylinder blind end. Discharge oil from the rod end passes through check Valve 4 and joins pump oil to gives regenerative action. As load pressure builds up to the cracking pressure of Valve 2, which has been adjusted for the desired changeover pressure, it opens Valve 2 and rod oil passes directly to tank. This gives a slower cylinder speed but at full force.

When Valve 1 is reversed to parallel arrow position, pump oil is connected through Valve 3 to the rod end for normal retraction.

Electrical Regeneration Circuit. Figure 7-54. Instead of going out of regeneration automatically when a certain load pressure is reached, the changeover is made in this circuit at a definite point in the cylinder stroke as determined by the location of limit Switch 3.

Valve 1 is the directional control and also takes care of pump unloading. The other one, Valve 2, controls regeneration. The center position of Valve 2 gives normal speed forward at full force; the straight arrow position gives fast forward speed in regeneration; the crossed arrow position is for reverse operation of the cylinder. With Valve 1 centered the cylinder is stopped.

Return oil from any 2:1 ratio cylinder is twice pump flow. A special feature of this circuit is its ability to dump a high volume of oil for retraction, flowing through two parallel paths; one through Valve 1, the other through Valve 2.

As in other press circuits, if there is a heavy gravity load on the cylinder, a counterbalance valve may have to be added at Position x. The chart under Figure 7-54 summarizes operation of that circuit, showing the particular solenoids to be energized for each cylinder function.

DIRECTIONAL CONTROL WITH 2-WAY POWER LOGIC VALVES

Up to this point we have studied directional control with 3-way and 4-way spool-type valves. A new series of 2-way, pilot-operated poppet-type valves has recently become available which can be used, with certain advantages on some applications, in place of spool-type valves. These valves, which have normally closed or normally open shut-off action, and which are called "power logic valves", will be described later. But first, a foundation should be laid for their use.

Directional Control With 2-Way Valves . . .

Double-Acting Cylinder. Figure 7-55. All possible cylinder functions can be controlled by opening and/or closing Valves 1, 2, 3, 4, and 5 in various combinations. For the purpose of studying circuit action, let these be considered as 1/4-turn, manually operated plug valves. In later circuits we will show these as power logic valves, opened and closed by pilot pressure under the control of miniature solenoid valves.

The first four valves are required in all control circuits for double-acting cylinders.

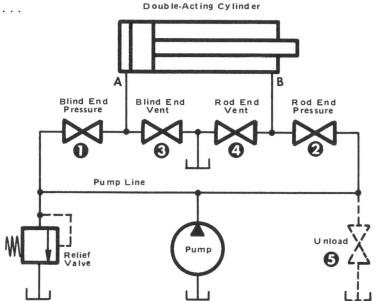

FIGURE 7-55. Directional Control With 2-Way Valves.

TABLE 7-1 – CYLINDER FUNCTIONS FOR FIGURE 7-55.					
Cylinder Functions	Valve 1	Valve 2	Valve 3	Valve 4	Valve 5
1. Moving, in normal extension speed.	Open	Closed	Closed	Open	Closed
2. Moving, at normal retraction speed.	Closed	Open	Open	Closed	Closed
3. Moving, at fast extension speed (regeneration)	Open	Open	Closed	Closed	Closed
4. Stopped, with pump unloaded and with all cylinder ports blocked (tandem center).	Closed	Closed	Closed	Closed	Open
5. Stopped, with pump line blocked and with both cylinder ports blocked (full closed center).	Closed	Closed	Closed	Closed	Closed
6. Stopped, with pump line blocked and with both cylinder ports vented to tank (float center).	Closed	Closed	Open	Open	Closed
7. Stopped, with pump line open to tank and with both cylinder ports also open to tank (full open center).	Closed	Closed	Open	Open	Open
8. Stopped, with pump line and cylinder Port A blocked and with cylinder Port B vented to tank.	Closed	Closed	Closed	Open	Closed
9. Stopped, with pump line and cylinder Port B blocked and with cylinder Port A vented to tank.	Closed	Closed	Open	Closed	Closed
10. Stopped, with pump line and cylinder Port B vented to tank and with cylinder Port A blocked.	Closed	Open	Closed	Open	Open
11. Stopped, with pump line and cylinder Port A vented to tank and with cylinder Port B blocked.	Open	Closed	Open	Closed	Open

131

Valve 5 is required only for functions, such as Numbers 4, 7, 10, and 11, where the pump line must be vented to tank, as for pump unloading.

Example: For Function No. 1, Valves 1 and 4 must be open, to connect pressure to the cylinder blind end and to vent the rod end. Valves 2, 3, and 5 must be closed to keep the pump flow from venting to tank. When all valves are placed in the condition described, the cylinder will move forward at normal speed and force.

Example: For Function No. 3 which is fast (regenerative) speed forward, Valves 1 and 2 must be open to connect pump flow to both ends of the cylinder. Valves 3, 4, and 5, must be closed to block the pressure line from venting to tank.

POWER LOGIC VALVES

Power logic valves are 2-position, 2-way, poppet-type valves, designed for handling large amounts of power – high pressure and/or high flow. They can be used singly or in groups for almost any function of directional or pressure control in hydraulic circuits. Those which are designed for directional control are normally closed.

Logic valves are built in cartridge form, to be installed in cavities bored in a manifold. They are retained in the manifold by a cover which is bolted to the manifold. Groups of logic valves, normally needed for directional control, are connected together by means of cross drilling in the manifold, to eliminate the maze of external plumbing which would be required. It is usually not practical to package them in individual bodies with threaded ports.

Figure 7-57. This is a simplified drawing showing the internal areas and working parts of a power logic valve. The normally closed version has the poppet spring loaded against its seat. A pilot port in the cover allows the valve action to be controlled either by pressurizing or venting the spring chamber. Industrial terminology designates the main port under the poppet as Port A, and the side main port as Port B. The pilot port is designated Port X.

FIGURE 7-56. *Cartridge-Type, 2-Way, 2-Position Power Logic Valve.*

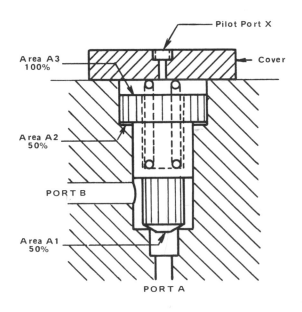

FIGURE 7-57. *Working Parts of a Poppet-Type, Normally Closed Power Logic Valve.*

Significant Areas. Operation of a logic valve depends on three significant areas inside the valve. The area under the poppet facing Port A is designated as Area A_1. On logic valves used for directional control, this area is 50% of the full area on top of the poppet designated as Area A_3. The net area on the internal shoulder is, therefore, 50% of the full Area A_3, and is equal to the bottom Area A_1. It is designated as Area A_2.

When pilot Port X is vented to tank, the valve will open for flow in either direction, A to B or B to A. Pressure coming to Port A works against Area A_1 and flows to Port B with only the restriction of the internal spring, very much like the free flow direction of a check valve. Likewise, pressure coming to Port B works against the net Area A_2, raising the poppet and flowing to Port A. When pilot Port X is pressurized, the poppet is held closed against the sum of pressures working at Ports A and B up to the level of the pilot pressure applied to Port X. For example, with a pilot pressure of 1000 PSI, the poppet will remain closed against the sum of pressures on Ports A and B up to 1000 PSI. Flow between Ports A and B can be started and stopped by using a miniature solenoid valve to either vent or pressurize Port X.

Advantages of Power Logic Valves. (1). They are virtually leaktight between Port A to either Port B or Port X, although there is leakage between Ports X and B when there is a pressure difference between them. In properly designed circuits there is less circuit leakage than with spool-type valves, with better circuit efficiency, less heating, and less tendency for cylinders to creep.

(2). Size for size there is less pressure drop at high flows through a logic valve than through a spool-type valve passage. This further improves efficiency, reduces oil heating, and permits handling larger flows through the same basic valve size.

(3). Logic valves have significantly faster response than spool-type valves.

(4). Logic valves are very versatile; When properly selected, they can almost entirely eliminate shock in the hydraulic system. This is especially important on fast-moving cylinder applications.

(5). Much of their control valving can be contained in the cover which retains them in the manifold. This may include solenoid valves, shuttle valves, orifices, needle valves, etc.

Limitations of Power Logic Valves. They are not for every system. Usually they are not available as separate components individually mounted in bodies with threaded ports. They are available as cartridges. They are for custom designed systems. Each system is different and requires a higher order of engineering skill than usually required for standard circuits using individual components. Those who plan to use them should consult with factory engineers from the company who will supply them, as there are many possibilities, variations, combinations, and effects which cannot be catalogued.

Most manufacturers now building power logic valves are following the German Standard DIN 24342, and these standards are expected to be issued as ANSI and NFPA American standards with only slight modifications.

Controlling Power Logic Valves. Figure 7-58. On industrial applications logic valves are usually controlled by a miniature 2-way or 3-way solenoid valve which is manifolded to the valve cover. Pilot pressure for holding the poppet on its seat can sometimes be taken from one of the main ports. Or it can be taken from another part of the circuit or from a separate pilot pump.

In Figure 7-58(A), the solenoid valve is connected so that when it is de-energized, the top area of the logic valve is vented. Pressure at either Port A or Port B can raise the poppet to provide a flow path. Flow can be in either direction through the main ports. When the solenoid is energized, pilot pressure on the top area holds the poppet seated. Flow is blocked in both directions through the main ports.

In Figure 7-58(B), the solenoid is connected to have the opposite effect. When de-energized, the poppet is held closed as long as sufficient pilot pressure is present.

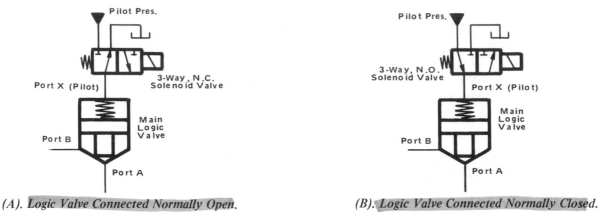

(A). Logic Valve Connected Normally Open.   *(B). Logic Valve Connected Normally Closed.*

FIGURE 7-58. *Controlling Power Logic Valves with Miniature Solenoid Valves.*

Internal and External Piloting. Figure 7-59. The preceding figure showed external piloting from an external source of pilot pressure. In this figure pilot pressure is taken from one of the main ports, whichever happens to be the pressure inlet. A solenoid valve controls the application of this pilot pressure either by bleed-off action with a 2-way valve as in Parts (A) and (B), or with a 3-way valve as shown in Part (C).

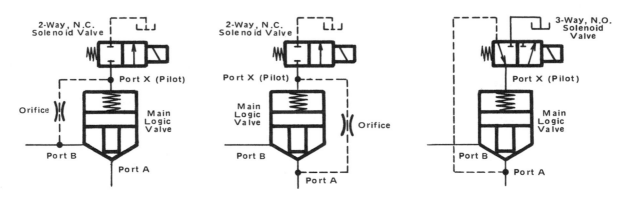

(A), Bleed-off, Pilot From Port B.   *(B). Bleed-off, Pilot From Port A.*   *(C). 3-Way Control, Pilot From Port A.*

FIGURE 7-59. *Methods of Piloting a 2-Way, Normally Closed Power Logic Valve.*

Internal Leakage of Power Logic Valves. Referring again to Figures 7-59(A) and 7-59(C), there is virtually zero leakage from Port A into Port B when the poppet is closed. There is a small amount of leakage between Ports X and B, but Port A is kept isolated from leakage into either of the other ports. In Figure 7-59(B) the leakage path is from Port B, up past the poppet into Port X, then down through the orifice into Port A. In order to minimize circuit leakage and cylinder drift it is important to have the circuit which is to be isolated from leakage connected to Port A.

Closed Pressure System for Double-Acting Cylinder. Figure 7-60. This circuit uses a variable displacement, pressure compensated pump. The main flow is handled through four power logic valves, and these are controlled by a miniature 4-way, 3-position solenoid valve having a "regenerative" spool center. See Page 99 for a description of this spool.

Circuit action is controlled as the 4-way solenoid valve pressurizes or vents the pilot ports of the logic valves. To extend the cylinder, Valves 1 and 3 must be closed by pilot pressure while Valves 2 and 4 must be opened by venting the top of their poppets. To retract the cylinder, logic Valves 1 and 3 must be opened by venting their poppets while Valves 2 and 4 must be closed by pilot pressure. To stop the cylinder, either at the end of its stroke or in mid stroke, both solenoids of the 4-way valve must be de-energized. When the spool of this valve is in center position, pilot pressure closes all logic valves. The cylinder is held in a fluid locked condition as long as the pump is producing pressure.

When the pump is shut down, it cannot furnish pilot pressure for the logic valves. The cylinder will hold position for a limited time against a reactionary force from the load because fluid pressure inside the cylinder, produced by load reaction, will pass through check Valve 6 or 7 and keep all logic valves closed. Eventually the cylinder may creep out of position because of leakage past the spool of the solenoid valve. If the solenoid valve has poppet action instead of spool action, circuit leakage can be reduced nearly to zero and the cylinder may hold against drift for a longer time. If a 4-way, 3-position poppet-type valve is not available, it may be replaced by two 3-way, 2-position poppet valves, or a lock valve or pilot-operated check valve(s) may be installed in the line(s) to the cylinder.

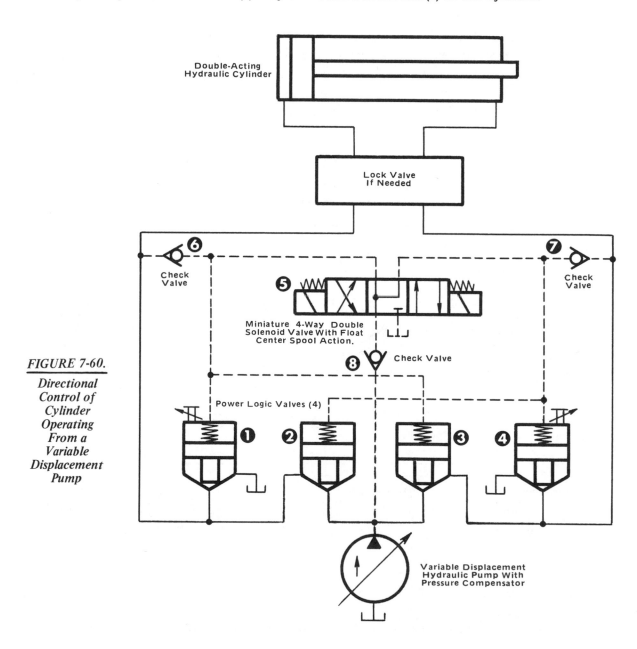

FIGURE 7-60.

Directional Control of Cylinder Operating From a Variable Displacement Pump

Open Pressure System for Double-Acting Cylinder. Figure 7-61. This system is designed for gear, vane, or other fixed displacement pumps which must be unloaded between cycles. It is similar to the 2-way valve control circuit of Figure 7-55, with logic valves replacing the manual 2-way valves.

The logic valves used in this circuit have covers containing a miniature 3-way solenoid valve to control action of the poppet. The logic valves (by energizing or de-energizing the solenoid control valves in various combinations) can be programmed into any of the cylinder functions shown in the first column of Table 7-2 on the next page. The more popular cylinder functions used with a gear pump, for example, would be Nos. 1 and 2 for extend and retract, and No. 4 to unload the pump. This would be equivalent to the action of a tandem center 4-way valve. In Figure 7-61, Solenoids 1, 2, 3, and 4 control flow to and from the cylinder. Solenoid 5 controls pump unloading.

Pilot pressure for the logic valves during extension and retraction of the cylinder comes from the pump. When the pump is shut off, or unloaded through Valve 5, the cylinder will hold for a short time against load reaction, in either direction, because pilot pressure to keep the logic valves closed comes from pressure build-up in the cylinder by load reaction. This pressure passes through shuttle Valve 6 or 7, to keep logic Valves 1 and 3, or 2 and 4, closed according to the direction of load reaction. Eventually the cylinder may drift out of position because of spool leakage in the solenoid valves. If poppet-type, instead of spool-type, solenoid valves are used, circuit leakage can be reduced to nearly zero and the cylinder will hold against drift for longer periods of time even with the pump shut off. Or, if necessary, a lock valve or pilot-operated check valve(s) can be installed in the line(s) to the cylinder.

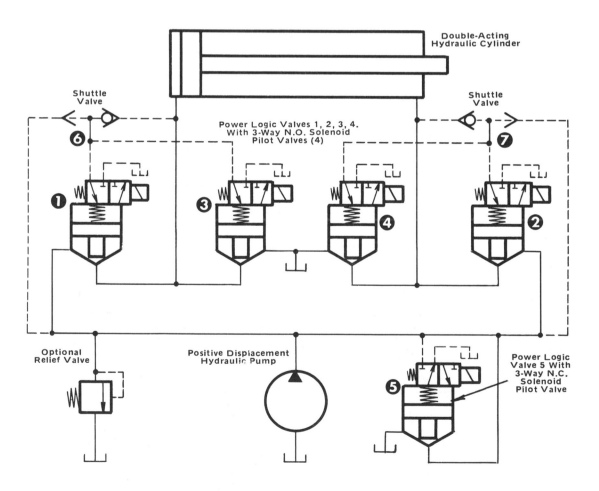

FIGURE 7-61. Directional Control of Cylinder Operating From a Fixed Displacement Pump.

TABLE 7-2 — POSSIBLE CYLINDER FUNCTIONS FOR FIGURE 7-61.

Cylinder Functions	Sol. 1	Sol. 2	Sol. 3	Sol. 4	Sol. 5
1. Moving, in normal extension speed	Ener.	De-ener.	De-ener.	Ener.	Ener.
2. Moving, at normal retraction speed	De-ener.	Ener.	Ener.	De-ener.	Ener.
3. Moving, at fast extension speed (regeneration)	Ener.	Ener.	De-ener.	De-ener.	Ener.
4. Stopped, with pump unloaded, both cylinder ports blocked	De-ener.	De-ener.	De-ener.	De-ener.	De-ener.
5. Stopped, pump line and cylinder ports blocked	De-ener.	De-ener.	De-ener.	De-ener.	Ener.
6. Stopped, pump line blocked, cylinder ports vented	De-ener.	De-ener.	Ener.	Ener.	Ener.
7. Stopped, pump and both cylinder ports vented	De-ener.	De-ener.	Ener.	Ener.	De-ener.
8. Stopped, pump line & Cyl A port blocked, Cyl B port vented	De-ener.	De-ener.	De-ener.	Ener.	Ener.
9. Stopped, pump line & Cyl B port blocked, Cyl A port vented	De-ener.	De-ener.	Ener..	De-ener.	Ener.
10. Stopped, pump line & Cyl B port vented, Cyl A port blocked	De-ener.	Ener.	De-ener.	Ener.	De-ener.
11. Stopped, pump line & Cyl A port vented, Cyl B port blocked	Ener.	De-ener.	Ener.	De-ener.	De-ener.

FIGURE 7-62. Plain Cover for Logic Valve.

FIGURE 7-63. Solenoid Valve Manifolded to Logic Valve Cover.

Logic Valve Covers. Depending on the manufacturer, a number of covers may be available. In addition to the plain cover, other covers may include a solenoid valve, relief valve, shuttle valve, flow control or needle valves, check valve, etc. Figure 7-63 is an example of cover with solenoid valve.

Capacity Range. Depending on the manufacturer, a range of sizes may be available to handle flows from 30 GPM to 1000 GPM or more.

Poppet Ratio. The 50% ratio of poppet areas as described here is generally best for logic valves used for directional control or where flow will be in both directions during the cycle. Other ratios are available. For example, a 7% ratio (shoulder Area A_2 is 7% of top Area A_3) may be used where flow will only be in the direction from Port A through to Port B, or where the area to be closed is exposed to the 7% area A_2. A 100% ratio is also available in which Area A_1 is equal to Area A_3 and there is no shoulder Area A_2. This model may be used where flow resistance from Port A to Port B must be a minimum.

Cracking Pressure. A choice of springs is usually available for various cracking pressures. For general use a cracking pressure of 25 PSI will usually work well. A higher cracking pressure of 50 PSI may be necessary when handling very high pressures on the main ports to prevent leakage through the valve. At low pressures the spring may be omitted and the valve mounted vertically.

Other Configurations. Most logic valves, especially those to handle flows in excess of 100 GPM, are furnished as cartridges. Individual valves in separate housings with threaded ports or for subplate mounting may be available from some manufacturers. They may be listed in catalogs as "pilot-to-close" check valves.

TWO-HAND CONTROL OF HYDRAULIC CYLINDERS

Where cylinders are operating mechanisms which might present a safety hazard to the operator, dual controls can be provided. These require the operator to use both hands to start the machine and to keep the cylinders in motion. Dual controls are also used on large presses which require two operators at different locations. Each operator has a start control. Neither operator, by himself, can reach both controls at the same time, and cannot start the press until the other operator is ready. Sometimes only one direction of cylinder motion is hazardous, and dual control circuits need to work in only one direction. If both directions of motion are hazardous, the control circuits will have to be designed so both hands of the operator must be on the controls in both directions of motion.

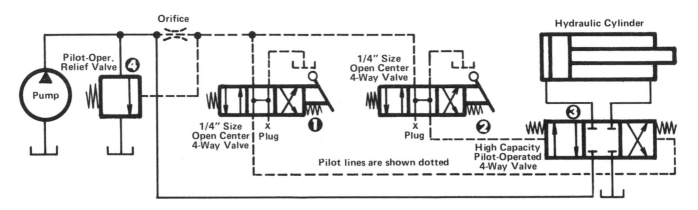

FIGURE 7-64. *Manually Operated 2-Hand Control for a Hydraulic Press.*

Figure 7-64. The main directional valve, 3, is sized as needed for the circuit. This is a 3-position, spring centered, double pilot valve with full closed center spool. It is controlled with a pair of miniature (1/8 or 1/4" size) manual valves, 1 and 2, which have full open center spools.

Pilot oil for shifting Valve 3 comes from the vent flow of the system relief valve, 4, which is of the pilot-operated type. When Valves 1 and 2 are centered, relief valve vent flow passes through the spool centers to tank, and the relief valve is unable to build up pump pressure. If one of the valves, 1 or 2, is shifted, vent flow is still passed to tank through the spool of the other valve. If both Valves 1 and 2 are shifted at the same time to the same side position, vent oil from the relief valve flows into one of the pilots on Valve 3. When the spool of Valve 3 has moved into a side position and has stalled, the relief vent line becomes blocked and the pump can build up system pressure.

The orifice is optional and may be omitted if not needed. Its purpose is to increase shifting speed of Valve 3 by adding a small supplemental pilot flow taken from the pump line. Normally it should be very small, perhaps 1/32 to 1/16" in diameter.

This circuit gives shockless shifting because the pump remains unloaded until the spool of Valve 3 has completely shifted to a side position. The circuit works in both forward and reverse directions of cylinder movement.

Two-hand circuits for compressed air are shown on Page 82. *Caution!* Two-hand circuits in this book are not offered as completely safe circuits to prevent operator injury by requiring both hands on the machine controls. They may not meet OSHA safety requirements. For more secure protection, please refer to the industry standard safety circuit shown in the Womack textbook "Electrical Control of Fluid Power". Unbeatable two-hand control packages are available from several manufacturers ready to integrate with your electrical control circuit.

8

PRESSURE CONTROL

...in Hydraulic Circuits

PRESSURE CONTROL WITH RELIEF VALVES

Valves used to control fluid power systems can be classified into three groups according to their general function. First, there are directional control valves whose purpose is to control direction of motion in cylinders, fluid motors, or other actuators. This group includes 2-way, 3-way, 4-way, and 5-way valves, and these have been covered in the preceding chapter and also in Volume 1. Then there is a group of pressure control valves which will be covered in this chapter. This group includes pressure relief, by-pass, and pressure reducing valves. Finally, there is a group of flow control valves whose primary function is to regulate the rate of fluid flow for the purpose of controlling speed of motion in cylinders and fluid motors. These valves will be described in a later chapter.

Pressure control valves are for the purpose of preventing the pressure either in the main pressure line or in a branch circuit from exceeding a maximum desired level. To understand why pressure control is needed, it is necessary to understand certain characteristics of a hydraulic pump. The primary function of a pump is to produce a FLOW of oil. Pressure is a secondary function. In the case of a pump supplying flow to move the piston of a hydraulic cylinder, a pressure gauge will read little or no pressure when there is no load against the piston rod. The pump will produce only enough pressure to make up friction resistance in the cylinder plus flow losses through the piping. As load is placed against the piston rod, the pressure gauge also reads load pressure, in direct proportion to the magnitude of the load. It is possible to place such a heavy load against the piston rod that the pump will exceed its safe pressure rating in trying to produce its normal flow. Relief valves are for the purpose of protecting the system against damagingly high pressure in case an overload is placed against the cylinder. Therefore, any pressure which can be read on a pressure gauge in the pump line is a direct indication of system losses plus load.

It is necessary to limit maximum pressure in a hydraulic system, first to protect all components in the system against structural damage; second, to obtain a reasonable life expectancy from the pump; and third, to limit the amount of horsepower in the system to prevent overloading of the electric motor or engine driving the system. These requirements are described in more detail:

(1). Safety Limiting of Pressure. All systems using a fixed displacement pump should include a pressure relief valve on the pump line and sometimes at other points in the system. This places a limit on pressure rise in event of an overload. Without a properly adjusted pump relief valve, if a cylinder should be subjected to an overload, either the pump would burst, the power source would stall, or some weak place in the system would fail. Pressure limiting is important not only to protect the equipment, but more important, to protect personnel. A hydraulic circuit without a pump relief valve is like an electrical circuit without a fuse or circuit breaker: an overload could cause line current to rise to a dangerous level.

While a relief valve is *always* recommended for fixed displacement pumps, it is usually recommended also for variable displacement pumps. These pumps, with pressure compensator, can sometimes be operated safely without a relief valve; it depends on the speed with which their compensator will act when an overload occurs, and also on the "tightness" of the oil circuit between the pump and the point in the circuit where a blockage of oil (due to an overload) occurs. A "tight" circuit is one in which there is a very small volume of oil in the pump line and where the oil compressibility will not absorb safely the momentary surge of oil volume before the compensator has time to reduce pump flow. As a matter of safety we recommend the use of a small size relief with all pressure compensated pumps. A small relief valve can discharge the very small amount of trapped oil until the compensator can act. In other words, a full size relief valve is not usually needed.

(2). Establishing a Working Level. While in some systems a relief valve is merely a safety device used only in emergencies, in others it is a regular working part of the total circuit. It can keep system pressure at a uniform level, perhaps by-passing part or all of the oil to tank during certain parts of the cycle. It can control the force output of the cylinder, to prevent damage to the workpieces from accidental excessive force, or crushing of a workpiece from clamping it with too much force.

(3). Establishing More Than One Working Level. Some systems are designed to work at a high force output for part of a cycle, then at a lower level for the remainder of the cycle. This can be readily accomplished with pilot-operated pressure relief or pressure reducing valves, and they can be remotely controlled with small manual or solenoid valves.

(4). Two Levels at the Same Time. A pilot-operated relief valve can establish a high level of pressure in the pump line while a pilot-operated pressure reducing valve can establish a lower level in a branch circuit. Each can be adjusted without affecting the other.

(5). Pump Unloading. Hydraulic systems which use fixed displacement pumps should be designed to allow the oil a free escape to tank during idle periods in the cycle when the cylinder is not moving. This is called "pump unloading" and there are several ways of designing this into the circuit which will be described later. These pumps should not be allowed to discharge across a relief valve during idling periods because of the power waste, the premature "wearing out" of the pump, and because of the heat produced in the hydraulic oil.

DIRECT-ACTING RELIEF VALVES

Direct-Acting Relief Valve. Figure 8-1. A direct-acting relief valve, as compared to a pilot-operated relief valve to be described later, is one in which a poppet is held closed by an adjustable (or fixed) spring to retain all of the pumped oil in the system until system pressure, as the load increases, rises to a level which will produce a force on the poppet area exactly equal to the spring force. This pressure is called the "cracking pressure". As used in this book, cracking pressure is defined as the threshold pressure just before discharge flow starts. From this point the relieving action takes place as follows:

With a small additional increase in system pressure above the cracking pressure, the spring is slightly compressed, opening up a very small flow passage to tank, and allowing a small part of the oil to escape to tank while retaining most of the oil in the system. As load against the system increases, the spring continues to be compressed, opening a wider passage to tank and allowing a larger part of the oil to escape. Finally, with a sufficiently large increase in system pressure, the spring is compressed enough to allow all the oil to escape. This is called the "full-flow" condition, and the system pressure may be substantially higher than at cracking pressure because it takes an additional rise in pressure to compress the spring far enough to pass the entire oil flow. The fluid starts to "bleed off" after the cracking point has been passed. If the operating point of the machine happens to fall in this "bleedoff" range between cracking and full flow by-pass, even for part of every cycle, there will be an excessive power waste which will reduce system efficiency and may cause the system to overheat.

Most direct-acting relief valves are limited to either low flow or low pressure operation. It is difficult to build one to

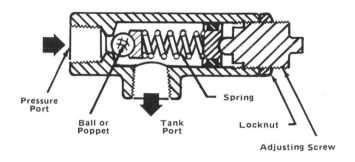

Cross Section of Typical Valve

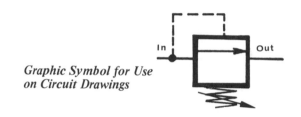

Graphic Symbol for Use on Circuit Drawings

FIGURE 8-1. Direct-Acting Relief Valve

operate at both high flow and high pressure at the same time. For example, if the poppet area were 1 square inch, sufficient to pass 80 GPM without excessive flow resistance, in order to operate and hold closed against 2000 PSI system pressure, the spring would have to exert 2000 lbs. of physical force to keep the poppet closed at cracking pressure. Of course, it is impractical to build an 80 GPM relief valve with a spring this size. At least one manufacturer has designed a direct-acting relief valve in which the flow forces assist the spring, making it possible to produce a valve which will operate at high pressure and flow while using a spring of reasonable size. These valves, however, are not widely available and on high pressure and/or high flow applications, pilot-operated relief valves, to be described later, are normally used.

The additional rise in pressure above cracking pressure is undesirable in a relief valve. It can be reduced by designing the valve to accommodate a longer spring. The longer the spring, the lower the spring rate and the better its performance with less rise in system pressure after the cracking pressure has been passed.

PRESSURE CONTROL WITH PILOT-OPERATED RELIEF VALVES

Pilot-Operated Relief Valve. A construction method is employed which improves the operating characteristics as described on preceding pages which also reduces the valve size. Hydraulic pressure taken from the line being controlled is used instead of a spring to hold the poppet on its seat.

Operating Principle. Figure 8-2. This is a 2-stage valve. The first stage contains the main poppet and light spring, and is sized to handle the maximum flow rating of the valve.

The second stage is usually mounted as a crosshead on the main body. It contains a miniature direct acting relief valve which controls the action of the poppet in the first stage.

The two main ports, inlet and outlet, are sized to carry the rated valve flow. The inlet is to be teed into the pump pressure line or other circuit to be protected. The outlet port must be connected to the system reservoir. Most valves have an RC (remote control) or vent port, of small size, located on the crosshead. At the option of the user, the RC port can be used for remote control of the action of the main poppet. Its use will be described later. If not used it should be plugged.

The main poppet, in the first stage, is normally closed and is held closed by hydraulic pressure from the pump line, through the control orifice, acting on the area above the poppet. The light main spring does not control the relieving pressure. Its function is to keep the main poppet in a closed position for starting, and to permit the valve to operate upside down or in any position. It has a cracking pressure of 75 to 100 PSI, and this is the lowest pressure to which the relief valve can be adjusted.

The second stage, the crosshead, with its miniature direct-acting relief valve, sets the maximum pressure permitted in the main spring chamber. If pump line pressure, due to a system overload, should exceed the maximum level permitted in the spring chamber, the higher pressure, working on the lower end of the poppet, will cause it to raise, opening a discharge path to tank through the valve.

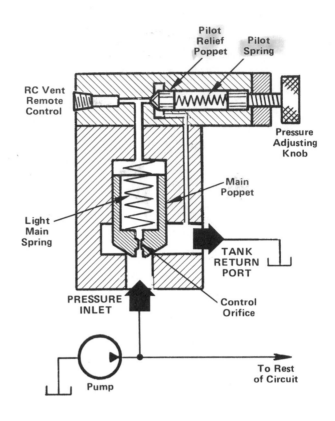

FIGURE 8-2. *Cross Section — Pilot-Operated Relief Valve.*

On a heavy overload, the main poppet will open enough to pass the entire pump flow to tank. On slight overloads, the poppet may partially open, just far enough to pass sufficient oil to tank to prevent any further rise in pressure in the pump line.

The control orifice is made small enough to limit the flow to the amount which can be handled by the miniature pilot relief poppet. If the orifice is too large, the pilot relief poppet would lose control of pressure; if too small, the response of the valve to overload pressure would be slow. When the relief valve starts to discharge flow to tank, the orifice provides an additional pressure drop.

To recap the valve action: The desired maximum system pressure is set on the control knob on the crosshead. This will set a limit on pressure rise inside the spring chamber from the pump line. If, then, pressure in the pump line, due to a system overload, should exceed the level in the spring chamber, the poppet cannot remain seated. It will open fully on a severe overload, or may open slightly on a light overload, just enough to keep pump line pressure from rising higher.

During operation of the valve, at all pressures below the cracking pressure, equal pressure from the pump line is acting on the lower and upper surfaces of the main poppet. Equal areas keep the poppet in balance and the light main spring keeps it seated. (Please see * note in the box on the next page). Now, if pump pressure rises higher than the cracking pressure set on the control knob, the higher pressure on the lower side of the poppet overcomes the limited pressure, plus spring force, in the spring chamber and the poppet starts to open.

The advantages of a pilot-operated relief valve given on Pages 144 and 145 are achieved primarily because controlling action is done in the pilot section (2nd stage) where there is very little oil flow. The pilot poppet opens only a few thousandths of an inch. So, the pilot spring can be designed with a nearly flat spring rate and with a better characteristic than is usually practical in a direct-acting relief valve. Characteristics of these two types will be compared on the next page in Figures 8-4 and 8-5.

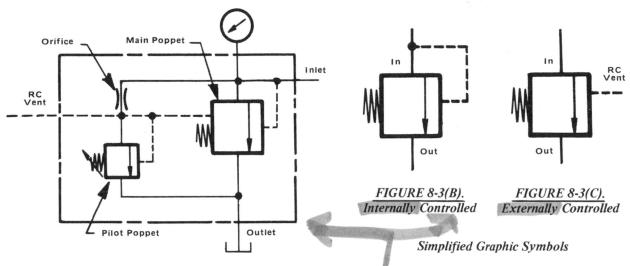

FIGURE 8-3(B).
Internally Controlled

FIGURE 8-3(C).
Externally Controlled

Simplified Graphic Symbols

FIGURE 8-3(A). Complete Graphic Symbol for Circuit Diagrams.

Graphic Symbol for a Pilot-Operated Relief Valve. Figure 8-3 (A). The complete graphic symbol shows that a pilot-operated relief valve is a composite assembly of several parts. The reader should be able to identify the parts in the above diagram by comparison with the schematic diagram on the preceding page.

The use of this complete symbol on schematic diagrams is obviously cumbersome, although for special circuits it may sometimes be necessary to show the complete symbol. On most diagrams a simplified symbol as shown in Parts (B) or (C) is satisfactory.

Part (B), Figure 8-3. Internal Control. This simplified symbol can be employed when the relief valve is used only for maximum pressure relief protection and is not to be remotely controlled. The RC vent port should be plugged. Indicate adjustable models with a slash arrow across the spring.

Part (C), Figure 8-3. External Control. If the valve is to be remotely controlled as for pump unloading, sequencing, etc., in addition to pump protection, the RC port is connected to the external device as shown on Page 147. Indicate adjustable models with a slash arrow through the spring.

---

*Note: Figure 8-2 is a conceptual drawing to illustrate the relieving action of a pilot-operated relief valve. Although the area on top of the main poppet appears to be larger than the area on the bottom, the physical construction must be such that these areas will be exactly equal for the valve to perform properly. During normal operation at pressures below the relieving level, there will be equal pressure working on top and bottom equal areas, so the poppet will be in balance and the main spring will keep it closed. A rise of pressure in the pump line of 75 to 100 PSI above the relieving level set on the control knob will cause the main poppet to start to open.

COMPARISON OF DIRECT-ACTING TO A PILOT-OPERATED RELIEF VALVE

Figures 8-4 and 8-5. The construction and operation of direct-acting relief valves has been described on Pages 140 and 141, and that of pilot-operated relief valves on Pages 142 and 143. The two graphs below compare performance of a typical valve of each type. The graphs do not represent any particular brand or model. Some direct-acting relief valves may have better performance than shown, and some pilot-operated relief valves may have poorer performance. The graphs represent performance on the same application, under the same conditions, using the same pump. They illustrate the possible rise in system pressure after cracking pressure has been reached, until the entire pump volume is able to pass across the relief valve to tank.

The pump used is rated for 20 GPM flow. The scale along the bottom of each graph shows what part of the 20 GPM is passing to tank across the relief valve at the pressure shown on the vertical

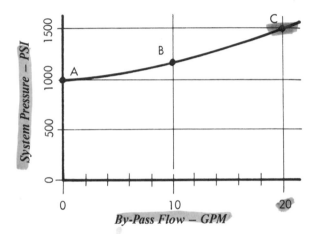

FIGURE 8-4. *Typical Direct-Acting Relief Valve.*

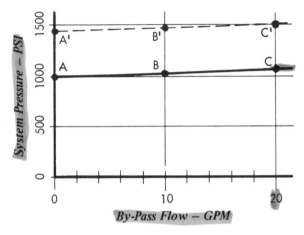

FIGURE 8-5. *Typical Pilot-Operated Relief Valve*

scale. On each graph, the valve has been adjusted to a cracking pressure of 1000 PSI. The rise in the line from Point A to Point B shows the increase in system pressure from the time the valve starts to crack open until it is opened far enough to discharge the full 20 GPM. Ideally, this line should be flat. The greater its rise between Points A and B, the poorer the performance of the relief valve.

Direct-Acting Relief Valve. Figure 8-4. The full 20 GPM is retained in the system until pump pressure, due to increase of the load, rises to 1000 PSI. At this pressure the relief valve is on the threshold of opening. Now, as the load further increases, say up to 1200 PSI, approximately represented by Point B on the graph, the relief valve has opened far enough to permit 10 GPM to discharge to tank. The cylinder is still moving but at a reduced speed. The relief valve is in a "bleed-off" condition. The 10 GPM still going to the cylinder is producing useful work output and no heat. The other 10 GPM going across the relief valve is producing no useful output; it is being converted into heat. The amount of power going into heat is 7 HP as calculated by the fluid HP formula. This may soon overheat the system if operation is maintained under this condition.

Now, when the load increases to the point where 1500 PSI is required to move it (Point C), the relief valve has opened far enough to permit discharge of the full 20 GPM from the pump. The cylinder now stalls against the load and the full 20 GPM discharges to tank and is converted to 14 HP of heat.

Pilot-Operated Relief Valve. Figure 8-5. Under the same conditions as before, the full 20 GPM is retained in the system until pressure, caused by an increase in load, rises to 1000 PSI, the cracking pressure. Now, as the load continues to increase, at Point C the relief valve has opened sufficiently to discharge the entire 20 GPM, with a rise in system pressure to only about 1100 PSI. This sharper characteristic, shown by the flatness of the graph line, indicates that there is less power waste and heat build-up as the system approaches its maximum capability. If the cylinder should remain in a stalled condition with the full 20 GPM going across the relief valve, the power waste will be only 12.8 HP instead of the 14 HP with a direct-acting relief valve. Cylinder performance is better and the efficiency is improved.

It is interesting to see what happens when the cracking pressure on the pilot-operated relief valve is raised to 1400 PSI (Figure 8-5). This is shown as Point A'. After cracking pressure has been exceeded, an additional load build-up of 100 PSI (Point C') is enough to open the relief valve sufficiently to discharge the full 20 GPM. System pressure at full cylinder stall (with 20 GPM across the relief) is only 1500 PSI, the same as with the direct-acting relief valve when set for a cracking pressure of 1000 PSI. When moving loads which build up near the end of the stroke, a pilot-operated relief offers a great advantage because there is less slow-down of the cylinder and less power waste until after the load pressure exceeds 1400 PSI, while on a direct-acting relief valve, the slow-down would start at 1000 PSI.

Additional Advantages of a Pilot-Operated Relief Valve. When used to control a large flow or high pressure, its main advantages over the direct-acting type are these:

(1). As described above, the rise of system pressure after the cracking pressure point has been passed, reduces cylinder slow-down and power waste and improves efficiency.

(2). It is adjustable over a very wide range of pressures. Usually, a direct-acting relief valve has only a 2:1 or 3:1 range of adjustment without changing springs. A pilot-operated relief valve is adjustable from a minimum of about 100 PSI up to 3000 PSI, and some valves can be adjusted even higher.

(3). It is more versatile; it has multiple uses. It can be used not only for maximum pressure protection; it can be controlled remotely for several levels of pressure and pump unloading.

(4). Operates quietly in the system; rarely chatters or produces undue noise. Because the poppet does not chatter on its seat, its service life may be longer and more trouble-free than an equivalent direct-acting relief valve.

(5). Since it permits less rise in system pressure after the cracking point has been passed, it offers better protection for all components in the circuit.

APPLICATIONS FOR DIRECT-ACTING RELIEF VALVES

Pump Relief. Figure 8-6. On low power systems, especially those with low flow, a direct-acting relief valve, because of its simplicity and lower cost, may be a good choice. But if used on high flow systems, it should be limited to emergency pump protection, and it may be adequate for some applications provided its cracking pressure can be set sufficiently high that under normal conditions the system will not be operating in the "bleed-off" range between cracking and full discharge flow. All circuits designed for intentional or continuous discharging during normal operation should use a pilot-operated relief in preference to a direct-acting type.

Cushioning. Figure 8-7. A pair of relief valves is often connected "back-to-back" across the ports of a reversible hydraulic motor or double-acting cylinder to absorb momentum energy of

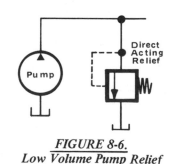

FIGURE 8-6.
Low Volume Pump Relief

145

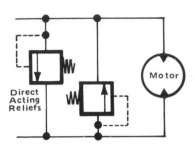

FIGURE 8-7.
Cushioning of a Hydraulic Motor

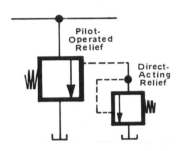

FIGURE 8-8.
Piloting a Larger Relief Valve

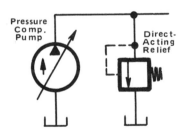

FIGURE 8-9.
Pressure Compensator Relief Valve

the load if the oil flow is suddenly blocked. Two individual relief valves may be used for protection in both directions of actuator movement, or a "cushion relief" valve, containing two relief valves in one housing, may be more convenient. The large pressure differential between cracking pressure and full flow pressure of a direct-acting relief valve makes it better for this purpose than a pilot-operated relief valve with its sharper characteristic. It can provide a smoother deceleration of inertia loads, and has a faster response. However, it does not take the place of a relief valve on the pump line.

For cushioning applications, each relief valve in the cushion valve assemby should be set substantially higher, say 500 PSI higher, than the pump relief valve to make sure none of the working oil escapes during normal operation of the system. A valve model which is non-adjustable should be used so no one in the field can tamper with its setting.

Piloting of a Larger Relief Valve. Figure 8-8. Pilot-operated relief valves will be described in the next few pages. Small direct-acting relief valves are used as pilot controls for them, and are incorporated in the same body. They are also used separately at a remote control point to control larger, pilot-operated relief valves. For these applications a very small, usually 1/4" size, direct-acting valve is sufficient.

Supplement to Pressure Compensator. Figure 8-9. On variable displacement, pressure compensated pumps, the compensator theoretically eliminates the need for a relief valve. However, a small time interval is required for the mechanical action of the compensator to reduce pump displacement. During this interval a high pressure spike can be generated and travel through the system. A small, direct-acting, relief valve should supplement the compensator by discharging the small volume of oil generated during the time interval. It will reduce the intensity of the pressure spike to protect components. A small size valve will do the job and act faster than a larger valve. Usually a 1/4 or 1/2" size is adequate except on very large systems where a larger size may be needed.

The relief valve should be non-adjustable, or should have a lock to prevent tampering. It should be set about 500 PSI higher than the "firing point" of the compensator.

Applications Not Suitable for a Direct-Acting Relief Valve . . .

(1). When discharging oil flow, some direct-acting relief valves become quite noisy at certain flows or pressures at which they are resonant. In addition to the annoyance of this noise, valve chatter may cause permanent damage to the poppet or its seat, causing it to "wear out" prematurely or to severely leak during normal operation of the system. Usually, it should only be used in those applications where it is intended to "pop open" momentarily in case of emergency. It should not be used on applications where the intent is to continually discharge a part of the oil flow during normal operation.

For example it should not be used as a pump relief when series-type flow controls are used to vary the actuator speed.

(2). Because of the relatively large difference between its cracking pressure and the full flow pressure at which it will discharge all of the pump flow to tank, a direct-acting relief valve on the pump line may not give adequate circuit protection. If it is set sufficiently high to avoid bleeding off a part of the oil during normal operation, it may allow the pressure to peak at too high a level when relieving the full pump flow. If its setting is reduced to give good circuit protection, it may "leak off" a part of the useful oil during normal operation, causing a slow-down in cylinder speed and causing overheating in the oil, as well as low efficiency because of the wasted power.

(3). Most direct-acting relief valves have a narrow range of adjustment, and a selection of several different springs may be required to cover a complete range of pressures up to 3000 PSI. If a wide range of adjustment will be required, a pilot-operated relief valve is preferred.

REMOTE CONTROL OF A PILOT-OPERATED RELIEF VALVE

Pilot-operated relief valves for use on mobile equipment usually do not have an RC vent port for remote control. Those models designed for industrial use do have this control port. To understand how and to what degree the main relief valve can be controlled through this port, the reader must first thoroughly understand the action of the relief valve. If necessary, review Pages 141, 142, and 143. Note that this port connects into the main spring chamber. At a remote point, the maximum pressure level in the spring chamber can be regulated independently of the setting on the knob of the relief valve itself, provided the setting on the knob is higher than any level set at the remote position. Since oil flows through the path of least resistance, the oil which passes from the pump, through the control orifice (Figure 8-2) will discharge to tank either through the pilot poppet in the relief valve crosshead, or out the RC port to the remote control, whichever path is easier to follow. See parallel relief valves, Page 152.

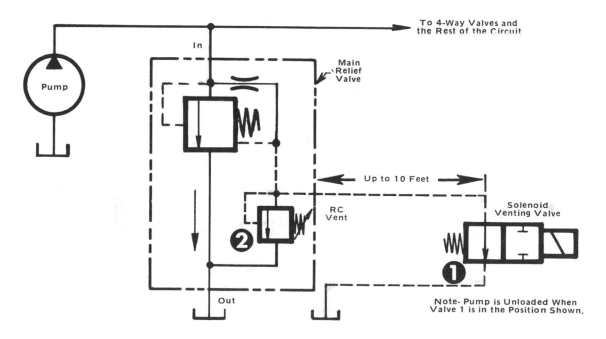

FIGURE 8-10. Externally "Venting" a Pilot-Operated Relief Valve to Unload a Pump.

Pump Unloading. Figure 8-10. If an external line from the RC vent port is connected to tank through a shut-off valve, the pump can be unloaded or "dumped" to tank when the valve is opened, or can be made to produce pressure up to the setting of the control knob when the venting valve is closed. An example of this type of circuitry is shown on Page 115.

The action of externally opening a free flow path to tank from the RC port is called "venting the relief valve". In this diagram, opening the remote shut-off valve provides an alternate free flow path to tank for oil in the main spring chamber. The main poppet cannot stay closed when hydraulic pressure in the spring chamber cannot build up. Pump flow lifts the main poppet from its seal and goes to tank through the outlet port of the relief valve.

Venting action does not reduce pump pressure completely to zero. There is a small residual pressure remaining, according to the strength of the main spring. This will usually be 75 to 100 PSI, and may be used as pilot pressure for any pilot-operated 4-way solenoid valves in the system.

In Figure 8-10, the remote venting valve, 1, is usually a 2-way, 2-position, normally open solenoid valve connected to the RC vent port of the system relief valve through a 3/8" diameter steel tubing line, and back to reservoir with a similar line. The separation distance should be kept as short as practical, preferably no greater than 10 feet. Longer separation causes the valve response to be sluggish. A normally open venting valve is preferred because a failure of the electrical control voltage will cause the solenoid valve to go to its open position which vents the relief valve and unloads the pump.

A manual plug valve of any type may be used on non-electrical systems. Any valve used for venting must have a pressure rating equal to the maximum system pressure.

Very little shock is generated when a pump is unloaded by venting the relief valve, and for high power systems (over 50 HP) it is a better unloading method than by using tandem or open center 4-way valves because of the high shifting shock created by those valves.

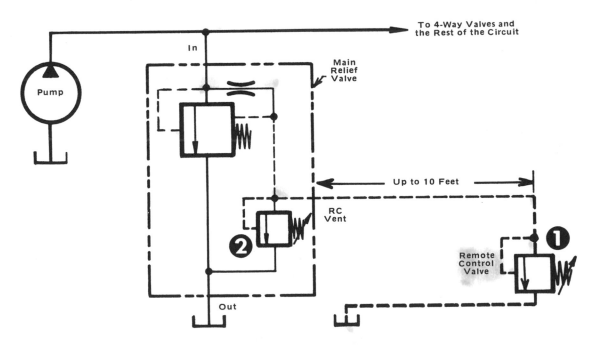

FIGURE 8-11. Reducing the Knob Setting of a Pilot-Operated Relief Valve From a Remote Control Location.

Remote Tonnage Control. Figure 8-11. The pressure relieving point of a pilot-operated relief valve can be adjusted from a remote point through the RC vent port, to override the pressure level set on the knob of the main valve. An example is a hydraulic press. Usually the hydraulic power unit and relief valve for the pump are located "upstairs" or at a distant point from the operator control panel. Since different jobs may require an adjustment in press tonnage, a small direct-acting relief valve may be installed on the operator control panel with a pair of 3/8" steel tubing lines connecting to the hydraulic power unit. The length of these lines should not exceed 10 feet or the relief valve action may become sluggish.

Note that the external remote control valve, 1, is in parallel with the pilot relief poppet inside the main relief valve. Since oil flow follows the path of least resistance, it will flow to tank either inside the main valve or through the RC port and the external relief valve, whichever path offers the least resistance. Therefore, the relief valve (either internal or external) with the lower setting will determine the relieving pressure of the main poppet. This rule then applies: maximum rated tonnage of the press must always be set on the knob of the main relief valve. Then, the tonnage can be reduced, when necessary, with the panel mounted remote relief valve. The remote relief can reduce the tonnage set on the main relief valve but can never raise it.

Multiple Reduced Pressures. There are many combinations of valving which can be installed in the RC Vent port of a pilot-operated relief valve to select several levels of system pressure or to unload the pump. It is important to keep length of plumbing as short as possible in the vent line; long lines cause the response to be slow. Optimum size of vent line plumbing seems to be about 5/16" I.D. To understand vent line action, remember that a stream of oil flows out of this port and is controlled by the external valving.

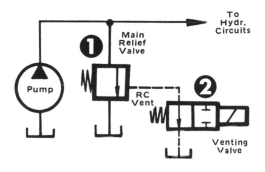

FIGURE 8-12.
Principle of Vent-Line Unloading

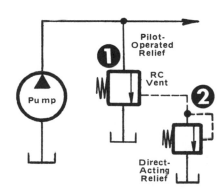

FIGURE 8-13.
Principle of Vent-Line Pressure Control.

Venting Principle. These diagrams show how a venting circuit would be drawn on a circuit diagram, with Valve 1 being the main relief valve of the pilot-operated type and miniature Valve 2 in each diagram being located externally to the main valve.

Figure 8-12. For unloading the hydraulic pump, miniature solenoid valves are preferred in electrically operated systems so the unloading solenoid can be tied in electrically with the directional valves.

While solenoid valve, 2, can be either normally closed or normally open, the normally open type is usually preferred to make the circuit "fail-safe". If the solenoid coil should burn out or if there should be a power failure in the electrical power supply, the system would return to a safe condition with the pump unloaded.

Figure 8-13. When designing venting systems for reduced pump pressures, remember that the highest system pressure must be set on the main relief valve, lower pressures being obtained with miniature direct-acting relief valves installed in the vent line.

In multiple step systems, two or more relief valves are often operated in series, and the rules on Page 152 for series relief valves will apply.

Step Control of Multiple Pressures. Figure 8-14. Valve 1 is the main relief valve of the pilot-operated type. Relief Valves 2 through 5 are miniature, non-adjustable relief valves for obtaining various pressure levels below the maximum of Valve 1. Oil flow coming from the RC port of Valve 1 can flow to tank through Valve 2, through Valves 2 and 3 in series, through Valves 2, 3, and 4 in series, or through Valves 2, 3, 4, and 5 in series, depending on the position of shut-off Valves A through E, whether open or closed.

Pressure levels on the pump available with this circuit are (1), the maximum as set on Valve 1, obtained with all shut-off valves closed; (2), four intermediate pressure levels obtained by opening one

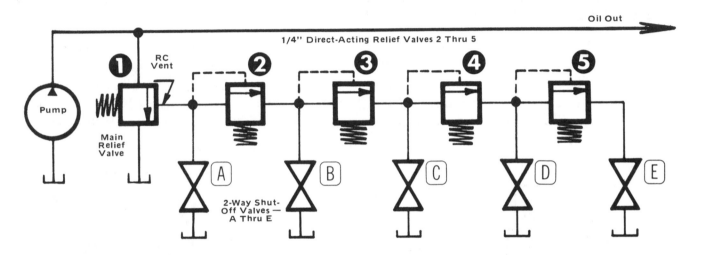

FIGURE 8-14. *Multiple Step-Level Control of System Pressure Level.*

of the valves, A through E. If more than one of these valves is opened, the pressure level is determined from the valve nearest the main relief valve; (3), pump unloading when Valve A is opened. When the relief setting of two or more of the miniature relief valves 2 through 5 add to higher than the setting on the main valve, the highest pressure obtainable is that set on the main valve.

Shut-off Valves A through E may be manual or solenoid, with a pressure rating at least equal to the sum of all control relief Valves 2 through 5 which are downstream from them. They may be very small in size since the flow from the RC port of Valve 1 is usually 1 GPM or less.

Figure 8-15. (See figure on next page). With a tandem center, double solenoid valve, 1, connected in the vent line to the main relief valve, 3, up to three pressure functions can be established for the pump. When connected as shown, the pump is unloaded when the solenoid valve is centered. When it is shifted to its left position, the vent line is blocked and the pump can produce pressures up to the adjustment setting of the control knob on the main valve, 3. In the right side position of the solenoid valve, the vent line is connected to a miniature relief valve, 2, which is set for a lower pressure than the main relief. The pump can produce pressures only up to the adjustment of Valve 2.

Figure 8-16. (See figure on next page). With a closed center double solenoid 4-way valve connected in the vent line to the main relief valve, 3, the center position of the spool blocks the RC vent and allows the pump to produce full pressure. The left position of the spool connects the vent line to auxiliary relief Valve 2, and the right position of the valve connects it to auxiliary relief Valve 3. This gives a choice of two intermediate pressures, both less than the main relief adjustment.

Control of Two Pumps. Figure 8-17.

This circuit may be used to control unloading of two pumps working on the same system. Both pumps may be unloaded or loaded at the same time; or one of them can be unloaded, leaving the other pump working for jogging, set-up, or in an electrically operated "high-low" circuit where one pump can use the entire input horsepower for operation at high pressure while the other pump is idling. Following the basic idea presented here, circuits can be designed for unloading control of three or more pumps.

Each pump, 1 and 2, has a pilot-operated relief valve, 3 and 4 respectively, for maximum pressure protection. The RC vent lines from these relief valves are controlled with a 3-position, closed center, solenoid valve, 5.

Center position of the solenoid valve blocks both vent lines and causes both pumps to be loaded and to produce pressure up to the adjusted setting of their control knobs. In the left side position of the solenoid valve, both vent lines are ported to tank and both pumps are unloaded. In this position, note that vent oil from relief Valve 4 passes through check Valve 9 and to tank through the solenoid valve. Also note that vent oil from relief Valve 3 passes through check Valves 8 and 7 and to tank through the solenoid valve. In the right side position of the solenoid valve, Pump 1 relief valve is vented while Pump 2 relief valve vent line is blocked. Pump 2 works at its normal maximum pressure while Pump 1 remains idling.

In Figures 8-15 through 8-17, manual valves of the ball or toggle type can be used in place of solenoid valves for vent

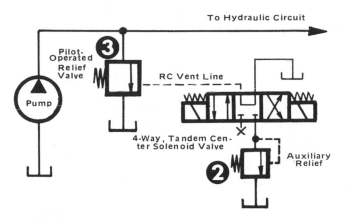

FIGURE 8-15. Pump Unload and Two Pressure Levels.

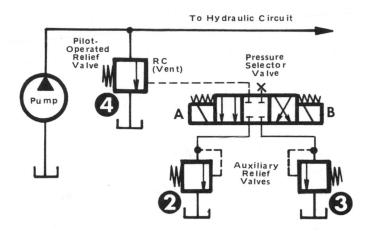

FIGURE 8-16. Three Pump Line Pressure Levels.

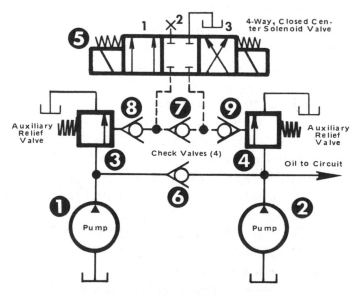

FIGURE 8-17. Vent-Line Control of Two Pumps.

control. Especially when using a solenoid valve, be sure its tank port is not exposed to pressure higher than its catalog rating.

CONNECTING RELIEF VALVES IN PARALLEL AND IN SERIES

Relief Valves in Parallel. Figure 8-18.

Relief Valves 1, 2, and 3 are connected in parallel across a deadheaded pump line. What will be the pressure which the pump must develop to discharge its flow to tank? The relief valves are adjusted to pressure settings of 2000, 1500, and 1000 PSI respectively.

Oil will flow through the path of least resistance. In this case the entire pump flow would pass through Valve 3, and the pump would put up a pressure of 1000 PSI.

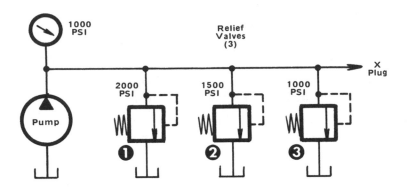

FIGURE 8-18. *Relief Valves Connected in Parallel.*

Relief Valves in Series. Figure 8-19.

This is a more difficult situation to visualize. The entire pump oil must pass through all three relief valves, and they are adjusted to pressure settings of 500, 500, and 1000 PSI respectively on Valves 1, 2, and 3.

Remember that pressure gauges are referenced to atmospheric pressure while relief valves are referenced to their own outlet which is not always at atmospheric pressure.

Starting at the tank port of Valve 3: Since both the relief valve and the gauge are referenced to atmospheric pressure, the gauge will read the same as the setting of the relief valve.

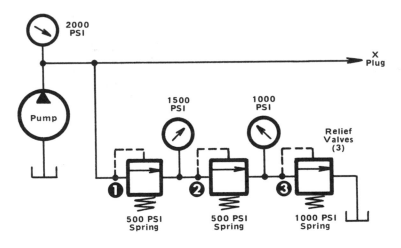

FIGURE 8-19. *Relief Valves Connected in Series.*

Continuing to Valve 2: Oil entering its inlet port must first have sufficient pressure to open its spring which is adjusted to 500 PSI. Then the oil must have enough additional pressure to open the spring on Valve 3 which is adjusted to 1000 PSI. Therefore, the oil must be at 1500 PSI or higher when entering Valve 2 or it cannot open both springs.

Finally, oil entering Valve 1 inlet must have sufficient pressure to open Valve 1 spring, then enough additional pressure to open Valve 2 spring, and finally must have enough pressure left to open Valve 3 spring. Therefore, the pressure required for relief valves in series is the sum of the individual spring settings of all valves through which the oil must pass to get to atmospheric pressure.

Note: The series arrangement of relief valves applies to the remote control of pressure reducing valves as detailed on Page 168. The parallel arrangement of relief valves applies to the remote control of pilot-operated relief valves as detailed on Page 148.

POWER LOGIC RELIEF VALVES

Please refer to Page 132 for general description of power logic valves and their use in directional control. They can also be used as pressure relief valves, and they are especially applicable to systems with very high oil flows in hundreds or thousands of GPM. As in the case of directional control, logic valves are not usually available as individual components installed in a body; they are manufactured as cartridges, to be installed in a manifold custom manufactured for the particular machine. Later, their use as by-pass and pressure reducing valves will be described.

Logic valve models suitable for relief valves are the normally closed type. The internal spring keeps the poppet in a normally closed position. It should have a cracking pressure of about 100 PSI. The valve must have a 1:1 poppet area ratio; that is, the area on the bottom of the poppet must be the same as the area on top. To use one as a relief valve, a small direct-acting relief valve and an orifice must be used in conjunction to establish the relieving pressure level.

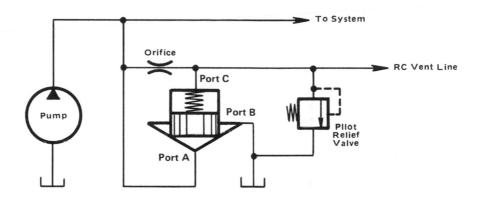

FIGURE 8-20. A Power Logic Valve Used as a Pressure Relief Valve.

Figure 8-20. The size of the logic valve is selected according to the volume of oil flow which must be discharged to tank at a minimum flow resistance. The pump pressure should be connected to the area under the poppet (Port A), and the tank return connection to Port B.

The relief valve which controls the relieving level, should be an adjustable miniature, direct-acting type, usually 1/4'' size for relieving flows up to 100 GPM and a 1/2'' size where larger flows are to be handled. It can be a separate unit mounted a short distance from the logic valve, but preferably it should be manifolded directly to the logic valve cover. It connects directly into the logic valve spring chamber to control the maximum fluid pressure which can build up on top of the poppet. When pump pressure under the poppet tends to be more than 100 PSI (spring cracking pressure) above relief valve setting, the poppet starts to open.

The orifice should have a sharp edge to make it insensitive to temperature changes in the oil. It, too, can be a separate unit mounted externally to the logic valve but preferably should be manufactured in the logic valve cover. It should be sized to pass from 1/2 up to several GPM, according to the flow volume of the system, at the maximum operating pressure.

The power logic valve and miniature relief valve are, in fact, a pilot-operated relief valve which can be remotely controlled as described for a conventional relief valve on Page 147. An RC venting connection should be taken directly from the internal spring chamber as shown on the diagram. The logic valve can be vented for pump unloading, or remotely adjusted for step level pressure control.

PRESSURE CONTROL WITH BY-PASS VALVES

<u>By-Pass Valves Defined.</u> The preceding pages have covered pressure control with relief valves of the direct-acting and pilot-operated types. We have seen how a pilot-operated relief valve is a multi-purpose valve used not only for maximum pressure limitation but for pump unloading and to establish various levels of pump pressure by remote control.

A by-pass valve is another multi-purpose valve which can serve various purposes of pressure control within the hydraulic circuit. Its general title is "by-pass valve", but it is sometimes called by other names according to its use in the circuit — such titles as "unloading valve", "sequence valve", "dump valve", "counterbalance valve", "relief valve", and others. The same valve can be used for all of these purposes if correctly connected for the source of its piloting pressure and for draining its spring chamber.

Basically the by-pass valve, like the pressure relief valve, is a 2-way, normally closed valve using either a sliding spool or a poppet. Since the valve is normally held in a closed position by an internal spring (or by hydraulic pressure), a pilot pressure signal is required to open it for flow.

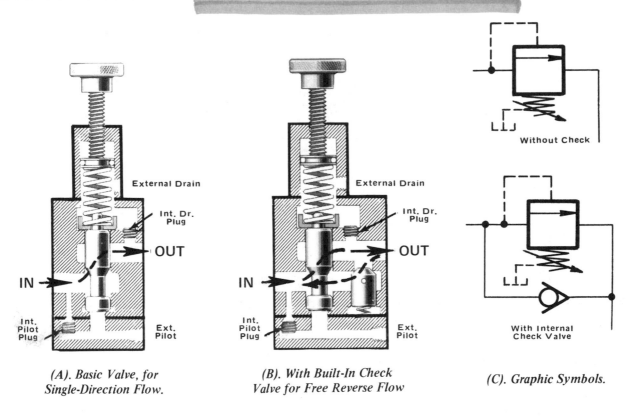

(A). Basic Valve, for Single-Direction Flow.

(B). With Built-In Check Valve for Free Reverse Flow

(C). Graphic Symbols.

FIGURE 8-21. *Direct-Acting By-Pass Valves, With and Without Return Flow Check Valve.*

<u>Direct-Acting By-Pass Valve.</u> **Figure 8-21.** A direct-acting by-pass valve is one in which an adjustable spring holds the spool or poppet closed and directly opposes the pilot pressure applied to open it. The spring force must be greater than the force from the pilot pressure, up to the "cracking pressure" of the valve. On the other hand, a pilot-operated by-pass valve makes use of hydraulic pressure, taken either from its own inlet or from another source, to hold the valve closed up to its cracking pressure. Pilot-operated by-pass valves will be described later.

On the direct-acting type, the spring tension is adjustable over a limited range, and the spring adjustment determines the threshold pressure on the opposite end of the spool or poppet which will start it to open. Like a relief valve, a further increase of pilot pressure is required to shift the spool far enough to pass the full system flow.

Flow through the spool or poppet of a by-pass valve is in one direction only, from the inlet to the outlet port. Flow cannot return through the poppet or spool in the opposite direction. When a by-pass valve is used in an application where return flow is required, a check valve must be placed around it. Usually a model can be purchased with a check valve built into the same housing, as in Figure 8-21(B).

Piloting a By-Pass Valve. Referring to Figure 8-21, Part (A), the main spool is pressure balanced, with equal area on each side of the groove, and no amount of pressure applied either to its inlet or outlet port will cause the spool to shift. According to the type of application, shifting (pilot) pressure must come either from the valve inlet port (as for sequencing, counterbalancing, or pressure relief applications), or from another part of the circuit (as for pump unloading, dumping, and by-pass applications), and must be applied to the end of the spool opposite the spring. Manufacturers usually build their by-pass valves so they can be converted in the field to receive pilot pressure from their inlet port (internal piloting) or from another source (external piloting). On the valve shown in Figure 8-21 an internal plug can be installed or removed by removing the bottom cap of the valve. When this plug is removed, and the external pilot port is plugged, pilot pressure to shift the spool comes from the valve inlet. When the internal plug is installed, pilot pressure from another part of the circuit can be connected to the external pilot port. Some brands use the internal plug method of changing from internal to external piloting; on other brands the bottom cap may be removed and rotated 90 or 180º to open up or close off internally drilled passages. Whether the valve should be internally or externally piloted depends on the nature of the application, and will be explained with circuit diagrams on the next few pages.

Draining a By-Pass Valve. Please refer again to Figure 8-21(A). In most circuitry, the spring chamber must be referenced to atmospheric pressure. It must always be drained; otherwise, internal leakage would soon fill up the spring chamber with trapped oil and the spool could not be shifted. The easiest way to drain the chamber is to connect it to the outlet port of the valve — provided the outlet port operates at tank or atmospheric pressure. In Figure 8-21 this is done by taking off the top cap of the valve and removing a small screw which plugs the hole which vents into the outlet port. On some brands the valve cover can be removed and rotated 90 or 180º to open up a drain hole.

However, on some applications like sequencing, the outlet port does not connect to tank; it goes to another circuit and may be at high pressure for a part of each cycle. In this case, the internal plug must be installed and a separate drain line run from the external drain port directly to tank. Whether or not the valve must be externally drained will be explained with diagrams to follow.

Note: It is very important when using an external drain line connected to the valve to make certain that the internal drain hole is plugged. If it happens to be open, part of the system oil will be short circuited to tank when the main outlet port is under pressure, and the valve will not work properly.

RULE FOR EXTERNAL DRAINING

If a by-pass valve is used on any application where back pressure may appear on its outlet port at any time during the cycle, its spring chamber must be separately connected to tank. If in doubt, the safe procedure is to externally drain it, making sure the internal drain passage is blocked.

REDDY FLUID

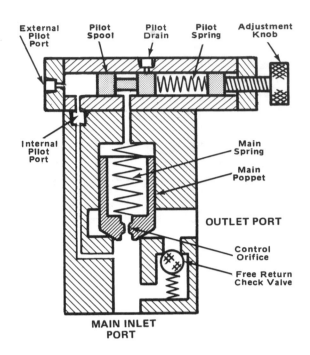

Figure 8-22. A Pilot-Operated Type of By-Pass Valve.

Pilot-Operated By-Pass Valves. Figure 8-22.

Most modern by-pass valves are of the pilot-operated type for many of the same reasons that make pilot-operated relief valves preferable over the direct-acting type — a wide range of adjustment, a very small rise in system pressure between cracking and full flow, efficient operation, and a more compact physical size. Instead of a large, heavy spring to hold the main poppet closed, pilot-operated valves use hydraulic pressure taken from the main valve inlet to hold the main poppet on its seat. In Figure 8-22, the crosshead blocks pressure inside the main spring chamber. When the 2-way valve spool in the crosshead shifts to the right by pilot pressure on its left end, trapped oil in the main spool chamber can vent out the pilot drain port to tank and system pressure can rise no higher. Pilot pressure can be obtained internally from the main inlet port or from an external source.

Note: Figure 8-22 is a conceptual view, oversimplified, to clearly explain operating principles. Exposed areas on top and bottom of the main poppet must be equal; the main spring keeps the poppet in a normally closed position until the 2-way spool in the crosshead shifts to the right, causing the top end of the main poppet to vent to tank. Also, the exposed area on the left end of the 2-way poppet spool must be very small so a light pilot spring can balance high pressure from the main inlet port. For applications where flow may return later in the cycle, use a model with built-in return flow check.

Piloting a Pilot-Operated By-Pass Valve. The nature of the application, as described in the following circuits, will determine whether pilot pressure for opening the valve should be obtained from the valve inlet port (internal piloting) or from another source independent of the valve (external piloting). For internal piloting, the external pilot port must be plugged and the internal pilot passage must be open. For external piloting, the pilot source must be connected to the external pilot port and the internal pilot passage must be plugged.

Draining a Pilot-Operated By-Pass Valve. Referring to Figure 8-22, the "Pilot Drain" port must be connected to tank for all applications. On some applications a separate tank drain line must be connected to this port. On other applications, the drain port can be combined with the main tank port. Some valve models have an internal passage so this can be done without a separate drain line. On the following circuits, the applications which require a separate drain line will be noted.

Pilot-Operated Relief and By-Pass Valves Compared. Both types are 2-way, normally closed, pilot-operated valves. By-pass valves can be used as relief valves by piloting them internally. But a relief valve is not suitable for most by-pass applications because there is no provision on a relief valve for external piloting nor for an external pilot drain.

Note: In the circuits to follow, either a pilot-operated or a direct-acting by-pass valve can be used. Important features to note on each of these circuits is whether that particular application requires internal or external piloting of the valve and whether a separate pilot drain line must be run to tank.

PRESSURE CONTROL CIRCUITS FOR BY-PASS VALVES

Pump Unloading. Figure 8-23. This two-pump circuit is the familiar "high-low" arrangement in which a high volume of oil from both pumps will cause a cylinder to rapidly traverse the distance to the work. As system pressure rises due to load resistance, a pressure level is reached which will cause one pump to unload, allowing the full input horsepower to be applied to the other pump to reach a very high pressure but, of course, at a lower speed.

In this circuit, Valve 2 is the system relief valve which protects both pumps. Valve 1 is the by-pass valve which unloads Pump PF-1 at a designated pressure level.

The idea of a high-low circuit is to cause a cylinder to move very rapidly during the forward traverse up to the work and on the return stroke. The input horsepower pro-

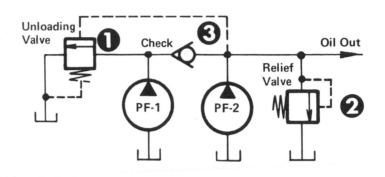

FIGURE 8-23.
Externally Piloted By-Pass Valve Used for Pump Unloading

vides a high flow of oil but is limited to fairly low pressure. Then, when heavy work resistance is encountered, one pump, PF-1, will unload so the entire horsepower can produce very high pressure from the other pump working alone. On most applications which have a relatively short working distance compared to the free travel distance forward and back, a great reduction in cost can be achieved with this circuit compared to one using one large pump for both high flow and high pressure. This would require a system of much greater horsepower capacity, but would be very little faster on the overall cycle.

Pump PF-1 becomes unloaded and placed in an idling condition when pressure from Pump PF-2 opens unloading Valve 1. The pressure at which PF-1 unloads can be adjusted on Valve 1, and is usually set at the pressure where the full input horsepower is being used. When it drops out, there is additional horsepower available to drive Pump PF-2 to a higher pressure.

On this application the by-pass valve, 1, can be internally drained because the main tank port is connected directly to tank. There is no need to run an extra drain return line.

By-pass Valve 1 must be piloted from a source of pressure other than its own inlet. When piloted from Pump PF-2 it will open wide when the unloading pressure is reached and will offer almost no restriction to the flow of Pump PF-1 to tank. If it were to be piloted from its own inlet (internal piloting), it would not unload; it would act like a relief valve when the unloading pressure point was reached.

Efficient Design of a High-Low System. Pump PF-1 is usually a high volume, lower pressure pump (like a gear pump for example), while Pump PF-2 is a lower volume pump (like a piston pump) which is capable of operating at the maximum pressure required in the system. The unloading pressure for Pump PF-1 should be set as high as possible, to the pressure level at which the full input horsepower is being used. A rule-of-thumb which often works out is to select the GPM ratings of the pumps, as near as possible, on a 4:1 ratio, PF-1 having 4 times the GPM output of PF-2. Then, the unloading pressure point can be set at about 1/5th the maximum system pressure.

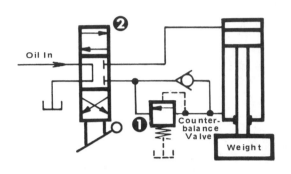

FIGURE 8-24.
Internally Piloted By-Pass Valve
Used for Counterbalancing

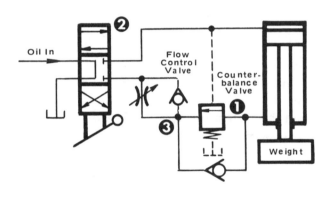

FIGURE 8-25.
Externally Piloted By-Pass Valve
Used for Counterbalancing

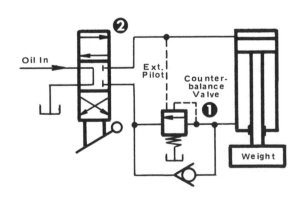

FIGURE 8-26.
Pilot-Assisted By-Pass Valve is
Ideal for Counterbalancing

Counterbalancing Circuits. A hydraulic cylinder which supports a heavy gravity load should be hydraulically counterbalanced for the safety of the machine and the operator. If operated without a counterbalance valve, when the 4-way valve is shifted for the cylinder to go down, the load will fall out of control. It will fall faster than oil can be supplied by the pump, and a vacuum will be created behind the piston. Aside from the safety consideration, a vacuum pulled in the rod area of a cylinder may draw air into the system through the rod seals.

There are several ways a by-pass valve can be used for counterbalancing a cylinder:

Internally Piloted Counterbalance Valve. Figure 8-24. The by-pass Valve 1 is connected to receive pilot pressure from its own inlet. Its pressure setting must be high enough to prevent its opening on fluid pressure produced in the rod end of the cylinder by weight of the load, but when the 4-way valve is shifted to go down, pump pressure on top of the piston will produce sufficient pressure in the rod end to open the by-pass valve.

This is a very smooth and shockless way of slowly lowering a very heavy load. But it may not be safe when the heavy load is lowered rapidly. If the 4-way valve were to be centered before the end of the downward stroke, momemtum energy would keep the cylinder in motion for a distance beyond the desired stopping point and would cause a vacuum on the top side of the piston. In other words, an operator would have difficulty in quickly stopping a fast-traveling massive load.

Another problem with this method is that pressure set on the by-pass valve is unavailable for producing force in the cylinder. This may not be a problem with moving loads, but on static loads, where the cylinder stalls against a load resistance, it would not have the full pump pressure for producing maximum force. Example: If the system relief valve is set for 3000 PSI, and if the by-pass valve, 1, is set for 500 PSI, the pressure available to the cylinder is 3000 PSI minus the by-pass valve setting of 500 PSI (after it has been reduced by the ratio of net area to full piston area.)

The by-pass model used should contain a check valve for free reverse flow. For maximum safety (in case a line to the cylinder should break), the by-pass valve should be installed as close as possible to the cylinder, plumbed to the cylinder with steel fittings, and its weight should be supported on a bracket. All unnecessary hose and fittings between by-pass valve and cylinder port should be eliminated.

The by-pass valve may be drained either internally or externally. However, external draining is preferred, especially if there are filters, heat exchangers, or other components in the tank return line which would cause back pressure. With external draining, the spool of the by-pass valve will already be in correct counterbalancing position at the moment the 4-way valve is shifted.

Externally Piloted Counterbalance Valve. Figure 8-25. The same by-pass valve, shown in the preceding diagram, may be used. In this case the valve receives its pilot pressure from the opposite cylinder port. External draining is preferred for the same reasons given before, although if necessary, internal draining can be employed.

The counterbalancing action is very abrupt, and shock may be generated if a fast-traveling load is suddenly stopped, or the cylinder may chatter as the load is lowered. If these problems should be encountered, a needle or flow control valve, 3, may be installed in series with the by-pass valve and adjusted for enough restriction to eliminate the shock or chatter.

An important advantage with external piloting is that none of the pump pressure is sacrificed in the counterbalance valve; all of it is available to work against static loads.

The pressure setting of the by-pass valve is not critical. The lower it can be set the less likely the cylinder is to chatter. However, it must be set lower than the system pressure relief valve.

Internally and Externally Piloted Counterbalance Valve. Figure 8-26. By-pass Valve 1 is especially built for counterbalancing applications. It is internally piloted but also has an external pilot port. It combines the desirable features of the preceding circuits. Internal piloting provides the smooth, shockless lowering of massive loads, and the external pilot, or "pilot assist" feature opens the poppet fully when working against a static load. Thus, full pump pressure is utilized if the cylinder should stall.

By-Pass Valves for Dumping. Figure 8-27. The term "dumping" as used here means discharging oil to tank with a minimum of back pressure. It means unloading of oil back to the reservoir.

Valve 1 is a standard by-pass valve connected to discharge oil from the cylinder blind end to tank in response to pilot pressure in the rod end of the cylinder. Its purpose in this circuit is to assist the 4-way valve in discharging the high flow of oil pushed out of the cylinder while it is retracting. This return oil is always greater than the pump flow, even up to twice the pump flow on 2:1 ratio cylinders. Since the by-pass valve can open only when piloted, it stays closed while the cylinder is extending.

The pressure setting of the by-pass

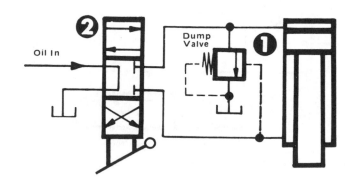

FIGURE 8-27.
Externally Piloted By-Pass Valve Used for Unloading Oil to Tank as the Cylinder Retracts.

159

valve is not critical, but it must be set higher than the back pressure from the cylinder rod port to tank so it cannot open while the cylinder is extending. It should be set fairly low, just high enough to open with the pressure required to retract the cylinder.

When handling high fluid flows on a large rod cylinder, it may be more economical to use a dump valve for handling peak return flow rather than to increase the size of the 4-way valve.

BY-PASS VALVES USED FOR SEQUENCING

In fluid power terminology, sequencing denotes a sequential action between two or more cylinders in which the cylinders extend and retract in a pre-programmed order on every cycle. By-pass valves are used for sequencing and when used in these applications are called "sequence valves". External draining is required on all by-pass valves used for sequencing; the spring chamber must always be referenced to atmosphere and not to the outlet port.

Sequence valves are non-electrical; they are not involved in, and do not add to the complexity of the electrical control circuit. They are the simplest and surest method for programming, step by step, the sequential action between cylinders. The priority action they give cannot be duplicated with limit switches and solenoid valves nor by cam valves. It can be duplicated after a fashion by pressure switches connected to solenoid 4-way valves, but at greater cost and complexity.

The classic example of programming with sequence valves is on a pressure casting machine, such as a die casting machine. There are two basic actions involved: First, the mold must be closed by a cylinder. Next, an injection cylinder pushes fluid material into the mold. It is of prime importance that the mold be held tightly closed by active hydraulic pressure during the injection part of the cycle. Any relaxation of mold closing pressure would cause the mold to momentarily open slightly, causing flash to appear on the product, resulting in rejected parts. A circuit designed with sequence valves will give priority to the mold closing cylinder, maintaining full active pressure on it during the injection.

Example of Sequencing. Figure 8-28. Let Cylinder 1 represent the mold closing cylinder of a die casting machine, and let Cylinder 2 represent the injection cylinder. When the 4-way valve is shifted to its left side position, oil flows to extend Cylinder 1 but is blocked from Cylinder 2 by the sequence valve. When Cylinder 1 reaches stall, with the mold tightly closed, pressure rises behind its piston to open the sequence valve and flow into Cylinder 2. While Cylinder 2 is

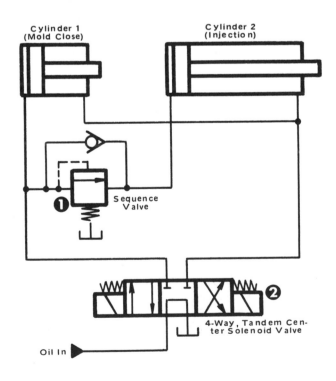

FIGURE 8-28.
Externally Drained By-Pass Valve Used for Sequencing

extending at whatever load it is moving, the sequence valve will throttle to keep full active pressure behind Cylinder 1. When both cylinders have reached stall, the sequence valve opens wide to put full system pressure on the pistons of both cylinders. When the 4-way valve is shifted to its right side position, the cylinders retract in random order, although their return strokes could be programmed with another sequence valve as shown in later circuits.

Sequence Valve Action Compared to Relief Valve Action. The graphic symbol for a by-pass valve when used as a sequence valve is illustrated in Figure 8-28 on the preceding page. It is similar to the relief valve symbol shown on Page 143. However, in the case of a sequence valve the source of pilot pressure can be either internal (from its own inlet), or external (from elsewhere in the circuit). But its spring chamber must *always* be externally drained to tank, never to its own outlet port. This is shown in Figure 8-28 and in other diagrams to follow.

On the other hand, a relief valve is *always* piloted from its own inlet and its spring chamber is *always* drained into its own outlet port. A relief valve cannot be used for sequencing as defined here because its spring chamber is not referenced to atmosphere. If it were to be used in place of a sequence valve, the pump pressure would divide between the two cylinders and full pressure would never be available to either of them. Study and compare the two diagrams below to understand the difference between a relief valve and a sequence valve.

Relief Valve. Figure 8-29. The spring chamber of a relief valve is always drained internally to its outlet port. In this example, the relief valve is adjusted for a cracking pressure of 500 PSI, and the valve inlet is connected to a source of 1250 PSI. The valve outlet port is deadended into a pressure

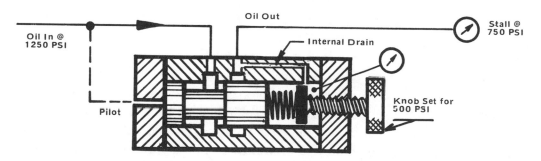

FIGURE 8-29. An internally drained by-pass valve acts like a relief valve. Back pressure on the valve outlet adds to the strength of the internal spring. Relief valves are always piloted from their own inlet and drained to their own outlet.

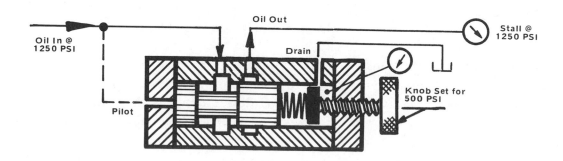

FIGURE 8-30. When a by-pass valve has its spring chamber externally drained to tank, any back pressure which may appear on its outlet will not affect the operation of the main spring.

gauge to measure the pressure level available downstream after the relief valve opens. When the inlet pressure is greater than the cracking pressure set on the valve, the valve spool opens. Then, any pressure appearing on the outlet port backs up through the drain passage and works against the spool, and adds to the strength of the spring. In this instance, with 1250 PSI on the inlet, the valve opens and pressure appears on the outlet port. When outlet pressure reaches 750 PSI, the total force trying to close the valve is 750 PSI fluid pressure in the spring chamber plus 500 PSI on the spring = 1250 PSI. This is the limit to which the pressure can rise on the outlet; any further increase on the outlet port would back up into the spring chamber and cause the valve to completely close.

Therefore, with a relief valve, the maximum pressure which can appear downstream is inlet pressure minus the PSI setting of the relief valve spring. The full inlet pressure is never available downstream.

Sequence Valve. Figure 8-30. The spring chamber of a by-pass valve used for sequencing must *always* be drained directly to tank through an external drain line to prevent fluid pressure from backing up in the spring chamber and increasing the closing force on the valve.

Under the same operating conditions of 500 PSI spring adjustment and an inlet pressure of 1250 PSI, as long as the inlet pressure is greater than the spring adjustment, the spool will be pushed to a full open position for free passage through the valve, and the full 1250 PSI will be available downstream.

Clamp and Work Sequencing Circuits. Several methods are available for extending and retracting two cylinders in a sequential program each cycle. In Chapter 3, the use of button bleeder valves, limit switches, solenoid valves, and flow control valves was described for operation of air cylinders. However, with hydraulic cylinders, most sequencing circuits use one of only two methods: sequence valves for non-electrical operation and solenoid valves for electrical operation. Programming involves the starting of a second cylinder when the first one reaches either a pre-set position in its stroke or when pressure in the fluid rises to a pre-set level.

"Clamp and work" is a term commonly used to describe a special arrangement of two cylinders in which the first one, the clamp, advances until it stalls against the workpiece, holding it firmly while the other one, the work cylinder, drives tooling for performing machining, pressing, forming, or shearing operations on the workpiece. One definite requirement for all clamp and work circuits is that the clamp, after stalling against the workpiece, shall continue to hold active force against the workpiece throughout the working part of the cycle regardless of whether the work resistance be light or heavy.

Sequence valve circuits are the best way of achieving this action. Pressure switches controlling 4-way solenoid valves might give acceptable results on some applications but the action is not as smooth and reliable as with sequence valves.

On some applications, clamping force need not be maintained while the work cylinder is retracting. The cylinders can retract randomly. But on other applications, full clamping force must continue during work cylinder retraction. The following circuits give examples of return action.

Random Retraction of Clamp and Work Cylinders. Figure 8-28 is a typical circuit for random retraction. When the solenoid valve is shifted to reverse the cylinders, the cylinder to retract first will be the one which presents the lesser resistance to oil flow. Circuitwise this is the simplest arrangement if clamping force is not required while the work cylinder is retracting. If it is, one of the following circuits may be used.

Sequential Retraction Without Active Clamping Force. Figure 8-31. If the clamp cylinder must remain in position during retraction of the work cylinder but does not need to maintain active clamping force, a second sequence valve, 2, may be added. It may also be necessary to add pilot-operated check valve, 4, to make sure "spring-back" from the load does not push the clamp cylinder back during retraction of the work cylinder. A description of pilot-operated check valves will be found on Page 200.

After the work cylinder has reached its home position, the pressure can build up and pass through sequence Valve 2. The pressure pilots Valve 4 open and retracts the clamp. When both cylinders have reached home positions, switches in the electrical control circuit should release the 4-way valve to its neutral, pump unloading, position.

Sequential Retraction With Full Clamping Force. Figure 8-32. On those applications where it is necessary to maintain full active clamping force until the work cylinder completely withdraws from the work, this circuit is suggested.

Each cylinder must have its own 4-way directional valve so the clamp can be maintained with forward force while the work cylinder is shifted to its reverse direction.

The electrical control circuit must be designed so Solenoids A and C are energized at the same time but through separate relay contacts. Energizing Solenoid A shifts Valve 2 to start the clamp forward. Energizing Solenoid C shifts Valve 3 to its crossed arrow block. However, the work cylinder does not

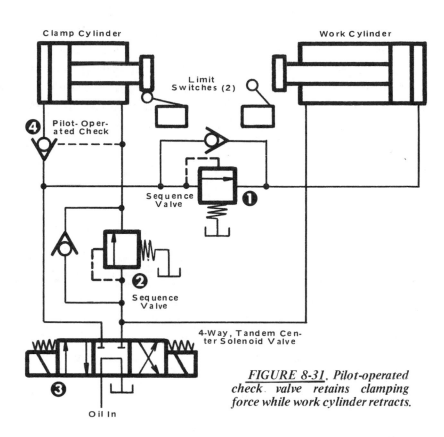

FIGURE 8-31. Pilot-operated check valve retains clamping force while work cylinder retracts.

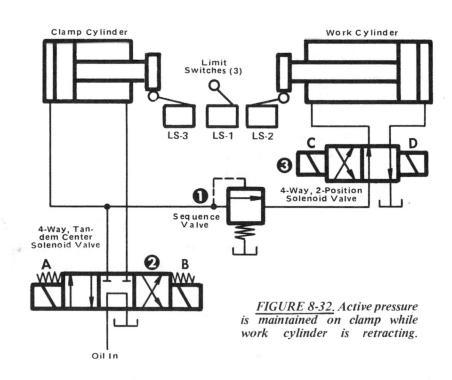

FIGURE 8-32. Active pressure is maintained on clamp while work cylinder is retracting.

move at this time because the sequence valve is closed and there is no oil supply for it. When the clamp stalls against the work, pressure rises to open the sequence valve and oil flows to Valve 3 and to the work cylinder to start it forward. It moves until limit Switch LS-1 is actuated. This energizes Solenoid D and reverses the work cylinder. During its retraction, Solenoid A remains energized and the clamp is still producing its full force against the workpiece because of the sequence valve action. At home position of the work cylinder limit Switch LS-2 is actuated. This signals the electrical circuit to shift Valve 2 to its opposite side position and starts the clamp to retract. At home position of the clamp, limit Switch LS-3 is actuated. This causes both solenoids of Valve 2 to become de-energized and the valve spool to center itself, unloading the pump.

Sequence Valve 1 is a by-pass valve connected for internal pilot and external draining. A model without return check valve can be used since there is never any reverse flow in this line.

PRESSURE CONTROL WITH REDUCING VALVES

Description of a Pressure Reducing Valve. A pressure reducing valve in hydraulics is the equivalent of a pressure regulator in a compressed air system, and the operating principle of the two is basically the same although they are constructed quite differently. A pressure reducing valve is for the purpose of establishing a lower pressure level in a branch circuit from that in the main pump line. The inlet is connected to a source of high pressure, usually the pump line, and the adjustment is set to deliver a constant maximum pressure at the outlet regardless of pressure fluctuations on the inlet, as long as the inlet pressure remains as high as or higher than the adjustment setting. Branch circuits operating from a reducing valve act the same way as any other hydraulic circuit. Pressure in the branch is directly proportional to the load resistance, and will be no higher than demanded by the load. But the maximum pressure available to the branch is limited by the adjustment setting. The load can move at full speed as long as it does not demand a pressure higher than the adjustment setting. But if the load should increase to the point that it demans more pressure than the adjustment setting, this in

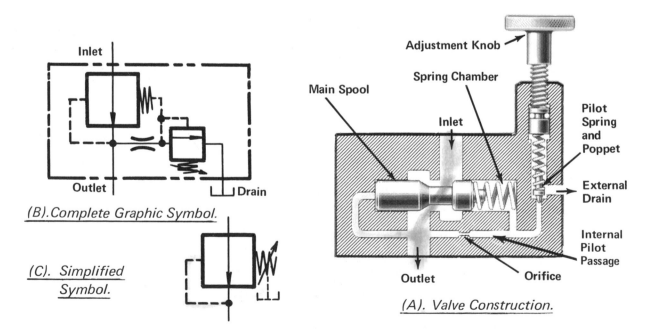

(B).Complete Graphic Symbol.

(C). Simplified Symbol.

(A). Valve Construction.

FIGURE 8-33. Pressure Reducing Valve of the Pilot-Operated Type.

effect is an overload. The load will either stall or will be forced to move at a slower speed — one which does not require pressure higher than the adjustment setting. A reducing valve offers very little flow resistance and power loss until the branch circuit demands pressure higher than the adjustment setting. At this point, the valve spool moves to a throttling position and will limit the flow through the valve to a value where the pressure will stay within the limit of the adjustment setting.

Applications for pressure reducing valves in hydraulics are limited but important. They cannot take the place of pressure relief valves in the pump line to protect the system from excessive pressure. They are normally used only in branch circuits in which the pressure must be limited to a level less than in the pump line. They are not needed in a system where all branch circuits are capable of operating at the same pressure level.

A pressure reducing valve does not add power loss and heating to the system except when the load demands pressure higher than the adjustment setting and the spool is in a throttling position.

Reducing valves are basically 2-way, 2-position, normally open valves. Being normally open, they are of the opposite configuration from pressure relief valves which are normally closed. The manufacture of direct-acting reducing valves for hydraulics has been virtually discontinued. Most, if not all, of the modern reducing valves are of the pilot-operated type, with a small relief valve mounted as a crosshead for limiting the pressure which can be delivered from the main outlet port.

Operating Principle, Pressure Reducing Valve. Figure 8-33. The main working parts can be seen in Part (A) of this figure. The spool is held in a normally wide open position by a light, non-adjustable, spring. A small, direct-acting, relief valve on the crosshead sets the maximum pressure level on the outlet port regardless of pressure fluctuations in pressure on the inlet (as long as inlet pressure is higher than the adjustment setting). The spool remains in the wide open position for full free flow to the outlet until an excessive demand for pressure from the load connected to the outlet causes the spool to move into a throttling position.

To visualize the action of the valve, start with zero pressure demand on the outlet. As outlet pressure increases (as the load increases), the rising pressure is sensed equally on both ends of the spool. Since area is the same on both ends, the spool remains pressure balanced and held in the open position by the spring bias. The spool remains open until outlet pressure rises to the cracking pressure of the relief valve on the crosshead. The adjustment of this relief valve puts a limit to the rise on the right end of the spool. As outlet pressure continues to rise, the pressure working on the left end of the spool becomes higher than the opposing pressure on the right end. When outlet pressure rises to become greater than the cracking pressure of the relief valve plus equivalent spring force, the spool starts moving to a throttling position to reduce flow through the valve. The spool will move as far as necessary to limit flow to prevent outlet pressure from exceeding the adjustment setting.

The higher the pressure demand on the outlet, the closer the spool will move toward its closed position. On a deadend condition, as for cylinder stall, the valve will close completely. On an open outlet condition, as for free travel of the cylinder, the spool will move to its wide open position.

There is no flow through the orifice at outlet pressures less than the adjustment setting, and its presence in no way affects valve performance. But at cracking pressure of the relief valve and beyond, the orifice limits flow into the spring chamber to a low value which can be handled efficiently by the tiny poppet in the relief valve.

There is a small loss of oil across the relief valve into the external drain when the spool is in a throttling position. On deadend service or with the outlet blocked, leakage across both ends of the spool escapes to tank across the relief valve, and the outlet pressure will not build up from leakage to a pressure higher than the adjustment setting.

A reducing valve will not pass flow very well in the reverse direction and in most applications is not required to do so. If reverse flow is required, a model with built-in reverse flow check valve should be used, or an external check valve may be plumbed around it.

The circuits in the section starting on Page 113, for operation of several branch circuits from one pump will, in general, apply to circuits where one or more branches operate from reducing valves.

Graphic Symbols for Pressure Reducing Valves. Figure 8-33. Part (B) shows the complete graphic symbol for a pilot-operated pressure reducing valve, and parts in the symbol can be identified by comparison of the symbol to the sectional view in Part (A). The complete symbol, because of its complexity, is rarely used, and the symbol of Part (C) is generally sufficient to indicate its operation in a hydraulic circuit.

Draining a Reducing Valve. All pressure reducing valves must have an external drain line to tank. The drain port must never be plugged. Since the main outlet port carries pressure during at least part of each cycle, the drain port cannot be teed into the outlet port. However, drain lines from several valves such as solenoid, sequence, or flow control valves can usually be teed together and run to tank without running separate drain lines from each valve which requires one.

CIRCUITS FOR PRESSURE REDUCING VALVES

Low Pressure Branch Circuit. Figure 8-34. By far the most common application for a pressure reducing valve is to provide a maximum lower pressure in a branch circuit either to protect components or other parts of a machine which are not rated for the full pump pressure, to provide a lower force level where this lower level is required for proper function, or to accurately control tool force in critical machining operations. Usually a reducing valve can be used in a branch circuit of any of the diagrams starting on Page 113 for operating several branch circuits from one pump.

The reducing valve cannot protect components on its inlet side against excessive pressure. A relief valve should always be used on the pump line. Usually it can protect all components on its outlet side except for situations where high mechanical reactive force against the piston rod of a cylinder, for example, might create a fluid pressure at the outlet port, during at least a part of the cycle, greater than the adjustment setting of the valve. On applications where this could occur, a reducing valve model containing a relief valve on its outlet port should be used, or a separate pilot-operated relief valve should be connected to the outlet port.

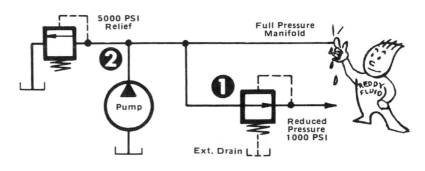

FIGURE 8-34. Pressure Reducing Valve in Low Pressure Branch Circuit.

Selective Reduced Pressure. Figure 8-35. The cylinder can extend with full system pressure, but on retraction the pressure is reduced to a lower value through a pressure reducing Valve 1. Some large cylinders, intended to produce high tonnage in the forward direction may not tolerate such high pressure on the rod end and on the packings.

Reducing Valve 1 should be a model with free reverse flow check valve built in. This valve remains wide open as the cylinder is retracting in free travel. But if the cylinder should stall at any point in its return stroke, the reducing valve would limit the pressure on the rod end. As with all pressure reducing valves, the external drain port must be run through a separate line to tank.

Reduced Clamping Force. Figure 8-36. In some applications the clamping force must be carefully controlled to prevent distortion or damage to the workpiece.

This is an example of two branch circuits working from one pump. It is a clamp and work application in which the clamp cylinder moves in to clamp the workpiece first. Then, when clamping force has been developed, the work cylinder is released through a sequence valve to move a cutting, forming, pressing, or shaping tool against the workpiece. The priority action of a sequence valve in programming two cylinders has been described on Page 160.

In this circuit, when the solenoid 4-way valve has been shifted to its left working block, oil can flow without restriction through reducing Valve 2 into the clamp cylinder. When the clamp has moved forward and stalled against the workpiece, pressure rises and closes the reducing valve when its pre-adjusted pressure setting has been reached. The purpose of the reducing valve is to protect the clamp cylinder from the full pump pressure.

With the clamp cylinder stalled and reducing Valve 2 closed, pressure against sequence Valve 1 inlet rises very rapidly. When circuit pressure equals or exceeds the adjusted setting of Valve 1, oil can flow through it to extend the work cylinder. The rest of the circuit action is the same as described in connection with Figure 8-28 on Page 160.

Both the reducing valve, 2, and the sequence valve, 1, should have built-in free reverse flow check valves, or external check valves should be used. Their external drain ports must be connected to tank. The drain ports of both valves and the drain port on the solenoid valve (if applicable), may feed into a common drain line direct to tank, but this drain line should not be combined with any return line carrying full pump flow.

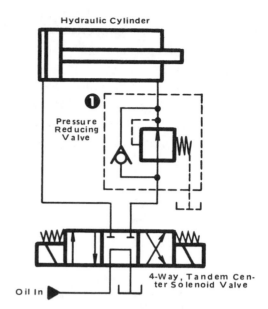

FIGURE 8-35. *Selective Reduced Pressure.*

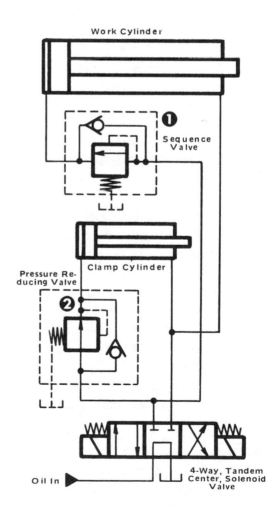

FIGURE 8-36. *Reduced Clamping Force.*

REMOTE CONTROL OF PRESSURE REDUCING VALVES

A pilot-operated reducing valve can be controlled from an operating position a short distance away, usually limited to 10 feet or less, but sometimes from a greater distance. Its outlet pressure can be remotely changed from knob setting up to full inlet pressure.

Unlike pilot-operated relief valves, reducing valves do not have an RC vent port. However, they can be controlled remotely by changing the flow resistance in their external drain line, in a manner which is similar to remote control of pilot-operated relief valves as described on Pages 147 and 148.

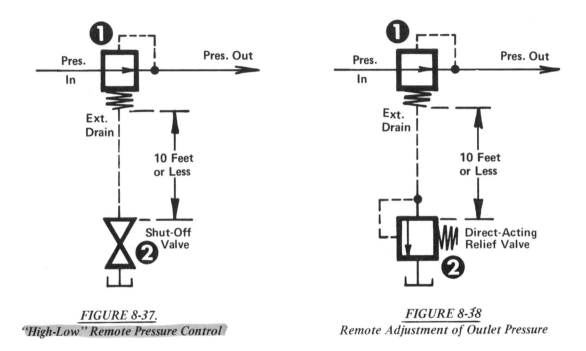

FIGURE 8-37.	FIGURE 8-38
"High-Low" Remote Pressure Control	*Remote Adjustment of Outlet Pressure*

Figure 8-37. A shut-off valve, either solenoid or manually operated, can be installed in the external drain line. If this valve is open, the reducing valve functions normally, delivering a maximum level of reduced pressure on its outlet according to the setting of its crosshead relief valve, and circuit operation is as previously described. If the drain shut-off valve is closed, or if the external drain port were to be plugged, the reducing valve spool would remain in its normal, wide open position, regardless of the adjustment setting on its knob; the spool would never move to a throttling position. This means that the pressure on the outlet port would be the same as that on its inlet.

If the drain shut-off valve is solenoid operated, full pressure or reduced pressure can be selected by electrical remote control.

The drain line should be kept as short and direct as possible, and of sufficient size that there will be no significant flow resistance to tank.

Figure 8-38. A miniature, direct-acting relief valve installed in the external drain line can be used to remotely change the outlet pressure delivered through the valve. Note from the sectional view of a reducing valve, Figure 8-33(A), that a relief valve in the drain line will be in series with the internal relief valve in the crosshead. The result of putting two relief valves in series is covered on Page 145. The two relief valve settings will be additive. In this case, adding a relief valve in the drain line has the same effect on outlet pressure as raising the setting of the crosshead relief valve: it raises the level of outlet pressure. The rule, then, for remote control of a reducing valve is to set the *lowest* desired outlet pressure on the crosshead relief. Then, this pressure can be raised remotely either by blocking the drain line, or adding flow resistance to it. This is the opposite procedure from remote control of a pressure relief valve where the *highest* desired pressure is set on the crosshead, then reduced remotely by valving placed in the RC vent line to tank.

Step Level Pressure Control. Figure 8-39. This circuit is an example of the numerous ways a reducing valve can be remotely controlled by miniature relief and/or shut-off valves.

Shut-off Valves 2, 3, and 4 may be solenoid or manually controlled to suit the application. Valve 1 is the main pressure reducing valve of a size to carry full flow required in the branch. The lowest pressure to be selected must be set on the crosshead of this valve. The drain line must be kept short and direct to minimize flow resistance. For remote control from distances greater than a few feet, the shut-off valves should be solenoid operated through an electrical circuit.

Miniature relief Valves A and B are adjusted for the intermediate pressure levels desired between the lowest, set on

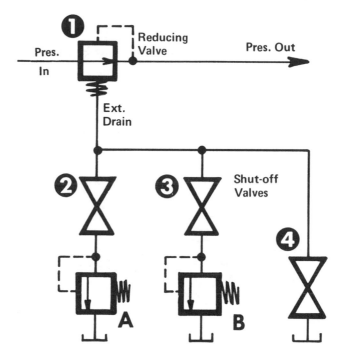

FIGURE 8-39. *Step-Level Pressure Control.*

the main valve, and full inlet pressure obtained when all shut-off valves are closed. Table 8-1 shows the condition of each shut-off valve to obtain various pressure levels on the outlet port of the main Valve 1.

TABLE 8-1 — STEP LEVEL ACTUATION CHART FOR FIGURE 8-39

Pressure Level	Condition of Valves 2, 3, and 4
Highest (Outlet = Inlet Pressure)	*Valves 2, 3, and 4 Closed*
Next Highest	*Valve 2 Open, Valves 3 and 4 Closed*
Next Highest	*Valve 3 Open, Valves 2 and 4 Closed*
Lowest (Outlet = Knob Setting on Main Valve 1)	*Valve 4 Open*

PRESSURE CONTROL WITH POWER LOGIC VALVES

Please refer to Page 132 for a description of power logic valves and their use in directional control applications, and to Page 153 for relief valve application. Certain models can be used for the relief, by-pass, sequence, counterbalance, and pressure reducing circuits shown earlier in this chapter.

They are furnished as cartridges to be installed in bored cavities in a manifold which has been custom designed and built by the user specifically for the application. They are held in the cavity by one of a choice of covers which usually contains auxiliary valving associated with the logic valves, such as miniature solenoid valves, relief valves, shuttle valves, and orifices. This method of mounting eliminates a great deal of external plumbing required for an equivalent system using standard valves.

Logic valve circuits should be approved by the factory which will furnish the valves. Their best applications are in the larger flow sizes from, say 50 GPM, up to hundreds or thousands of GPM.

Logic valve models for by-pass and pressure relief are normally closed and have an area ratio of 100% – equal area on top and bottom with no net area on the side at Port B. The internal spring should have a closing force equivalent to 50 to 100 PSI. The pressure level at which the poppet will open is adjusted with a miniature relief valve, usually mounted on the cover.

Models used for pressure reducing must be normally open, and with an area ratio of 100% –bottom area equal to top area. The internal spring should be equivalent to 50 to 100 PSI.

Logic valves, like the by-pass valves described earlier, must receive their pilot pressure (internal or external) from the proper source for the application, and the pilot control relief valve (mounted on the valve cover) must be separately drained to tank on applications where the main outlet (Port A) does not connect directly to tank. Logic valves can be used in previously described pressure control circuits as noted in the following diagrams:

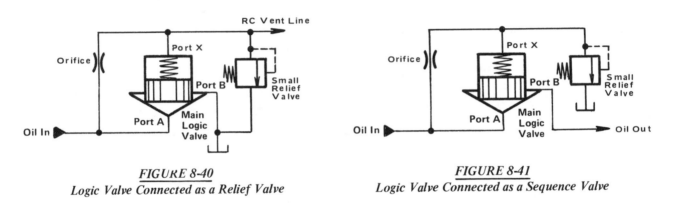

FIGURE 8-40
Logic Valve Connected as a Relief Valve

FIGURE 8-41
Logic Valve Connected as a Sequence Valve

Figure 8-40. Internal Pilot, Internal Drain, Normally Closed. Used primarily as a pressure relief valve as illustrated in the circuit on Page 153. It may be remotely controlled by restricting or venting the line indicated, to change pressure levels or to unload a pump.

Figure 8-41. Internal Pilot, External Drain, Normally Closed. May be used for sequencing as shown in Figures 8-28, 8-31, and 8-32, and as an internally piloted counterbalance valve in Figure 8-24. The pilot drain (from the relief valve control) must not be connected with Port A (main outlet) on any application where Port A is not connected directly to tank.

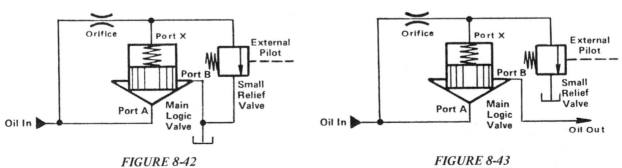

FIGURE 8-42
Logic Valve Connected as an Unloading Valve

FIGURE 8-43
Logic Valve Connected as a Counterbalance Valve

<u>Figure 8-42. External Pilot, Internal Drain, Normally Closed.</u> This connection applies to the unloading valve in the high-low circuit of Figure 8-23 and in the dumping (unloading) circuit of Figure 8-27.

<u>Figure 8-43. External Pilot, External Drain, Normally Closed.</u> This is typical of externally piloted counterbalancing as in Figure 8-25. Also used in some sequence valve circuits with meter-in speed control as shown in Figure 9-12.

<u>Figure 8-44. Internal Pilot, External Drain, Normally Open.</u> This is a pressure reducing valve and can be used in any of the circuits of Figures 8-34 through 8-39. Pressure on the bottom of the poppet (from Port A inlet) is balanced by equal pressure taken from outlet Port B through an orifice to work on an equal area, so the poppet remains open. The miniature relief valve determines the maximum pressure which can build up on top of the poppet. When relief valve cracking pressure has

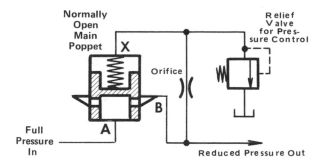

FIGURE 8-44. *Pressure Reducing Valve.*

been reached (by pressure build-up on the outlet Port B), pressure in the spring chamber can rise no higher. Therefore, any further increase in outlet pressure above spring force, will move the poppet toward a throttling position or completely closed to prevent any further pressure rise on outlet Port B.

PUMP UNLOADING METHODS

The Importance of Unloading a Pump. Most hydraulic machines operate through a cycle in which the system is working for part of the time and is idle for other periods. It makes good sense to let the pump rest during those periods when fluid power is not needed. This prolongs the working life of the pump, reduces the average power input to the machine, and reduces heating in the hydraulic oil. Most hydraulic circuits should be designed with a means for allowing the pump to unload or rest during non working periods in the cycle.

The most straightforward way of unloading a pump is to turn it off, either by shutting off the prime mover (electric motor or engine) which drives it, or by de-clutching it from the prime mover. This is rarely done, however, except on small systems such as tailgate lifts, because of the delay and difficulty in re-starting the pump when power is needed. On larger systems the pump is allowed to continue its rotation but as much of the pressure load against its outlet port is removed as practical. When allowed

to pump freely against little or no resistance, its life expectancy extends toward infinity, it consumes very little power from its driving source, and it produces little or no heating of the oil. This section is a summary of various methods of unloading a pump, some of which have been covered elsewhere in these textbooks and will be referenced as they are presented.

Unloading Principle. Figure 8-45. A manually operated shut-off valve can be opened to present a free and relatively unrestricted flow path to tank even though the pump continues to run and to

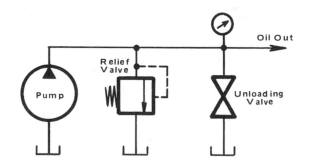

FIGURE 8-45. *Direct Unloading of a Pump.*

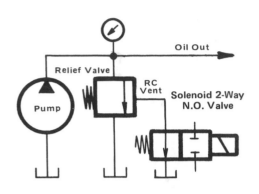

FIGURE 8-46
Unloading by Relief Valve Venting

produce a full flow of oil. A pressure gauge on the pump line should show near zero pressure when the unloading valve is open. Even though this method is simple it is seldom employed because it is not automatic; it requires extra attention from an operator, and therefore, through possible operator neglect, the pump may not operate unloaded when it should, and the system might overheat. Nearly all systems use one of the automatic unloading methods to be described.

Relief Valve Vent Unloading. Figure 8-46. Venting the pump relief valve is the preferred way of unloading the pump in an electrically operated system. The unloading valve is usually of miniature size, 2-way, normally open. A 3-branch circuit using this system is shown on Page 115.

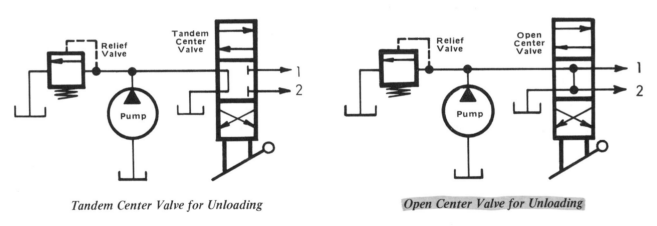

Tandem Center Valve for Unloading   *Open Center Valve for Unloading*

FIGURE 8-47. Unloading of a Pump Through the Spool of a 4-Way Directional Valve.

Unloading Through the 4-Way Valve Center. Figure 8-47. Because of its simplicity, unloading the pump through the spool of a 4-way tandem or open center valve is often used on low power systems which are either electrically or manually operated. Tandem center valves are recommended for cylinder operation and open center valves for hydraulic motor operation. This method is not recommended on higher power (over 50 HP) systems because of the high system shock produced when the valve is shifted. Caution: When using an open center valve to operate a cylinder, the cylinder may drift out of position when the valve is centered if there is appreciable back pressure in the tank return line. Please refer to the text on Page 100.

Bank Valve Unloading. For manually operated hydraulic systems, particularly mobile systems, sectional or bank valves provide an ideal way to unload the pump when none of the branch circuits is working. See Page 119 for description of these valves.

Accumulator Systems. Those systems which charge an accumulator during periods when branch circuits are idle use either a pilot-operated special by-pass valve or a pressure switch and solenoid valve to unload the pump after the accumulator has received a full charge. A sample circuit is on Page 90.

Pressure Compensated Pump. Some models of piston and vane-type pumps use a built-in pressure compensator to unload the hydraulic pump when all branch circuits are closed to oil flow. See circuit and description on Page 91.

Partial Unloading Circuits. The two circuits on this page do not completely unload a hydraulic pump, but they do reduce the load on the pump to a low level which may be acceptable on certain carefully considered applications. If one of these circuits can be used, it will simplify the hydraulic circuit and the electrical control system. Of course these partial unloading circuits are practical mainly on low power systems and those where the pressure to retract the cylinder is quite low.

Figure 8-48. Two full size relief Valves 1 and 2 are used to control maximum pressure for extension and maximum pressure for retraction respectively. Valve 1, of course, is set for rated system tonnage in the working (extension direction). Valve 2 is set for the least pressure which will cause the cylinder to retract, presumably very low. When the cylinder has fully retracted, it stalls against a positive stop or against its own end cap and allows the full pump flow to discharge across relief Valve 2. If Valve 2 can be set very low, the amount of power waste and heat generation may not be significant.

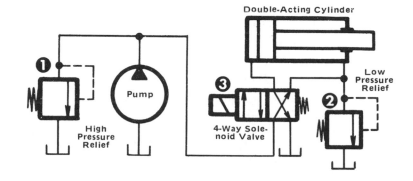

FIGURE 8-48. Dual Relief Valve Partial Pump Unloading.

The advantage of not unloading the pump at home position of the cylinder is that a simple 2-position model can be used for 4-way Valve 3. The electrical circuit can also be relatively simple because it does not require the use of holding relays.

Since relief Valve 2 is connected to the rod port of the cylinder, its presence will not affect the tonnage output of the cylinder while it is extending.

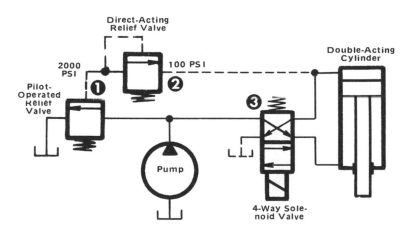

FIGURE 8-49.
Partial Unloading by Relief Valve Venting

Figure 8-49. This circuit operates in exactly the same way as the preceding one, but the action is accomplished in a slightly different way. Relief valve 1 in this circuit is a pilot-operated model with RC vent port for remote control. It is adjusted for the maximum pressure of the system in the extension direction of the cylinder. Valve 2 in this circuit is a miniature model connected in the vent line of the main relief valve. The key to the uniqueness of this circuit is that Valve 2, instead of discharging to tank, discharges to the blind end port of the cylinder. It should be adjusted for the least pressure which will cause the cylinder to retract.

While the cylinder is extending, relief Valve 2 is non-functional because its outlet port is blocked by being connected to high pressure on the blind port of the cylinder. So the cylinder will extend at the maximum pressure adjusted on the main relief Valve 1. But while the cylinder is retracting, relief Valve 2 is discharging to tank through the 4-way valve. This causes the pressure setting of the main relief Valve 1 to be lowered to the setting of the relief valve in the vent line. At home position the cylinder is allowed to stall. The full pump flow will discharge across relief Valve 1 but at a pressure low enough to avoid excessive power waste and heating.

This circuit offers the same advantages as the preceding one – simpler and cheaper components, and a very simple electrical control circuit without the use of holding relays. Of course it can only be used on applications where the cylinder can be retracted with a very low pressure.

Cam Valve Unloading. Cam valves may provide a convenient means of unloading the pump when the cylinder has retracted to home position. However, they are generally unavailable in larger than 1/4 or 3/8" size, which limits their use either to unloading small oil flows or to acting as a pilot control on a larger valve to handle higher flows.

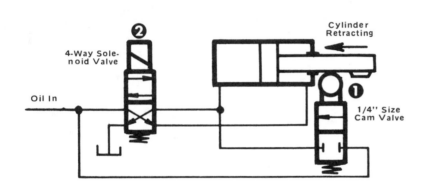

FIGURE 8-50. Cam Valve Pump Unloading.

Figure 8-50. Cam Valve 1 is placed where it will be actuated by the cylinder or by a moving part of the machine when the cylinder reaches its home position. Normally it is closed, but when actuated it provides a free flow path for pump oil to be vented to tank through 4-way solenoid Valve 2. When the next cycle is started by energizing Valve 2, pump oil is disconnected from its vent path to tank and is connected directly to the blind end port of the cylinder. The cylinder can now operate forward and return with normal pressure until the cam valve is again actuated.

This circuit can only be used on oil flows which can be handled by the size of cam valve which is available. Its main advantage over other unloading circuits is the simplicity of the electrical circuit needed to control it. Holding relays can be eliminated by using either single solenoid valves or 2-position double solenoid valves.

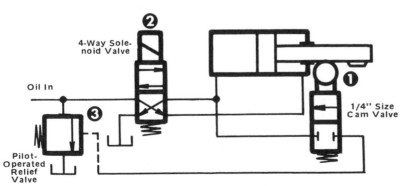

FIGURE 8-51. Cam Valve Unloading of a Large Pump.

Figure 8-51. Alternate cam valve unloading circuit which unloads the pump when the cylinder retracts to home position and actuates cam Valve 1. The 1/4" size cam valve controls a large pilot-operated relief Valve 3. Pump oil does not unload directly through the cam valve; it goes to tank through the relief

174

valve when that valve is vented by the cam valve. In other respects this circuit has the same action described for the preceding one, Figure 8-50.

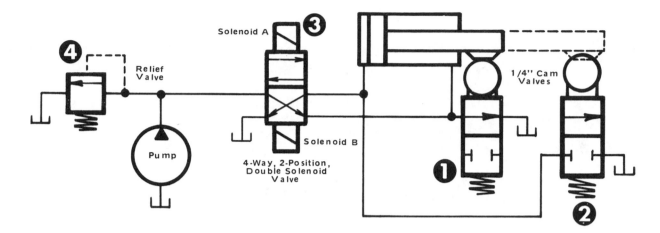

FIGURE 8-52. Cam Valve Unloading at Each End of the Stroke — at Each Half cycle of the Cylinder.

Figure 8-52. The cylinder can make a half cycle and stop with the pump unloaded. Extension and retraction limiting positions are established by positioning the cam Valves 1 and 2. When Solenoid A on the 4-way valve is energized, the cylinder extends until cam Valve 2 is actuated. The pump flow then is unloaded to tank which stops the cylinder. To retract the cylinder, Solenoid B must be energized. At retracted home position, cam Valve 1 is actuated. This unloads the pump and stops the cylinder.

This circuit will handle only small flows, according to the flow rating of the cam valves. To handle higher flows, the basic idea in the preceding circuit can be used — the cam valves can control a pilot-operated relief valve to unload a much higher flow than could be handled directly by the cam valves.

9

SPEED CONTROL
...in Hydraulic Circuits

METHODS OF SPEED CONTROL

The travel speed of a hydraulic cylinder can be regulated in just one way — by regulating the volume of fluid metered into or out of it. There are several ways to do this.

(1). By metering the rate of flow to or from the cylinder with throttle-type valves including needle valves, pressure compensated or non-compensated flow control valves, or modulating the flow with a manually operated directional control valve.

(2). By metering off to tank a part of the flow so the full pump volume does not reach the cylinder. The same valving can be used as for paragraph (1).

(3). Regulating the shaft speed of the pump with a variable speed drive. There are many types of drives including D-C motors, variable frequency A-C motors, engines, two-speed A-C motors, variable pulleys, step pulleys, mechanical variable speed drives, and others. The flow output of a pump is directly proportional to its shaft RPM.

(4). By using a variable displacement pump. Some piston pumps are built with an external control by which an operator can change pump displacement to control hydraulic motor or cylinder speed. The flow output of such a pump is directly proportional to its cubic inch per revolution displacement.

The most commonly used, and therefore the most important, speed control methods are those in Paragraphs (1) and (2), and are those which will primarily be covered in this chapter.

FLOW CONTROL VALVES

Needle Valve. Figure 9-1. Although any type of 2-way valve can be used for throttling flow in fluids, including ball, gate, globe, and plug valve types, a needle valve is built especially for this purpose with a tapered needle for fine adjustment of flow rate. It can be used on air, water, hydraulic oil, and

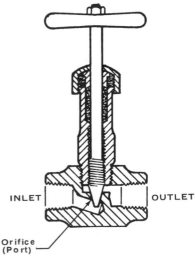

INLET   OUTLET

Orifice
(Port)

FIGURE 9-1. Simple Needle Valve.

other fluids. Needle valves are relatively inexpensive and will operate at either low or high pressure. A summary of the more important characteristics of 2-way valves, including needle valves, will be found in Chapter 3 of Volume 1.

Since a needle valve is designed for restricting flow, its internal orifice is usually smaller than its rated connection size. This gives a finer degree of control than if it had a large orifice. It can control flow equally well in either direction, but is usually marked for the preferred direction of flow to keep inlet pressure off of the stem seal when the valve is closed.

The main objection to a needle valve for controlling speed of a cylinder is that, like any other orifice, if any circuit condition should cause a change in the pressure drop across it from inlet to outlet, the flow through it would also change. If the pressure difference should increase, the flow would increase and vice versa. These two illustrations show the effect on speed as the load against the cylinder changes:

Figure 9-2. When a needle valve is used as a speed control in series with the oil flow to a cylinder, whatever pressure is not needed to overcome the load will appear as pressure drop across the needle valve. In this case, with a light load, only 500 PSI is enough to move the load. Therefore, with the relief valve set at 2300 PSI, 1800 PSI will appear across the needle valve and will force a certain flow through the valve. The operator controls cylinder speed by adjusting the orifice on the needle valve. As long as the load on the cylinder remains constant, cylinder speed will also remain constant.

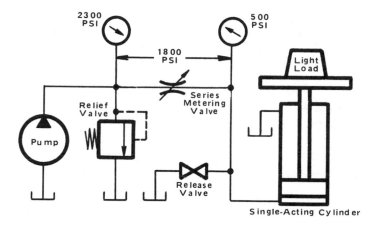

FIGURE 9-2. *Pressure Distribution in a Series Metering Circuit When Circuit is Lightly Loaded.*

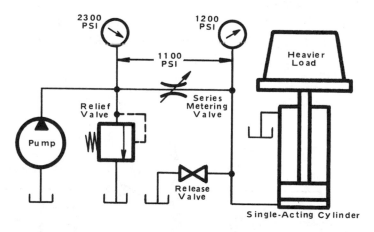

FIGURE 9-3. *Pressure Distribution in a Series Metering Circuit Changes if the Load is increased.*

Figure 9-3. If the load on the cylinder is increased so it requires 1200 PSI to move it, the pressure drop across the needle valve drops from 1800 to 1100 PSI. At this reduced differential pressure, less oil is able to pass through the needle valve at the same orifice setting as before and the cylinder slows down. To restore the original cylinder speed, the needle valve must be opened to a larger orifice.

A viscosity change in the oil will also affect the flow rate through a needle valve. As oil temperature increases, the oil becomes "thinner" and will flow more readily through the same size orifice. At a given orifice size, cylinder speed will increase as the oil temperature rises.

In view of the change in speed with change in load, a needle valve may or may not give satisfactory control of cylinder speed. Later in this chapter the advantages and disadvantages of a needle valve will be compared with those of a pressure compensated flow control valve which will maintain a constant flow regardless of variations in load against a cylinder.

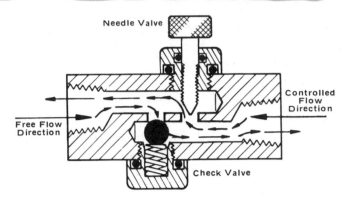

FIGURE 9-4. A flow control valve includes both a metering needle plus a free-return check valve in one housing.

Non-Compensated Flow Control Valve. Figure 9-4.

This is basically a needle valve with a built-in check valve to pass fluid freely in the reverse direction, and is used at points in the circuit where flow must be metered in one direction and must return later in the cycle without being restricted. The use of flow control valves for speed control of single-acting and double-acting cylinders is covered in Chapter 3 of Volume 1.

Simple flow control valves of this type can be used for any fluid compatible with the materials of their construction; brass construction is preferred for air or water, steel for hydraulic oil, and stainless steel for certain corrosive fluids. Some manufacturers offer models with micrometer read-out allowing the needle to be returned to a previously recorded setting.

Pressure Compensated Flow Control Valve.

A pressure compensated flow control valve is designed to accurately control the speed of a hydraulic cylinder (or motor), eliminating the variations in speed caused by changes in load. It has a needle-type metering orifice plus a compensating spool, both usually housed in a common body. An external drain may or may not be required depending on the construction method employed. The compensating spool measures the rate of flow through the metering orifice and automatically moves to one side or the other, as needed, to add to or remove restriction to maintain a constant flow through the valve according to the orifice diameter adjustment. In Figures 9-2 and 9-3 on the preceding page, if a pressure compensated flow control valve instead of a needle valve had been used, the valve compensator would have automatically enlarged the orifice as the load increased to avoid a loss of speed.

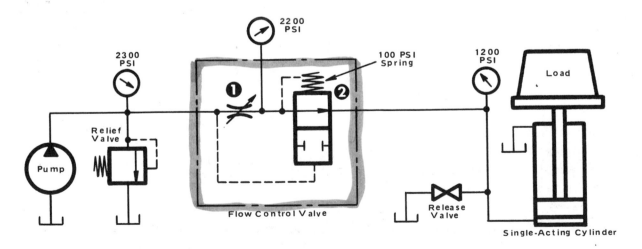

FIGURE 9-5. A pressure compensated flow control valve includes a metering orifice plus a pressure reducing type of compensating spool.

In the following description, the strength of the non-adjustable internal spring which keeps the compensator spool open, is designed so that with the area on the end of the compensator spool, about 100 PSI on the opposite end of the spool will move the spool into a throttling position. This design figure may vary slightly between valve manufacturers.

Pressure Compensating Principle. Figure 9-5.

This is the same circuit as Figure 9-3 but with a pressure compensated flow control valve to replace the needle valve for speed control. This diagram will serve to show how the compensating spool in the flow control valve acts automatically to maintain a constant flow to the cylinder. The valve works on the principle that if the pressure drop across an orifice can be maintained at a constant level, the flow through the valve will also remain constant.

The compensating spool, as shown later in Figure 9-7, is a 2-way valve, normally open, with spring bias equivalent to 100 PSI to keep it normally open. Opposite ends of the spool are equal in area and are connected to the upstream and downstream sides of the metering orifice. Any difference in pressure across the orifice causes a force unbalance on the compensating spool.

On starting from zero flow, as the flow increases through the metering orifice, the compensator spool will remain wide open until the pressure drop across the metering orifice (due to flow) reaches 100 PSI. At this point, opposite ends of the spool are in a force balance condition. Any further increase in flow will create a higher pressure drop across the orifice and will increase the force on the bottom end of the spool relative to the force on top. The spool will start closing to limit any further increase in flow. On an application where the cylinder continues to move, the pressure from the pump will divide between the load, the metering orifice and the compensating spool as shown by the gauge readings in this diagram. Any pump relief valve pressure which is not used by the load will appear as pressure drop across the flow control valve, with 100 PSI across the orifice and the remainder across the compensating spool. In the diagram, 1200 PSI is needed for the load. The remaining 1100 PSI appears across the flow control valve, 100 PSI across the orifice and 1000 PSI across the spool.

On a flow control valve with adjustable orifice, as the orifice size is changed, the maximum flow which the valve will pass is reached when the pressure drop across the metering orifice reaches 100 PSI.

In effect, a pressure compensated flow control valve adds an artificial load to the system to absorb the extra horsepower output of the pump which the cylinder does not use. If load on the cylinder should increase, more of the pump fluid horsepower would be transferred to the cylinder and less of it would be dissipated on the artificial load created by the compensator spool. The artificial load produces heat in the oil which is undesirable. A system using this type of speed control operates at highest efficiency and minimum heating when the flow control valve is wide open and the cylinder traveling at maximum speed. It operates at less efficiency and with greater heating at reduced cylinder speeds.

Fixed Flow Control Valve. Figure 9-6.

This is a non-adjustable version of a pressure compensated flow control valve. Often used in sets of 2, 3, 4, or more matched valves to keep several cylinders moving in synchronization at the same speed. Figure 10-1 shows their application in such a circuit. Also used in branch circuits to split off a small constant flow from the pump for a special purpose. Some models may also include an internal check valve for free reverse flow.

These valves have the same basic elements — metering orifice and compensator spool — as described in connection with Figure 9-5, and they operate in the same way. The metering orifice primarily determines the flow rating and is accurately sized at the factory for the specified flow. The spring strength has a small effect on flow rating, and in some models is adjustable to obtain a very limited range of flow adjustment.

The control orifice diameter is sized to obtain a 100 PSI pressure drop at the rated flow. It is made with a knife edge so it will be relatively insensitive to oil viscosity. The compensating spool is exposed

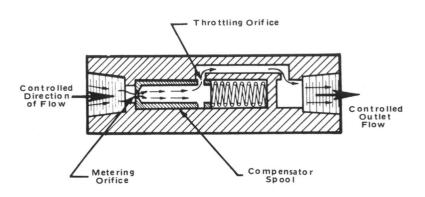

FIGURE 9-6. *A fixed rate flow control of the pressure compensated type includes a non-adjustable metering orifice plus a spring loaded compensating spool of the pressure reducing type.*

on its left end to pressure upstream of the orifice, and on its right end to pressure downstream of the orifice. If pressure drop across the orifice (due to excessive oil flow) should exceed 100 PSI, the spool will move toward the right, compressing the spring. Oil which flows through the orifice then flows through a series of radial metering holes in the compensating spool into a collecting groove in the valve body. Metering starts when flow causes a pressure drop in excess of 100 PSI across the orifice. Metering rate is proportional to the distance moved by the compensator spool.

Oil can flow through the valve in the return direction but is uncontrolled. Return flow at the rated GPM will suffer about a 100 PSI pressure loss through the orifice. If the return rate is higher than the valve rating, pressure loss will be approximately proportional to the square of the flow. To reduce high loss in the return flow, a check valve should be connected in parallel with the valve, or a model with built-in free return check valve should be used.

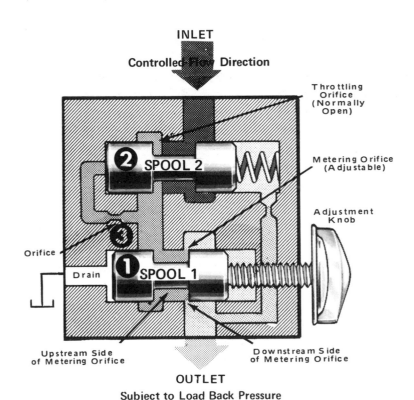

Pressure Compensated Flow Control Valve, 2-Port Type. Figure 9-7. This illustration depicts an adjustable flow control valve of the 2-port type. It has an inlet port and an outlet port. Some brands may also have a small external drain port. The metering orifice is in series with flow between the two main ports. Oil which cannot pass through the metering orifice backs up and usually is discharged across the pump relief valve. Another type, the 3-port model, will be described later.

In this particular model, oil entering the valve inlet flows first through the compensating spool, 2, then through the metering orifice, 1, which in this case is shown as a spool in order to obtain the full adjustable flow range in one turn of the adjustment knob. The use of a spool for metering requires an external drain to carry away spool

FIGURE 9-7. *A pressure compensated flow control valve has a reducing-type compensating spool preceeding the metering orifice to keep a constant pressure difference across the orifice.*

leakage. Other brands which use a needle valve as a metering orifice do not require the external drain.

As in the valves previously described, the compensator spool is biased to its normally open position with a 100 PSI spring. The left end of this spool is piloted from pressure taken upstream of the metering orifice, Spool 1, and the right end is piloted from pressure on the downstream side of Spool 1.

When the adjustment knob has been set for a certain GPM, flow through the valve can increase until the pressure difference across the metering orifice, Spool 1, reaches 100 PSI. Spool 2 is now in exact force balance. Any additional flow will create a pressure difference of more than 100 PSI. The force on the left end of the spool, now greater than the opposing force on the right end, would cause the compensator spool, 2, to move toward the right to prevent a further increase in flow.

Internal orificing, as at 3, may be employed to stabilize the valve action to prevent hunting or erratic operation of the compensator spool. The drain port must be piped directly to tank and not combined with the outlet port.

Characteristics of Needle-Type Flow Control Valves. (1). Compared to pressure compensated flow controls they can meter the flow more accurately at very low feed rates. They have no internal slippage to cause inaccuracy at low flows. At low feed rates, if the cylinder load is fairly constant throughout the stroke, a needle valve may well outperform the more expensive type. However, at medium to fast cylinder speeds, a change in cylinder load may cause an unacceptable change in cylinder speed.

(2). They work best on low pressure hydraulics and compressed air. At high pressure the change in cylinder speed on a changing load may be so great as to be unacceptable.

(3). If the cylinder load is constant throughout the stroke, a needle valve will work just as well as a pressure compensated flow control.

Characteristics of 2-Port Pressure Compensated Flow Control Valves. (1). Compared to needle valve speed controls, pressure compensated flow control valves are designed to meter a constant flow of oil to a cylinder (or hydraulic motor) during changes in the load. If the load remains constant throughout the stroke, these valves have no important advantage over needle valves.

(2). On a system pressure of 500 PSI or less, the pressure compensating principle does not work well. Too much (100 PSI) of the pump pressure is consumed in operating the compensator. They are inefficient and may become erratic. They are completely impractical on pressures of 150 PSI and less; only a needle valve flow control should be used at these low pressures.

(3). They become highly inaccurate when metering very slow cylinder feeds. Spool slippage in the compensator becomes too high a percentage of metered flow.

(4). Their best accuracy is at the high end of their flow range, and in the mid range of pressures. Increased slippage at high pressure decreases their accuracy. Their best overall performance is in the pressure range of 1000 to 3000 PSI.

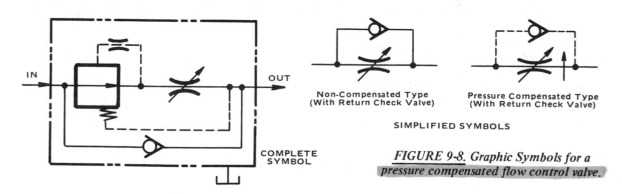

SIMPLIFIED SYMBOLS

FIGURE 9-8. Graphic Symbols for a pressure compensated flow control valve.

Graphic Symbols for 2-Port Flow Control Valves. Figure 9-8. The complete symbol on the left is for a 2-port pressure compensated flow control. The drain symbol is optional depending on whether the model used has a leakage drain. A check valve could be added to the symbol, inside the dash enclosure, for models designed for reverse flow. The complete symbol, however, is too cumbersome for general use on circuit drawings, and one of the simplified symbols is usually preferred.

The non-compensated symbol is typical of a plain needle valve with reverse flow check valve. The same simplified symbol is used for a pressure compensated valve with a small vertical arrow added to indicate pressure compensation.

Three-Port Pressure Compensated Flow Control. This is a less commonly used type of flow control valve, and may offer advantages on certain applications. On single-branch circuits — only one cylinder operating from the pump — it may improve efficiency or may reduce oil heating when the cylinder is being operated at reduced speed.

It has three full size ports: inlet, outlet, and tank return. It has an adjustable metering orifice and a compensating spool. It, too, is designed to maintain a constant flow to the cylinder under changes in load against the cylinder. To understand how it is different from the 2-port control, please review:

All pressure compensated flow control valves operate on the principle that if the pressure difference across an orifice can be kept constant, the flow through the orifice will remain constant. As oil flows through a metering orifice, the pressure drop across the orifice is measured and used as motive force to shift a compensating spool to limit the maximum flow. On the 2-port flow control which has previously been described, a pressure reducing type spool (normally closed) is used as the compensator. The pressure ahead of the metering orifice is reduced to a level just 100 PSI above the outlet pressure as reflected back to the outlet by the load resistance. As the load changes, the reducing valve spool automatically changes, as necessary, to always keep the metering orifice inlet 100 PSI above whatever pressure is reflected by the load. The complete graphic symbol, Figure 9-8, shows the compensating spool as a pressure reducing valve with its spring connected not to tank but to the load pressure which appears on the valve outlet.

A 3-port flow control valve uses a relief valve (normally closed) compensating spool, with the relief setting able to "float" up or down but always to be 100 PSI higher than the outlet port. This valve, instead of blocking excess flow which cannot pass through the metering orifice, discharges the excess flow to tank through the tank return port, at a pressure just 100 PSI higher than load pressure instead of at full relief valve pressure. This can sometimes save a considerable amount of wasted power and reduce the oil heating.

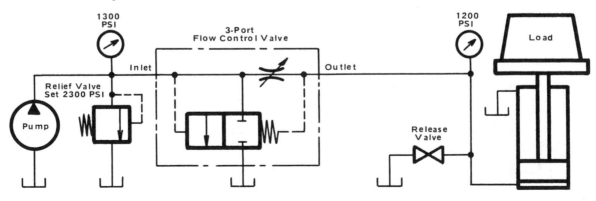

FIGURE 9-9. Speed control of a single-acting cylinder with a 3-port flow control valve.

Three-Port Valve Speed Control Circuit. Figure 9-9. A 3-port flow control valve is used for speed control. Circuit conditions are the same as for Figure 9-5, with a load requiring 1200 PSI and with the pump relief valve adjusted for 2300 PSI. Compare with Figure 9-3 needle valve speed control and with Figure 9-5 in which a 2-port pressure compensated flow control valve is used.

The metering orifice is set to the desired maximum GPM. As flow increases through the orifice, the pressure drop across the orifice also increases. When the flow has increased to a level where the pressure drop is 100 PSI, the two ends of the compensating spool are in a force balanced condition. Any further increase in flow will cause a relatively higher force to appear on the left end of the spool, and the spool will move far enough to the right to prevent any further increase in the flow. When this condition has been reached, the surplus pump oil which cannot pass through the orifice can discharge to tank across the compensating spool at a pressure lower than the pump relief valve adjustment. Notice the gauge readings in this diagram compared to those in Figure 9-5. A unique feature of the 3-port control is that pump oil which cannot pass through the metering orifice is allowed to discharge to tank at a pressure of 1300 PSI rather than full relief valve pressure of 2300 PSI as in Figure 9-5.

Characteristics of 3-Port Pressure Compensated Flow Control Valves. (1). Compared to 2-port flow control valves, since the metering orifice is in series with the flow from pump to cylinder, this valve offers the same good speed regulation as the 2-port model. This is especially important when operating a hydraulic motor instead of a cylinder. (See Volume 3 for information on speed regulation).

(2). When the cylinder is operating at reduced speed, the 3-port control is more efficient, wasting less power and creating less heat in the oil. But at full speed it offers none of these advantages.

(3). Only one branch circuit can be operated from the pump with 3-port flow control valves. If the system has more than one branch circuit, the by-pass path in one flow control valve would prevent full pressure from building up in any other branch. For multiple branch applications, the 2-port model must be used.

(4). Compared to needle valve speed control, the information on 2-port valves on Page 181 will generally apply to 3-port models.

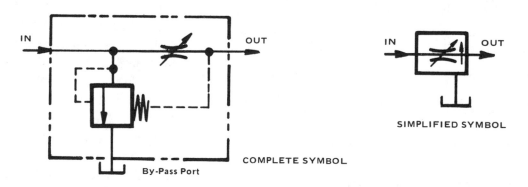

FIGURE 9-10. Graphic symbols for 3-Port Pressure Compensated Flow Control Valves.

Graphic Symbols for a 3-Port Flow Control Valve. Figure 9-10. The complete symbol shows that this is basically a self-adjusting pressure relief valve placed in front of a metering orifice. The relief valve is the compensator for the orifice and is different from an ordinary relief valve because it is referenced to the downstream side of the orifice instead of to atmosphere. It sets the relieving pressure in front of the orifice to always be just 100 PSI higher than the outlet pressure to the load. For most diagrams, however, the simplified symbol is more convenient to use.

HYDRAULIC SPEED CONTROL METHODS

Information in this section is limited to double-acting hydraulic cylinder operation. Speed control of hydraulic motors is a more involved matter and is covered in detail in Volume 3. Speed control of air cylinders and single-acting hydraulic cylinders has been well covered in Volume 1. Speed control of combination air/oil systems is covered in Chapter 12 of this volume.

Pressure compensated flow control valves, either 2-port or 3-port type, are shown in the diagrams. Needle-type flow control valves could also be used in place of 2-port controls if their performance is acceptable on the particular application.

On compressed air cylinders, speed must usually be controlled in both directions of movement because the cylinder is working from an unlimited supply of fluid. But a hydraulic cylinder, working from a limited supply of fluid, usually requires speed control only in the working direction, with return speed at the full volume of the pump. On most applications the speed control valve is located in one or both lines which connect the cylinder to its directional valve.

As will be shown in the following diagrams, there are several ways to install a speed control valve and there are several valve types from which to choose. In these diagrams, the working direction is assumed to be the extension stroke, with uncontrolled return speed. These circuits can be modified for speed control in both directions or only on the return stroke. Some of them are arranged with the speed control in "series" with the oil flow to the cylinder. Others have the control valve in "parallel" or "by-pass" with the cylinder. Some are arranged for "meter-out" control and others for "meter-in" control. The relative merits or disadvantages of each of these arrangements will be discussed following presentation of the circuits.

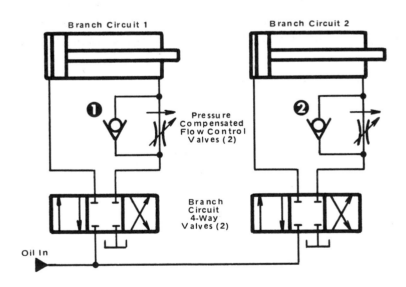

FIGURE 9-11. Meter-Out Speed Control for Cylinder Advance.

Meter-Out Speed Control. Figure 9-11. This is shown as a 2-branch system but the speed control method can be extended to any number of branches operating from the same pump. The flow control valves, 1 and 2, are in series with the oil supply to their cylinders. Cylinder speed is controlled by metering fluid out of the rod end as the cylinder extends. Any surplus of pump flow which cannot pass through the flow control valves backs up against the pump relief valve (not shown) and discharges to tank at full relief valve setting even though the cylinders may be operating with light loads. The flow control valves should have a built-in return check valve or one should be added externally. The amount of heating in the oil depends on cylinder speed as noted later in the comparisons.

Meter-In Speed Control. Figure 9-12. The same flow control valves have been re-positioned in the circuit to meter the oil rate into the blind side of the cylinder. Surplus oil which is not used to move the cylinders is discharged to tank at full relief valve setting. Performance with this method will be compared later to the meter-out arrangement of the preceding figure.

By-Pass Speed Control. Figure 9-13. The speed control valves used here (without return check valves) are connected in parallel with the cylinder. Oil from the pump has two alternate flow paths. A part of it will be discharged directly to tank by the flow control valve. The remainder will be forced into the cylinder and will determine the cylinder travel speed.

One very important limitation of by-pass speed control is that it is usually applicable to systems having only one branch operating from the pump. If used in multiple branch systems the flow control valves must be kept isolated from one another. They must be used with 4-way valves having closed cylinder ports and they must be installed in connecting lines between 4-way valve and cylinder. Only one branch may be operated at a time, while the 4-way valves in other branches must remain closed. In this figure, tandem center valves are shown since only one branch may ever be operated at a time. If two 4-way valves were to be shifted at the same time, the flow control valves would interact. Any change of speed or load in one branch would affect the speed of the other branches. The branch operating at

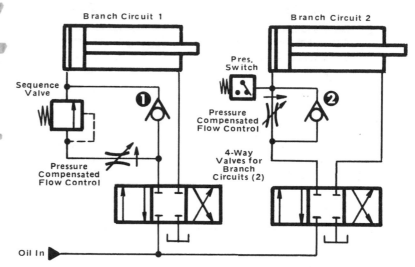

FIGURE 9-12. *Meter-In Speed Control for Cylinder Advance.*

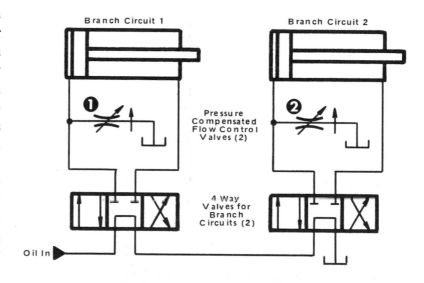

FIGURE 9-13. *By-Pass Meter-In Speed Control for Cylinder Advance.*

the lowest pressure would prevent other branches from receiving the higher pressure they need.

Another possible limitation to consider before selecting this type of control is that it can never be used to meter oil out of a cylinder. It is always a meter-in form of control. Therefore, it does not have the potential "hold-back" against an overrunning load which is needed in some applications.

Still another limitation, if a pressure compensated type of flow control valve is used, is that a cylinder cannot be completely stopped with the flow control valve even at wide open position, unless the mechanical load against its piston takes more pressure to move it than the pressure drop of the full volume of pump oil flowing across the metering orifice to tank. Otherwise the cylinder may not stop or may drift out of position. Nor can its speed be accurately controlled anywhere in the speed range at load pressures less than 100 PSI. The compensator in the flow control valve would not work. Speed control would be entirely with the metering orifice and would be uncompensated. At these low pressures a non-compensated, needle valve flow control valve should be used.

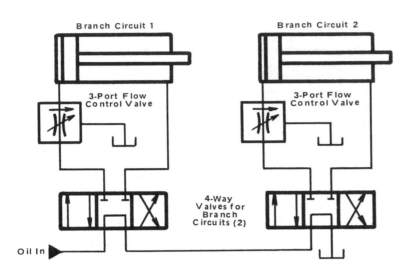

FIGURE 9-14. 3-Port Meter-In Speed Control for Cylinder Advance

Three-Port Flow Control. Figure 9-14.

The 3-port flow control valve, in some circuits, combines the accuracy of 2-port series control with the lower power waste of the 2-port by-pass control. Since the metering orifice is in series with oil flow from pump to cylinder, there is excellent speed regulation compared to parallel metering. And the surplus oil not metered to the cylinder can discharge to tank at a pressure less than the pump relief valve setting when the cylinder is operating at less than maximum pressure, thus minimizing power waste and oil heating.

While the 3-port method of speed control offers advantages which are useful on some applications, it has some of the limitations previously mentioned for by-pass speed control: it is usually applicable only in single-branch circuits. If used in multiple-branch applications, the flow control valves must be kept isolated from one another. In Figure 9-14 this is done by installing them in connecting lines between 4-way valve and cylinder and by using 4-way valves with closed cylinder ports in neutral. Only one branch at a time can be operated. If both 4-way valves were to be shifted at the same time, the two flow control valves would interact. The branch circuit operating at the lower pressure would prevent a cylinder in the other branch from receiving a higher pressure. The second cylinder would stall and all of the oil which should go to it would discharge to tank through the tank port of the first flow control valve.

Like the 2-port by-pass speed control, this method can never be used for meter-out control, and this might make it unsuitable for certain applications.

COMPARISON OF SPEED CONTROL METHODS

Series Compared to Parallel or By-Pass Methods. Series methods are shown in Figures 9-11 and 9-12 for meter-out and meter-in control respectively. By metering the oil in series through a metering orifice, speed control is more accurate because internal slippage in the pump is isolated from the cylinder and cannot affect its speed. With by-pass methods, pump slippage is not isolated. When load resistance increases, there is a drop in the pump flow output due to increased slippage. This causes a slow-down in cylinder speed no matter how accurate the flow control valve may be. The amount of slow-down with respect to cylinder no-load speed is called "speed regulation". Because it is a much more important factor on speed control of hydraulic motors, it has been placed in Volume 3. For more detailed information on speed regulation, the reader is referred to Chapter 7 of Volume 3.

One of the series speed control methods, either meter-out or meter-in must be used in systems where there is more than one branch circuit operating from one pump, and where two or more branch circuits must be operated at the same time. The reasons for this requirement are given in connection with Figures 9-13 and 9-14.

Disadvantage of a series speed control circuit is that all oil which cannot pass through the flow control valve (or valves) must discharge across the pump relief valve. This, of course, wastes input power which turns into heat in the oil. One of the by-pass methods in Figures 9-13 and 9-14 will conserve more of the input power and cause less heating because surplus oil can discharge to tank at a lower pressure. However, this is true only when the cylinder is working against a load which requires less than relief valve pressure.

Three-Port Compared to Two-Port Flow Control Valve. The 3-port valve gives good speed regulation because the metering orifice is in series with oil flow to the cylinder. This isolates the cylinder from pump slippage. On loads which do not require full relief valve pressure, it will conserve input power and generate less heat than a series control 2-port valve.

Meter-In Compared to Meter-Out Speed Control. As a rule, meter-in control as shown in Figure 9-12 is preferred over meter-out control or either of the by-pass control methods shown in Figures 9-13 and 9-14. It avoids pressure intensification in the rod end of the cylinder which could damage the cylinder if operating, for example, with rod end down on a hoisting application. Meter-in should always be used in circuits where a pressure switch or sequence valve is teed into the blind port of a cylinder to detect a sharp pressure rise to trip the switch or open the sequence valve when the piston "bottoms out".

Meter-out control may work slightly better on those applications where the speed of the cylinder is to be reduced to a very low feed rate. The cylinder piston moves while being held between two relatively high pressure columns of oil, so it may have less tendency to move erratically. If load is suddenly removed from the cylinder, the meter-out speed method will not entirely eliminate "lunge" but in some cases it may reduce it.

CHOOSING A SPEED CONTROL CIRCUIT

The best choice for type of speed control valve and circuit for a given application depends on which circuit characteristics are more important. If there is more than one important requirement and they conflict with one another, it may be necessary to carefully consider the circuits presented in preceding pages and to pick a valve and system which seems to offer the best compromise. Some requirements are listed here, with recommendations for the most suitable speed control system:

(1). **Accuracy.** If accuracy is more important than any other requirement, use one of the series metering systems of Figures 9-11 or 9-12. Avoid the use of by-pass control as in Figure 9-13. However, if it should be necessary to use by-pass speed control for some other important reason, select a pump, like a piston pump, which has low internal slippage from no load to full load pressure.

If possible, design the system to operate at a moderate pressure and at the highest flow consistent with the moderate pressure. Accuracy will vary over the range of the flow control valve, being best at the high flow end of the range. And, of course, use a high quality or precision type flow control valve.

(2). **Minimum Oil Heating.** Avoid series type speed circuits. Use the by-pass circuit of Figure 9-13, or better yet the 3-port flow control of Figure 9-14. Remember, however, that only one branch circuit can be operated at any one time from a pump with these circuits. If necessary to use a series speed circuit, be sure the pump relief valve is set as low as possible.

(3). **Holdback.** If important to restrain the cylinder from lunge when a cutting tool is breaking through the work, use only the series meter-out circuit of Figure 9-11. If this circuit does not give enough restraint, consider the addition of a hydraulic checking cylinder (shock absorber) to catch the cylinder as the tool breaks through.

(4). **Several Branch Circuits.** If more than one branch at a time will be operating, use meter-out or meter-in control, Figures 9-11 or 9-12. If no more than one at a time will be operated, use any of the circuits shown.

(5). Low Pressure System. At pressures less than 500 to 1000 PSI, pressure compensated flow control valves do not work well. A needle valve type control will probably give better results.

(6). Constant Load System. If the load against the cylinder will be reasonably constant throughout the working stroke, any of the speed control circuits may be used. Actually, a needle valve type control will work just as well, perhaps better, than a pressure compensated control.

(7). Pressure Switch or Sequence Valve. If one of these pressure sensitive devices is used, avoid the use of the series meter-out circuit of Figure 9-11. Any of the other circuits are suitable.

(8). Wide Range of Speed Adjustment. Widest range is obtained with a needle valve type flow control valve. Of the pressure compensated valve circuits, the series metering of Figures 9-11 and 9-12 will give a wider adjustable range with acceptable cylinder performance than will the other circuits.

OPERATIONAL NOTES ON SPEED CONTROL SYSTEMS

Efficiency of Pressure Compensated Flow Control Valves. A certain amount of pressure, about 100 PSI, is consumed in the operation of the compensator spool, and is unavailable in the system for producing output from a cylinder. This figure is average and may vary a little one way or the other according to the manufacturer.

These flow control valves are more efficient when a certain horsepower level in the fluid system can be obtained by using a high pressure to low flow ratio rather than the other way around. They consume pressure, and the higher the system pressure, the smaller the percentage of pressure which has to be tapped off for control operation. For example, on a 3000 PSI system, 100 PSI to operate the compensator is about a 3% pressure loss. But on a 500 PSI system it would be a 20% loss. Horsepower loss, through the flow control is proportional, not to actual PSI loss but to its percentage of the total pressure of the system. If the pressure loss is 20%, the power waste is also 20%.

Accuracy of Pressure Compensated Flow Control Valves. Definite figures on accuracy or repeatibility are difficult to quote. This varies with accuracy of manufacture, the flow rating of the control, and the pressure drop across it. It also depends on the purpose for which the control was primarily intended — for mobile use where cylinder speed can be visually monitored, or for industrial use where accuracy may be much more important. In general, accuracy of any flow control valve is best when the valve is operating near its maximum rated flow and when operating at medium pressures. Accuracy is poorer on flows near its minimum rating because the internal slip across the compensator spool is a higher percentage of total flow through the main ports. Accuracy is poorer at very high pressure because of increased slippage across the compensator spool at high pressure. An accuracy of ±15% can usually be obtained, and an accuracy of better than ±5% is difficult to obtain even under the best of conditions.

SPEED CONTROL OF VERY LOW FEED RATES

When attempting to control the movement of a hydraulic cylinder at a very low feed rate, several factors, including choice of components and type of feed circuit, should be carefully considered.

(1). Select a cylinder with very tight piston seals. Multiple-V or chevron seals are probably the best, if available. O-rings usually make very tight seals if their grooves are designed and machined properly. Other types of soft seals, while usually not as tight, may be acceptable. Leakage across the piston seals will cause inaccuracy, the cylinder slowing down due to increased seal leakage, if the intensity of the load should increase.

(2). Consider the choice of an oversize cylinder — one with a bore larger than necessary to produce the feed force. An oversize cylinder can be fed with a higher oil flow for the same speed. A higher flow can be metered more accurately than a smaller flow. Also, the lower pressure required for a larger bore cylinder will reduce the amount of piston seal leakage and may allow a pressure compensated flow control valve to operate in a more accurate part of its range. Incidentally, increasing the bore diameter of a cylinder does not increase the "springiness" of the cylinder. Only the length of oil column under compression affects the rigidity in the system.

(3). Order the cylinder with the shortest possible stroke. Lay out the plumbing to have the shortest possible oil column under compression. The shorter the oil column the greater the rigidity in the fluid part of the system. On most systems the oil column length would be measured from the pump, through the 4-way valve, to the piston of the cylinder. On servo systems the oil column is measured from the 4-way (servo) valve to the cylinder piston. The effect of having a long oil column is to reduce the rigidity of a cutting tool driven by the cylinder, causing possible chatter or tool breakage.

(4). Always use hydraulics; compressed air and air-over-oil are not usually successful on slow feed applications. Compressibility of the air will nearly always cause erratic operation, tool chatter, and inconsistent feeding.

(5). The best speed control method for slow feed rates may be the series meter-out method shown in Figure 9-11 except on heavy overhung loads (where the load is pulling the cylinder during parts of the cycle). The cylinder may have less tendency to "chatter" because its piston is held between two columns of pressurized oil. If the cylinder is pushing a cutting tool there may be less "lunge" when the load is suddenly removed, as by a drill breaking through the work.

(6). Use care in selecting the flow control valve. A non-compensated, needle-type, flow control valve is always preferred over a pressure compensated control, if the application is such that one can be used. Refer to Page 177 for limitations of this type valve. However, if the load resistance will vary widely during the stroke it may be necessary to use a pressure compensated flow control. It has one other advantage over a needle valve: most of them have calibrated dials and can very easily be re-set to a desired position.

Some designers make the mistake of automatically specifying pressure compensated control valves without considering their limitations.

OTHER METHODS OF SPEED CONTROL

To vary the travel speed of a hydraulic cylinder it is necessary to be able to regulate the volume of oil flow into or out of it. In most cases this is done with a flow control valve in one of the circuits previously described.

On industrial fluid power machines (those operating inside a plant, usually with solenoid directional control valves), the flow control valve is usually adjusted to a selected speed on the working part of the cylinder stroke and remains in this position cycle after cycle. To reduce power waste, piston pumps are sometimes used which have a "load sensing" control. This type of control is briefly described in Chapter 6 of Volume 3. By means of a pressure feedback line from the flow control valve to the pump, the pump displacement is automatically changed, as the flow control valve setting is changed, to permit the pump to produce only the volume which can pass through the flow control valve. This eliminates the waste of discharging surplus oil across a relief valve.

But on mobile equipment, controlling cylinder speed is a normal part of the operator's function, and cylinder speed may be varied several times during its stroke. Directional control is by means of manual valves, and the function of speed control is combined with directional control by modulating handle position to restrict the flow of oil into and out of the cylinder which is controlled by that handle. Sectional valves for directional control have been covered in Chapter 7 of this book

Since flow control valves waste input power and create heat in the oil, variable displacement, pressure compensated pumps are often used to save some of this wasted power. The pump will automatically reduce its displacement and its output flow to avoid discharging surplus oil across a relief valve. Piston pumps are available from some sources with electric control of pump displacement. The operator uses a rheostat to control a servo-type torque motor on the pump to position the pump swash plate to give exactly the pump flow needed without waste.

SUMMARY OF SPEED CONTROL

For satisfactory operation of a speed control system on a hydraulic cylinder, and to minimize power waste and heat generation, the speed control method should be chosen with care. The important points covered in preceding pages of this chapter may be summarized as follows:

(1). Series meter-in control, Figure 9-12, is more often used for hydraulic cylinders to avoid the danger of damage from pressure intensification in the cylinder rod end if meter-out control is used. This is a particular danger on overhung loads. Also, it must be used in circuits where a pressure switch or sequence valve is teed into the cylinder blind end to sense a pressure rise behind the cylinder piston.

(2). Series meter-out control, Figure 9-11, so popular with air cylinders, is not often used on hydraulic cylinders. On some applications it may tend to stabilize the cylinder against chatter or to minimize "lunge" as a cutting tool breaks through the work.

(3). By-pass control, Figure 9-13, is more efficient than series control and generates less heat when the cylinder is operating at reduced speed, but its use should be limited to systems having only one branch operating from the pump. It gives no protection against cylinder "lunge".

(4). Three-port control, Figure 9-14, is an improved variation of by-pass control. It has greater accuracy than straight by-pass control and generates less heat than the series control circuits. But it, too, is limited to systems having only one branch operating from the pump.

(5). On series meter-out or meter-in circuits, use a pilot-operated type of relief valve on the pump, and carefully adjust its pressure setting to the lowest pressure which will give full cylinder speed. Setting it higher than necessary will add to the power losses during periods when the cylinder is operating at reduced speed. If possible, lock or seal the adjustment to prevent unauthorized personnel from tampering with the adjustment and setting it so high that the system may overheat.

(6). When specifying a design, and assuming maximum cylinder speed will be with the flow control valve wide open (or completely closed for by-pass control), do not extend the lower end of the speed range lower than necessary. When using a fixed displacement pump, heat is generated and power is wasted in direct proportion to the amount the speed is reduced from maximum. Performance (speed regulation) also becomes increasingly poor as the speed is reduced.

(7). Select the hydraulic pump with no more flow than needed at maximum speed with the flow control valve wide open. Any surplus flow which must be discharged to tank produces heat. On belt-driven pumps choose the sheave ratio to produce the pump speed which will deliver no more than the GPM needed for maximum cylinder speed with the flow control valve wide open (or completely closed if by-pass speed control is used).

(8). Use a flow control valve in only one direction of cylinder motion. Using a control in both directions is poor design on most systems and generates unwanted heat.

10

HYDRAULIC CIRCUITS

...Miscellaneous

SYNCHRONIZING SEVERAL HYDRAULIC CYLINDERS

One of the problems which plagues hydraulic circuit designers is how to make several cylinders, operating in parallel from the same pump and through the same 4-way control valve, move at the same speed throughout their stroke. The natural tendency of oil is to flow toward least resistance. Therefore, the cylinder having the least load will tend to take all or most of the oil, leaving little or none for other cylinders. Of course this problem can be solved with complicated and expensive electrical servo systems with a device on each cylinder to feed back position information to a "black box" containing electronic circuits. But on ordinary hydraulic systems, the accuracy of this type of synchronization is not necessary and is far too costly. There are several simpler methods which synchronize the cylinders, not perfectly, but well enough for many applications.

Synchronization With Flow Control Valves. Figure 10-1. Simple needle-type flow control valves may sometimes be used on identical cylinders having the same stroke if the load on both cylinders is not necessarily equal, but remains constant throughout the stroke. Both cylinders must be allowed to "bottom out" at both ends of the stroke so on the next stroke they will start in synchronization.

Where more accurate synchronization is necessary, and especially where the load distribution between the cylinders may change during the stroke, pressure compensated type flow control valves should be used subject to the limitations stated in Chapter 9. If possible, they should be connected in a series meter-out arrangement as in Figure 9-11.

In Figure 10-1 flow control Valves 1 and 3 meter equal flows out of the cylinders while they are extending, and Valves 2 and 4 meter equal flows out of the cylinders while they are retracting.

Pressure loss through the metering orifice of a pressure compensated flow control valve is approximately 100 PSI per valve. Return oil flowing back through the orifice will be uncontrolled. The pressure drop of returning oil will be proportional to the square of the flow. For example, if return flow is 2 times the metered flow rating of the valve, pressure drop through the orifice will be about 400 PSI. The flow control valves should have built-in check valves, or external check valves should be used to by-pass return flow around the valve to avoid pressure loss through the metering orifice.

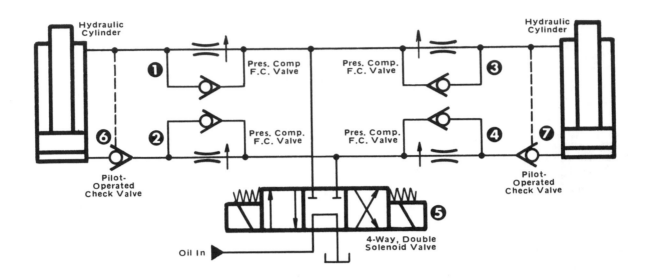

FIGURE 10-1. Cylinder Synchronization With Pressure Compensated Flow Control Valves.

Important! For a pair of flow control valves (such as Valves 1 and 3 or 2 and 4 in Figure 10-1) to divide the flow with maximum accuracy, the combined rating of the pair should be less than the total flow available to them, with a surplus being discharged to tank through the pump relief valve. If the oil flow to be split is less than their combined ratings they cannot divide it equally. For this reason, flow control valves are recommended only for electric motor driven systems where pump flow remains constant. They should not be used on systems driven by an engine because when the engine speed is reduced the pump will not produce sufficient flow to be divided equally.

The flow control valves may be purchased in matched pairs for greater accuracy. Notice in Figure 10-1 that the outlet flow from both cylinders to be metered through Valves 1 and 3 while the cylinder extends, is substantially less than pump flow because of the difference in area on opposite sides of the cylinder piston. Likewise, the flow from the cylinder blind ends to be metered as the cylinders retract is considerably greater than pump flow. This must be taken into account when ordering the flow control valves. The amount of undersizing or oversizing with respect to the pump flow will depend on the area relationships in the cylinder.

Example: Assume we are working with a pump flow of 20 GPM. Half of this, or 10 GPM must be split off for each cylinder. Suppose, with the cylinder used, the area ratio is 4:3 between full piston area and the net area around the piston rod. As the cylinders extend, oil flows freely through the check valves in Valves 2 and 4. A 10 GPM flow into each cylinder would push out 7½ GPM from the rod port. Flow control Valves 1 and 3 should be ordered with 7½ GPM flow rating. As the cylinders retract, a 10 GPM flow into each cylinder would push out 13.33 GPM from the blind end port. Flow control Valves 2 and 4 should be ordered with 13.33 GPM flow rating. Note in this example that 20 GPM is not the free flow pump rating; it is the minimum flow which the pump can deliver when under full pressure.

Pilot-operated check Valves 6 and 7 prevent transfer of oil from the more heavily loaded cylinder to the more lightly loaded one if the 4-way valve is centered with cylinders partly extended. Oil transfer would cause one cylinder to sag, causing a bind in the mechanism which might damage the machine. To restore the machine to normal operation might be difficult. Pilot-operated check valves are almost leaktight and would hold the cylinders for a long time without sag assuming the cylinders had leaktight piston seals. Counterbalance valves could be used either in place of, or in conjunction with Valves 6 and 7, although their leakage rate is usually higher than that of pilot-operated check valves.

Synchronization With Flow Divider Valves.
Figure 10-2. Flow divider valves will accept a flow on their inlet port and will split the flow into two equal parts. Circuits using spool-type dividers and combiners and rotary-type dividers are shown in Volume 3.

Flow dividers are preferred over flow control valves on systems powered by an engine. They will equally divide either high or low flow from the pump. Their accuracy is about the same as that of flow control valves although they may be more efficient because there is no surplus oil to discharge across the pump relief valve.

Pilot-operated check valves, as shown in the preceding diagram may be necessary with flow dividing circuits to prevent oil transfer by leakage from one cylinder to the other through the flow dividing valves when the cylinders are stopped in a partly extended position. The cylinders should be "bottomed out" at both ends of their stroke to bring them back into synchronization for the next stroke.

Synchronization With Separate Pumps.
Figure 10-3. There are various ways of using separate pumps and separate 4-way solenoid valves for each cylinder. This could be a more accurate way of synchronizing two or more cylinders if low slippage pumps (piston-type) are employed and if the cylinders have leaktight seals. There will always be slight differences in pump displacement and cylinder displacement. And there will be differences in pump, valve, and cylinder slippages. Total circuit accuracy depends on careful selection of high quality components.

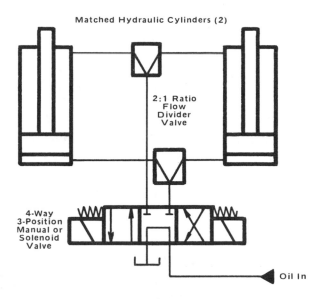

FIGURE 10-2. Flow Divider Valves.

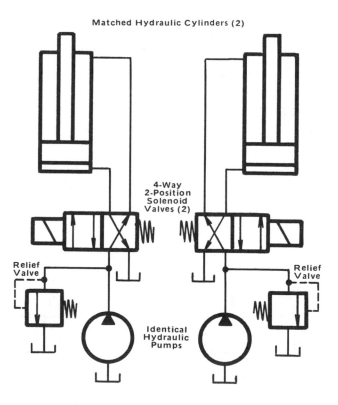

FIGURE 10-3. Dual Pumps for Cylinder Synchronization.

193

Dual or tandem (2-section) pumps of equal displacement may work out conveniently in the circuit of Figure 10-3.

See also the synchronizing circuit on Page 228 using two double-end-rod hydraulic cylinders connected in series.

OPERATION FROM TWO OR MORE ALTERNATE LOCATIONS

Some hydraulic systems must be controlled from two or more locations. Examples are a gate or door; or a hydraulic machine designed for two operators at different locations. On many of these systems the best solution is to use solenoid 4-way valve control, and to design multiple controls in the electrical system. We have shown examples in the Womack book "Electrical Control of Fluid Power". In the present book we will limit our study to circuits which can be manually controlled or to those which can be solenoid controlled with very basic and uncomplicated circuits.

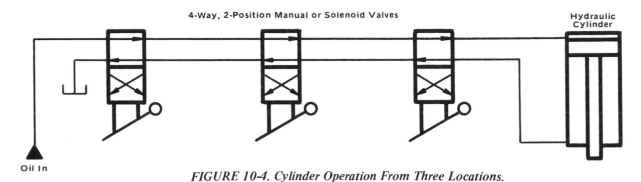

FIGURE 10-4. Cylinder Operation From Three Locations.

Simple System. Figure 10-4. Either manual valves or double solenoid valves can be used. They should be 4-way type, 2-position, with friction or detent positioning. Control may be established at any number of locations; the diagram shows three. Shifting any one of these valves will cause the cylinder to move a half cycle from its existing position. Normally this system is used for "eyeball" control, where the cylinder position, forward or back, can be seen by the operator. Spring return

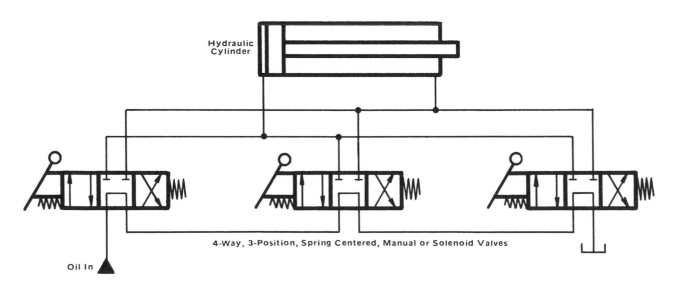

FIGURE 10-5. Control of Cylinder From Three Locations Using 3-Position Manual or Solenoid Valves.

valves should not be used; the spool should stay in the position to which it was shifted. One disadvantage to this simple circuit is that an operator cannot tell in advance which way the cylinder will go unless he can see it.

Blind Control. Figure 10-5. The cylinder can be out of view of the operator and he knows which way to shift one of the valves to extend it. The valves can be either manual or solenoid type. They should be 3-position, spring centered, 4-way valves with either tandem center or closed center spools. This circuit is primarily for control of small hydraulic cylinders with miniature 4-way valves. The following circuit is designed for control of very large cylinders.

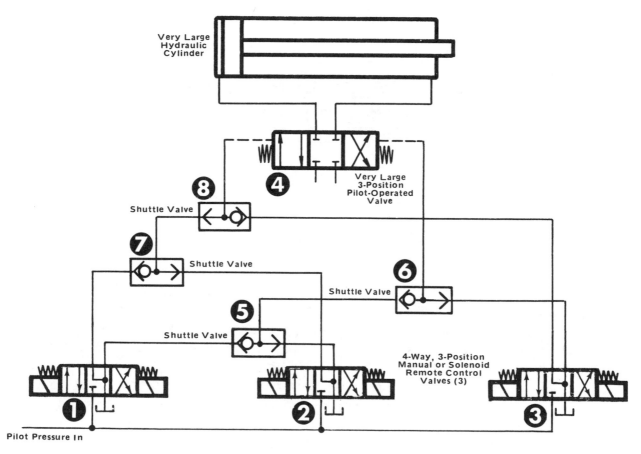

FIGURE 10-6. Pilot Control of a Very Large Cylinder From Three Locations.

Control of Large Cylinders. Figure 10-6. Valve 4 is sized to handle full flow to the cylinder. It is a double piloted 4-way valve, 3-position, spring centered. Remote control Valves 1, 2, and 3, can be miniature manual or solenoid valves controlling pilot pressure to Valve 4. They should have float center spools to allow the spool of Valve 4 to center when all remote valves are centered. Separation distance between main valve and remote control valves should generally be less than 50 feet.

Shuttle Valves 5, 6, 7, and 8 must be used to isolate the control valves from one another; to keep pilot pressure initiated at one of these valves from by-passing to tank at another location. Refer to Volume 1 for description of shuttle valves.

The control principle, shown here for three remote stations can be extended to additional stations by adding two more shuttle valves for each additional location.

TWO-HAND CONTROL CIRCUITS

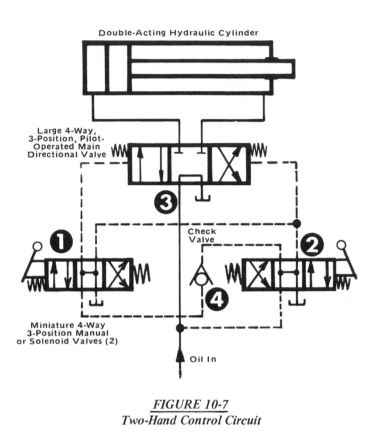

Double-Acting Hydraulic Cylinder

Large 4-Way, 3-Position, Pilot- Operated Main Directional Valve

❸

Check Valve

❶

❷

❹

Miniature 4-Way 3-Position Manual or Solenoid Valves (2)

Oil In

FIGURE 10-7
Two-Hand Control Circuit

Two-hand control circuits require two valves or switches to be actuated at the same time to start the machine. They remind and encourage an operator to keep his hands clear of the machine when it is started. They are also useful as a safety precaution on machines controlled by two (or more) operators at different locations. The machine cannot start until both operators have activated their starting controls. The reader is referred to 2-hand controls for air cylinders on Pages 82 and 83, and to electrical 2-hand circuits in the Womack book "Electrical Control of Fluid Power".

Figure 10-7. This circuit supplements the other circuits mentioned above. The main oil flow to the cylinder is handled through 4-way Valve 3 which is a double piloted, 3-position, spring centered model with tandem center for unloading the hydraulic pump. It is sized according to the main flow. Operator controls are Valves 1 and 2, which can be 3-position, spring centered manual or solenoid valves with open center spools. They can be miniature models, 1/8" or 1/4" size since they handle only pilot oil to shift the main 4-way valve. Both of these valves must be actuated at the same time and to the same side of center position before main Valve 3 can receive pilot pressure to shift it out of center neutral. During cylinder travel, either extending or retracting, if one of the control valves is released or shifted to its opposite side position, Valve 3 immediately drops into neutral position to stop the cylinder. When one of the control valves is released, it by-passes to tank the pilot pressure initiated by the other control valve.

Caution! *Two-hand control circuits in this book are not offered as completely safe circuits to prevent accidental injury to personnel. Other precautions as required by OSHA should be observed.*

CUSHIONING OF HYDRAULIC CYLINDERS

Figure 10-8. If a hydraulic cylinder is moving and then is suddenly stopped by shifting the 4-way control valve to neutral with a blocked cylinder port, the momentum energy of the cylinder and its load creates a shock wave in the line connecting cylinder port to the 4-way valve. This is a very brief pressure spike which appears during the time the cylinder is coming to a stop. It can damage the cylinder, the valve, or the connecting line.

We usually employ pressure relief valves to discharge these shock waves to tank. While this does not entirely eliminate them it does reduce their intensity. Valve 3, shown in dash lines, is the usual position for installing cushion relief valves. In order not to interfere with normal cylinder operation, Valve 3 should be set to a cracking pressure about 500 PSI higher than the setting of the pump relief valve. It should be a non-adjustable model to prevent anyone from tampering with its setting. A second relief valve could be installed alongside it and facing the other way to protect against shock waves while the cylinder is retracting.

While this type of cushioning is effective to a degree, the circuit of Figure 10-8 gives a more satisfactory control of these shock waves. The cushioning relief valve, 2, is connected to discharge into the pump line

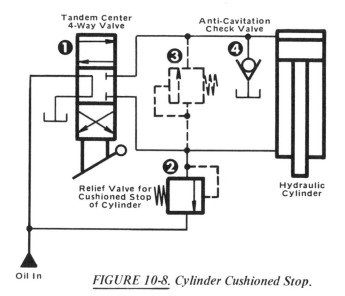

FIGURE 10-8. *Cylinder Cushioned Stop.*

instead of directly to tank. It should be an adjustable model and should be set for the best cushioning effect. It can be set higher than the pump relief valve and will give a very abrupt action, or it can be set to any pressure lower than the system pressure for a softer action. Its setting, whether high or low, has no effect on normal cylinder operation. But when 4-way Valve 1 is shifted to center neutral, the relief valve can discharge shock wave oil into the pump line where it can vent off to tank through the tandem valve spool center. Check Valve 4, selected for a very low cracking pressure, will prevent severe cavitation back of the piston, by pulling in oil from the reservoir while the cylinder is coming to a stop. By experimentally adjusting relief Valve 2, a pressure setting can be found which will decelerate the cylinder rapidly, but without creating excessive shock.

For best cushioning effect, Valve 3 can be a miniature size (1/4" or 1/2") because it handles a very small oil discharge. It should be a direct-acting, rather than a pilot-operated type, for a faster response.

For adding cushion protection while the cylinder is retracting, a second relief valve could be added on the blind end cylinder port and a second anti-cavitation check valve added on the rod port.

HOW TO MINIMIZE OVERHEATING IN HYDRAULIC SYSTEMS

Information in this section summarizes similar information scattered through this book and in the other Womack fluid power textbooks. It may be used as a checklist or as a design aid for new systems.

Maintaining a moderate but not excessive oil temperature is important for all hydraulic systems. Machine tool manufacturers recommend their hydraulic systems be operated in the range of 110 to 130°F. This temperature range is best for heavy duty systems designed for maximum reliability and which are required to work steadily 8 to 24 hours every day. First of all, this temperature will keep the oil viscosity at the right value for best performance and highest efficiency of the machine. Secondly, many undesirable chemical compounds such as acids and sludge are formed with oxygen from the air or from water condensate under the influence of heat. These impurities cause accelerated wear in moving parts, and will most certainly hasten the day when a breakdown will occur, perhaps at a critical and costly time.

Excessively high temperature is always detrimental both to the machine and to its hydraulic oil. But for practical reasons, where the machine duty is not as severe as for machine tool service, many hydraulic systems are permitted to operate in the range 130 to 160°F. However, the machine designer should be aware that temperatures even this much above the optimum range will definitely reduce life of some components and will increase the chances of a system breakdown. Practical reasons for allowing the oil to run at a higher temperature are the extra cost of the equipment, space limitations, light-duty service, infrequent operation, and cost and operation of cooling equipment. Mobile equipment, in particular, presents some very real difficulties in keeping oil temperature below 160°F., and these systems, out of necessity, are usually allowed to run at temperatures which would be highly undesirable for industrial equipment because there is no really practical way to keep the oil cool. Rubber seals will rapidly deteriorate and components with moving parts such as pumps and hydraulic motors will rapidly wear out at temperatures over 200°F. However, due to technical improvements in oil additives and in seal compounds, high temperature operation is not as destructive as it once was.

If unavoidable to operate a heavy-duty system at elevated temperatures, the oil should be discarded and completely replaced at intervals depending on severity of usage. These intervals might be at the start of every season, after (so many) hours of operation, or on some other basis appropriate for the severity of the service.

Checkpoints for Reduction of Oil Heating . . .

(A). **Pump Unloading.** On new designs the circuit should by all means provide for letting the pump run free (idle) between cycles and even during intervals in the cycle if pressure is not needed. Various ways of unloading a pump have been shown in this and other volumes in this series. All systems should have a provision for pump unloading with the possible exception of fractional horsepower applications operating at low pressure, or on systems where the heat generated by not unloading the pump can be radiated from surfaces of reservoir, cylinders, and plumbing, or by heat exchangers.

On existing systems, if overheating suddenly becomes a problem, check to be sure the pressure in the pump line falls to a low value when the cylinders are not moving. Make any repairs needed.

(B). **Speed Controls.** All flow control valves placed in the circuit for controlling cylinder speed will generate heat. With series-type speed control, heating will be worse in proportion to the amount cylinder speed is reduced. Excess power available to, but not used by, the cylinder will convert into heat. Do not operate a cylinder slower than necessary. Make sure the pump relief valve is at the lowest pressure setting which will operate the cylinder satisfactorily.

Pressure compensated flow control valves usually generate more heat than simple needle-type flow control valves. Do not use a pressure compensated flow control valve if a needle valve will give satisfactory control. Refer to Chapter 9 for detailed information on the use of needle valves.

(C). **Pressure Relief Valves.** All oil which discharges across a pressure relief valve produces heat in direct proportion to the volume of flow and to the relieving pressure. The amount of heat produced can be calculated with the fluid power horsepower formula: $HP = PSI \times GPM \div 1714$. Any circuit in which the pressure gradually rises to a peak before cutting out, such as a baling press, should most definitely employ a pilot-operated in preference to a direct-acting relief valve, to avoid the power waste in the area between cracking pressure and full relieving pressure. This is covered on Page 143.

In fluid power, if the pump flow or any part of the pump flow discharges from a higher to a lower pressure level without producing work, the energy in this part of the flow is converted into an equivalent amount of heat. Flow to a cylinder is converted into useful output and does not produce heat. Any discharge across a relief valve or an orifice should be avoided except where this is purposely planned for a useful purpose.

(D). Pressure Reducing Valves. These valves produce heat when the output pressure is up to the adjustment level and oil is flowing through the valve. Principally the heat is produced by the compensator spool having moved into a throttling position. A little heat is produced across the metering orifice. These valves do not produce appreciable heat during periods when the output port is delivering flow at less than the adjusted setting. The use of these valves should be avoided if there is a more efficient way of accomplishing the same result.

Heating may be produced indirectly by a pressure reducing valve when its compensator spool is in throttling position because of pump oil upstream which is not allowed to pass through the valve and must discharge across the pump relief valve.

(E). Reservoir. If properly placed baffle plates are used inside the reservoir they will keep the oil circulating properly, and will keep it from stagnating in corners. This will make the reservoir a more efficient heat radiator and will help to keep oil temperature down. Reservoir design is covered in Volume 1, Chapter 6.

A medium size reservoir seems to be most efficient. A good rule is to contain an oil volume equal to about 3 times the pump circulation rate. For example, when using a 50 GPM pump, optimum reservoir size is about 150 gallons for an open loop system. For a closed loop system, reservoir capacity should be about 3 times the circulation rate of the charge pump. Increasing reservoir volume beyond this 3 to 1 rule may be of some benefit but the system size and weight may be unnecessarily increased. On unusually large reservoirs it is difficult to maintain a good internal oil circulation which tends to offset the additional advantage of the larger external surface area for heat radiation.

A great deal of useful information on reservoirs will be found in the Womack book "Fluid Power in Plant and Field".

The reservoir should be located where it is exposed to vertical (convection) air currents on all sides. It should be elevated to allow air to circulate underneath. Forced air circulation against the side of the reservoir will often improve heat radiation. It should never be enclosed in a console or cabinet unless there is a forced air cooling system inside the console.

(F). Sunlight Radiation. If possible, always shield the reservoir and all other hydraulic components from direct rays of the sun. But shielding should be done in a manner which will not interfere with convection air circulation around the outside and under the reservoir. Sunlight radiation is a factor often overlooked by designers of mobile equipment. Using a hollow frame menber for the oil reservoir, while saving space and weight, is very poor design from the standpoint of oil cooling, filter replacement, reservoir access, and proper circulation of oil inside the reservoir.

(G). Heat Exchanger. It is hard to give a specific rule as to whether a heat exchanger may be needed when a new design is being worked out. A great deal depends on the circuit used, especially the method of speed control, and on how many other heat producing devices are in the circuit. The matter of possible excessive heat should always be considered on a new design. On all systems of more than 10 horsepower, access connections for running return oil through a heat exchanger should be provided. These connections can be initially plugged, but are available later, if needed, without re-plumbing the tank return line. Usually, systems of less than 10 horsepower, providing they have a properly built reservoir with 3:1 ratio capacity mentioned before, and providing they are installed with good air circulation around them, will seldom need additional cooling. Even larger systems, if operated for an hour or two then allowed to cool, may operate satisfactorily with no additional cooling. Systems operating outdoors may overheat in the summertime. Systems operating indoors, in an air conditioned environment, may not overheat even though handling more than 10 horsepower.

If additional cooling should be needed, and if a sufficient supply of cooling water is available at a reasonable cost, a shell and tube heat exchanger is more efficient, is lower in cost, and much smaller in size than an air blast type. Just as a rule-of-thumb, the horsepower rating of the heat exchanger, if required, should not exceed 25% of the motor or engine horsepower driving the system, and usually an even smaller size is adequate. Systems mounted on an engine-driven vehicle frequently make use of the engine radiator fan by mounting an oil radiator core in front of the engine radiator.

Although not highly recommended, in marginal cases where a small amount of additional cooling capacity is needed, coiled copper tubing can be immersed under the oil in the reservoir, with cooling water through it. Caution! Water connections to the cooling coil must be made underneath the minimum oil level and must be brought outside the tank below the oil level. Cold tubing inside the reservoir would condense atmospheric water and contaminate the oil.

HOLDING AGAINST STATIC LOADS

On hydraulic presses used for applications such as bonding, laminating, curing, or molding, static force must be held on the work during a curing period. Whether this force can be maintained simply by blocking a port on the cylinder(s) after the curing force has been reached, or whether the hydraulic pump must be left running under pressure during the curing period to make up for leakage losses in various parts of the circuit, depends on several factors: How long or over what period of time does the pressure have to be held? How much leakage will there be in the cylinder and other components? How much pressure drop in the cylinder is acceptable during the curing time? Several circuits for maintaining holding pressure under static conditions will be considered:

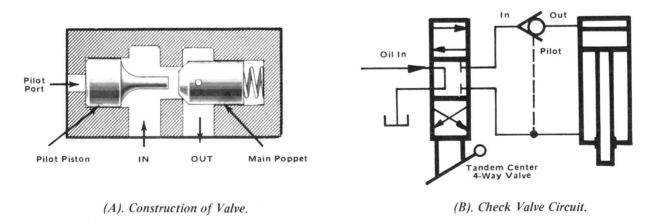

(A). Construction of Valve.   *(B). Check Valve Circuit.*

FIGURE 10-9. Holding Circuit Using a Pilot-Operated Check Valve.

Pilot-Operated Check Valve. Figure 10-9. A pilot-operated check valve has a main poppet which is biased to a normally closed position with a light spring. It will pass oil flow in one direction and will block return flow. In this respect its action is the same as a standard check valve. It also has a pilot piston which, when activated by pilot pressure, will push the main poppet off of its seat, allowing free flow in both directions. The graphic symbol used on circuit diagrams for this valve can be seen in Part (B) of this figure.

In this circuit, the working direction of the cylinder is downward. A pilot-operated check valve is placed in the connecting line to its blind end port. When the 4-way valve is shifted to its upper block, the cylinder travels downward and stalls against the work with full pump pressure. When the 4-way valve is centered, the hydraulic pump can unload and the check valve prevents trapped oil in the cylinder from leaking off through clearances in the 4-way valve spool. The check valve itself is a poppet-type valve and is almost leaktight. It will hold pressure in the cylinder for a reasonable length of time provided the cylinder piston seals are leaktight. A reasonable time might be from a few minutes to a half hour. Over this time period there may be a slight drop in pressure because neither the pilot-operated check valve nor the cylinder seals are completely leaktight.

When the cylinder is to be retracted, the 4-way valve is shifted from center position to the lower block. Pump pressure on the rod end port acts as pilot pressure to open the check valve for free return flow, and the cylinder can retract.

Accumulator. Figure 10-10. An accumulator teed into the blind end port of the cylinder can store a supply of oil during the downstroke of the cylinder. This stored oil can feed back into the cylinder during the holding period to make up oil which may escape through clearances in the 4-way valve spool and past the piston seals.

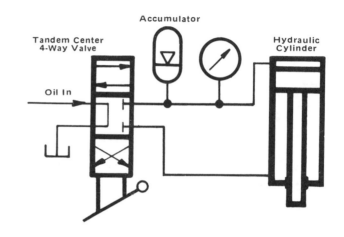

FIGURE 10-10. *Accumulator Holding Circuit.*

To extend the cylinder, the 4-way valve must be shifted to the upper block. When the cylinder stalls against the work, the 4-way valve must be left in shifted position for a brief period until the accumulator is filled with oil up to the system pressure level. Then the 4-way valve can be centered to unload the pump.

During the holding period the pressure behind the cylinder piston will slowly drop as oil is fed out of the accumulator to make up circuit leakage. The greater the capacity of the accumulator the less the pressure will drop, but using too large an accumulator will require a longer time to re-charge it before the 4-way valve can be centered. It will also be less energy efficient because all of the stored oil must be discharged every time the cylinder retracts.

When the cylinder is to be retracted, the 4-way valve is shifted to the lower block. The remaining oil in the accumulator is discharged to tank.

Better results are always obtained when using a cylinder with soft, leaktight seals. A rotary shear-type 4-way valve usually has a lower internal leakage than a spool-type valve and will hold the pressure at a higher level for a longer time. A pressure gauge should be installed on the cylinder port to make sure that the cylinder and accumulator have reached full pressure before the 4-way valve is centered.

The reader is referred to Volume 1, Chapter 6 for a more complete description of accumulators and their operation.

Air-Driven Pressure Intensifier. When active pressure from a hydraulic pump must be maintained on the cylinder throughout the holding period because of high circuit leakage, an extra long holding period, or to keep the pressure on the cylinder piston from dropping during the holding period, an air-driven pressure intensifier is one of the best solutions. This application is described in more detail in Chapter 11.

The pressure intensifier is a single-piston pump driven with an air cylinder. Many models are available to suit the pressure range to be covered. The intensifier will deliver a small flow of oil, sufficient to make up all oil lost through component clearances. It will remain in an active pumping condition, delivering only the oil flow required for make-up. It consumes power from the air line, and does not add heat to the hydraulic oil.

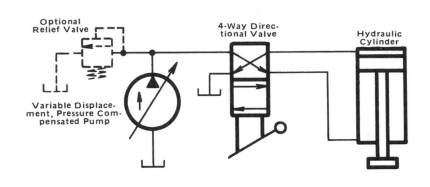

Optional Relief Valve

4-Way Directional Valve

Hydraulic Cylinder

Variable Displacement, Pressure Compensated Pump

FIGURE 10-11.
Holding Circuit With Pressure Compensated Pump.

Pressure Compensated Hydraulic Pump. Figure 10-11. The hydraulic pump in this diagram is a variable displacement type with pressure compensator. The compensator is adjustable, and can be set to the pressure level desired on the cylinder during a curing period. It will permit full pump delivery until the cylinder stalls against the work. Then it will reduce pump displacement to almost zero, just enough flow to make up slippage losses in circuit components — cylinder piston seals, 4-way valve spool, and slippage in the pump itself. The cylinder can be held in a stalled condition against the work during curing time and very little input power will be required and there will be very little heat generated in the oil. The working principle of a pressure compensator is illustrated in the next diagram. More information on applications for this type pump will be found on Pages 91 and 114.

Pressure Compensating Principle. Figure 10-12. A compensator is built into the pump at the factory and usually cannot be added to a pump in the field. Not all pumps can be built with a pressure compensator — only piston pumps and a certain type of vane pump having a single-lobe cam. Most other types of positive displacement pumps including internal and external gear, balanced (double lobe) vane, gerotor, and screw types cannot be built with variable displacement, so they cannot be pressure compensated to a lower than maximum displacement.

Figure 10-12 is a schematic of a check valve axial piston pump, variable displacement, controlled with a pressure compensator. The pistons, usually 5, 7, or 9 in number, are stroking inside a piston block which is keyed to, and is rotatating with the shaft. The left ends of the pistons are swiveled to piston shoes which bear against and slide around on the swash plate as the piston block rotates. The swash plate in this particular pump does not rotate; it is mounted on a pair of trunnions (hinge pins) so it can swivel from neutral (straight up and down) to a maximum tilt angle. When the swash plate is at an angle to the vertical, the pistons stroke as they are carried around by the piston block. Their length of stroke (and the GPM output of the pump) is proportional to the tilt of the swash plate. At less than maximum pressure, the swash plate remains at maximum tilt angle, held there by spring force, by hydraulic pressure, by the dynamics (mostly centrifugal force) of pump construction, or by a combination of these forces, and pump flow remains at maximum. When the pump is operating against light loads which do not require the maximum pressure, hydraulic pilot pressure (obtained from the pump outlet port), is not sufficient to compress the compensator spring. When pressure on the outlet port rises high enough to overcome the adjustable compensator spring, the "firing" pressure has been reached. The compensator piston starts to pull the swash plate back toward neutral, reducing

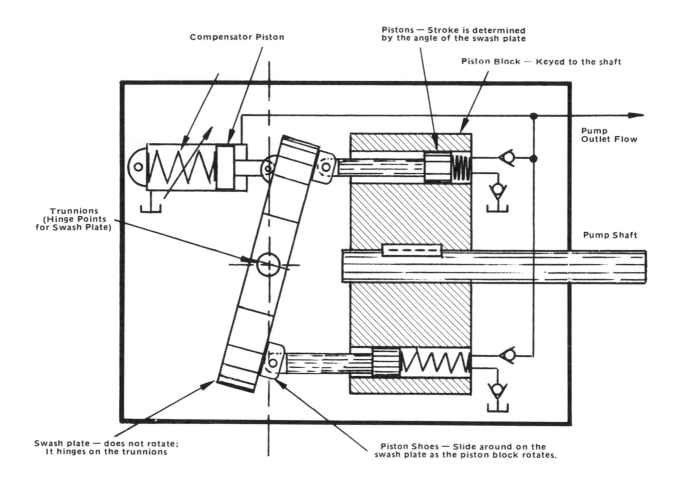

FIGURE 10-12. Schematic of a variable displacement, check valve piston pump, with built-in pressure compensator.

pump displacement (piston stroke) and output flow. The tension of the compensator spring can be adjusted for the desired maximum or "firing" pressure.

Some pump models are available with built-in or externally adjustable internal stops to limit the tilt angle of the swash plate. These stops limit the maximum flow and limit the HP demand on the prime mover (electric motor or engine) which drives the pump.

The pump in Figure 10-12 does not have a control lever for the operator to change the displacement. The compensator spring usually has a wrench adjustment. The pump always operates "wide open" until outlet pressure rises to a maximum. Piston pumps used in closed loop hydraulic transmissions usually have an external control lever to allow the operator to control displacement at pressures less than the compensator setting. The compensator then acts to "override" the operator. It will automatically reduce displacement when compensator setting has been reached even though the operator still has his control lever set for maximum displacement.

Dual-Pump Holding Circuit. Figure 10-13. This is a 2-pump "high-low" circuit as described on Page 157. In this case, Pump 1 is a high volume pump operating at low pressure to extend and retract the cylinder when it is running in free travel in both directions. Pump 2 is a very small displacement pump of any kind, sized for enough flow to make up leakages in the cylinder and valve while the

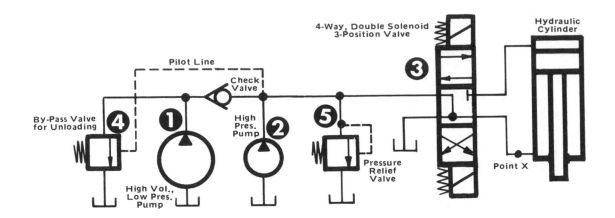

FIGURE 10-13. Dual Pump Holding Circuit of the "High-Low" Type.

cylinder is holding against the work during the curing cycle. Relief Valve 5 is set for cylinder force during the curing period. By-pass Valve 4 unloads Pump 1 during the curing period. It should be adjusted for a moderate pressure level, sufficient to move the cylinder in free travel. This is usually 200 to 400 PSI.

When the 4-way valve is shifted to its upper block the cylinder extends, using the combined flow from both pumps. When it stalls against the work, pressure rises behind its piston, causing by-pass Valve 4 to open to unload Pump 1. Pump 2 then builds up system pressure to the setting of relief Valve 5, and continues to discharge across Valve 5 during the curing cycle. The 4-way valve must be left in this side position during the curing period. To retract the cylinder, the 4-way valve is shifted to its lower block. When the cylinder reaches home position, the 4-way valve can be released to its center neutral position to stop the cylinder and unload the pump.

Valve 3 should have a spool in which the rod end cylinder port is vented to tank in neutral. This is to prevent the possibility of building up pressure under the piston by circuit leakage. Any build-up of pressure on the rod end of the cylinder would offset a part of the downward force of the cylinder during the curing period. If necessary to counterbalance the cylinder during its downward travel, a counterbalance valve may be installed at Point x and adjusted to the lowest counterbalancing pressure which will keep the cylinder from running ahead of the pump.

Isolated Piston. There are several brands of check valve type piston pumps in which one or more individual pistons can be isolated and their flow brought out on a separate port. In effect this is a dual pump and can be used in the "high-low" circuit shown above. The isolated piston would be Pump 2, and would provide an active oil supply to make up circuit leakage during the curing period while the remainder of the pistons, represented as Pump 1, would be unloaded by by-pass Valve 4.

11
PRESSURE INTENSIFICATION
...Useful Applications in Hydraulics

Intensification Principle. At certain places in a hydraulic or pneumatic system an unusually high pressure may be needed for operations like pressure testing, high pressure injection, or crimping which require a very small volume of the high pressure fluid. Pressure intensifiers may be ideal for these operations. There are several types of pressure intensifiers available, but information in this section will be limited to the rotary type

Air-Driven, Piston-Type Pressure Intensifiers

—From Haskel

for hydraulics and to the reciprocating piston type for both hydraulics and pneumatics.

Intensifiers receive pressure (or power) from a source connected to their inlet port. They can deliver an amplified pressure at their outlet port either static force or force with a small movement. The amplified pressure can be delivered either in the same or in a different fluid medium. For example, compressed air pressure on the inlet can produce an amplified hydraulic pressure at the outlet.

An intensifier cannot amplify power. It is a constant horsepower device, delivering output at a higher pressure level than the inlet but at a lower flow, and at the same horsepower minus inherent losses in the intensifier due to mechanical friction and fluid flow. An intensifier in a fluid system is the equivalent of a transformer in electrical power transmission which operates at a constant horsepower but converts input voltage on the primary winding to a higher voltage on the secondary winding but at a reduced current flow. It is also the equivalent of a gear train in mechanical transmission which is also a constant horsepower device converting a lower into a higher torque but at a corresponding reduction in speed.

Intensifiers offer a convenient and economical means of amplifying shop air pressure or low hydraulic pressure into high pressure air or hydraulic pressure without the use of a rotary power pump.

Sometimes they offer the only practical means of generating extremely high pressure. For example, 100 PSI from a shop air line can produce hydraulic pressure at 15,000 PSI or higher.

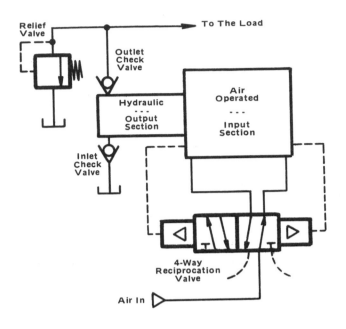

FIGURE 11-2. Automatically Reciprocating Pressure Intensifier.

Air-Driven, Hydraulic Intensifier.
Figure 11-2. One of the most common applications for intensifiers is to produce a small flow of high pressure oil using shop air pressure as a power source. A piston-type intensifier includes an input section with a large diameter piston working from shop air pressure, and an output section with a small diameter piston producing intensified hydraulic pressure. The two pistons are joined together. Mechanical force from the large air piston drives the small piston which acts as a single-piston check valve type pump. The hydraulic pressure produced will be in the same ratio as the area ratio between the two pistons.

Example: If the large piston is 4" in diameter with an area of 12.56 square inches and is exposed to 100 PSI air pressure, and if the small piston is 1/2" in diameter with an area of .196 square inches, the ratio between areas is 12.56 ÷ .196 = 64. The hydraulic pressure produced with 100 PSI inlet pressure will be 100 x 64 = 6400 PSI.

In this figure the 4-way reciprocation valve (button bleeder type) and the two check valves are built into, and are an integral part of the intensifier assembly. The relief valve is optional and may be omitted. Maximum pressure on the output section can be limited by controlling the inlet air pressure.

When air is connected into the intensifier it will reciprocate between its stroke limits. The volume of oil delivered on each stroke will be proportional to output piston diameter, stroke length, and number of strokes per minute. An air-activated intensifier of this type will stroke very rapidly when producing a low output pressure, and will stroke more slowly when required to produce a higher pressure. Finally, when the maximum intensified pressure level is reached, it will stroke very slowly, pumping only enough oil to maintain this pressure level against any leakage loss in the output circuit.

Important! If the intensifier is used in the same circuit with another pump, a rotary power pump for example, both pumps must draw oil from, and discharge back into, a common reservoir. If more than one reservoir is used in a system, all reservoirs must be on the same elevation and must be interconnected. If they are on different elevations, there will be an oil transfer, with one reservoir eventually running dry while the other one overflows.

Intensifier. Construction. Figure 11-3. Simplified construction diagram illustrating principle of operation. Actual construction would include a means to prevent valve spool from stalling in mid-travel.

The air-driven input section is on the left side, and the intensified hydraulic pressure is pumped by the output section on the right side. The intensified pressure level is determined by the ratio of piston areas in these two sections.

When compressed air is connected to the air inlet port on the 4-way reciprocation valve, the air piston will start into continuous reciprocation. At each end of its stroke, when it pushes the actuator pin, a signal is transmitted to the 4-way valve, which in this case is a button bleeder valve, and the valve reverses the travel direction of the air piston. A manual bleed button on each end of the 4-way valve provides manual overrides for the internal actuator pins.

Continuously reciprocating intensifiers, similar to the one described, are

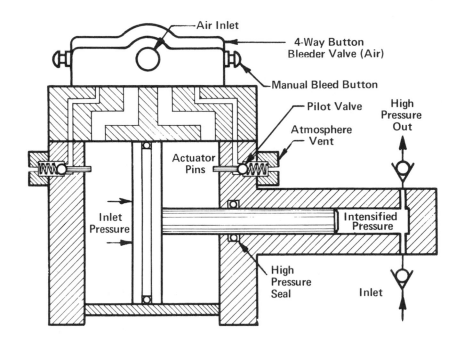

FIGURE 11-3. *Typical Construction of a Piston-Type Intensifier.*

usually built with a relatively short stroke for economy of construction and minimum mounting space. For reasons of economy, those designed for compressed air on the input section are built for single-ended operation; that is, with an output section only on one end of the input section as shown later in Figure 11-4. The air piston returns very rapidly and very little pumping time is lost. Those designed for low pressure hydraulics on the input section have a return speed the same as the forward speed. They are usually built with two output pistons, one on each end of the input section to increase the total output flow. This construction is illustrated later in Figure 11-6.

Note that in Figure 11-3 the check valves and the 4-way reciprocation valve are built as an integral part of the pressure intensifier.

Fluid Media. Although the air-driven hydraulic intensifier described above is used more often than any other kind, intensification can be used on many kinds of liquid and gas applications. The same or different fluids can be handled on the input and output sections of the same system. For example, air to water, hydraulic to hydraulic, air to air, and air to nitrogen operation are used with models designed to handle these fluids. Stainless steel models are also available for pumping corrosive liquids.

Maximum Ratio of Intensification. With liquids, the pressure ratio between input and output is limited by congealing of the fluid at high pressure. Petroleum oil can be pumped to 50,000 PSI or higher. Some fluids can be pumped to 200,000 PSI or higher with special intensifiers.

When pumping gas, the ratio is limited by the compression efficiency. In practice, an intensifier of the type shown in Figure 11-2 is limited to about 3:1 ratio. If carefully designed for high volumetric efficiency, with a minimum piston clearance and minimum dead space at the end of the piston travel, ratios of 7:1 can be obtained. The critical maximum ratio for air is about 13:1. At this ratio the heat of compression may be sufficiently high to ignite any mineral oil that may be mixed in with the air, resulting in an explosion.

207

TYPICAL APPLICATIONS FOR PRESSURE INTENSIFIERS

The applications described here are for reciprocating intensifiers but may also be suitable for either "1-Shot" or rotary mechanical intensifiers which are yet to be described. Most of them involve pressure intensification but some are primarily concerned with handling liquids which cannot be satisfactorily pumped with rotary pumps; pressure intensification may or may not be involved.

Note: Air-driven pressure intensifiers should not be used on applications which can be handled satisfactorily with ordinary rotary pumps. As with all compressed air equipment, efficiency of air-driven intensifiers is quite low. They consume a large volume of air relative to the hydraulic horsepower they can produce. The low efficiency is due partly to the compressible nature of air and to the power lost in heat when it is compressed. They are intended as a means of obtaining a low flow of high pressure fluid when more conventional methods seem to be impractical, or where there appears to be an important advantage in spite of the low efficiency. Before purchasing one, be sure your plant air supply has sufficient capacity to run it over the time period involved.

(1). Component Testing. Occasional testing or production testing of manufactured parts for physical strength, burst rating, or for leaks — such parts as sand castings or die castings, forgings, welded or seamless pipe, welded containers or assemblies, or pipe lines, to name a few. Figure 11-8 shows a typical testing set-up for vessels. Leak testing of parts can be done either by immersing the part in water while air pressure is applied, or by applying high pressure inside the part and observing any drop-off in gauge pressure under static conditions over a measured time period.

(2). Fluid Transfer. This is one of those applications where transfer of corrosive fluids rather than pressure intensification is the main objective. Low ratio intensifiers (or even 1:1 models) constructed with stainless steel parts are used to transfer acids and other chemicals which are difficult or impossible to handle with conventional pumps. Solid particles should be filtered out before the fluid is permitted to enter the pump.

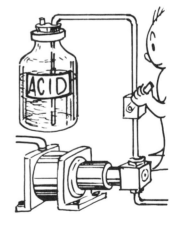

(3). High Pressure. Very high pressure, impossible with rotary pumps can be produced, up to 15,000 PSI with ease by an intensifier having the proper piston ratio. Up to 200,000 PSI can be produced by using special assemblies, special high pressure port connections, and with suitable fluids. Intensification level is limited by congealing of the fluid at high pressure, strength of pump barrel, piping, and port connections.

(4). Gas Storage. Storage of gas in pressure bottles or recovery of residue gas remaining in such bottles after terminal pressure in the bottle has dropped below a usable level.

(5). Long Holding of Pressure on Presses. An air-driven intensifier, delivering a small volume of hydraulic oil at high pressure is an accurate and inexpensive way of maintaining active pressure under static conditions for an indefinite time without generating heat or consuming power, except to make up for leakage losses in the circuit. It will maintain active pressure on bonding, laminating, or curing presses for long curing periods while the main pump is turned off. See further details on Page 212.

(6). Liquid Injection. Air-driven intensifiers may be used for injecting liquids such as odorants, anti-freeze, and lubricants into high pressure gas or liquid lines.

(7). Replace Hand Pump. Some installations such as hand-operated shop presses can be upgraded by replacing the hand pump with either a 1-shot or a continuous cycling intensifier. Pumping speed can be increased by 5 to 20 times and operator fatigue greatly reduced.

(8). Charging Accumulators. High pressure nitrogen bottles are used for charging accumulators. However, most of the nitrogen becomes unusable when bottle pressure falls below 1500 PSI which is the most common charging level. A pressure intensifier, used in the cascading circuit of Figure 11-10, can recover most of the remaining gas. For production work where quantities of accumulators are to be charged, intensifiers will be a money-saving addition to the operation. This use of intensifiers is especially useful for salvaging rare and costly gases in certain processes.

(9). Miscellaneous. Numerous small jobs which require only a small movement at high pressure are done with intensifiers. Some of these applications include riveting, spot welding clamps, bending, crimping, forming, piercing, punching, stamping, shearing, marking, etc. In many cases a 1-shot model driven with shop air or low pressure hydraulics is ideal for the job.

TYPES OF PISTON-OPERATED PRESSURE INTENSIFIERS

Single-Ended. Figure 11-4. The input section is powered with compressed air or other gas. The output section will pump oil, water, gas, or other chemicals compatible with materials of construction. The power stroke will be slow or fast depending on the pressure pumped against. The return stoke is very fast so little pumping time is lost. Single-ended construction is inefficient when the input is powered from a hydraulic supply because of time lost on the return stroke

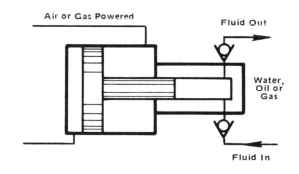

FIGURE 11-4. Single-Ended Intensifier.

Pumped fluid is delivered in "shots" with a short break each time the piston retracts. For this reason it may be unsuitable for certain jobs unless pumped into an accumulator.

Caution! Air-driven intensifiers should not be used to replace standard rotary hydraulic pumps if those pumps are suitable for the job. As in all air operated circuits, air consumption is high for the amount of fluid horsepower produced. The capacity of the plant air supply must be sufficient for the additional demand.

Tandem Type. Figure 11-5. Has two or more input pistons operating in parallel from the driving source. This is a single-ended type and works more efficiently when powered by compressed air. It will pump air, gas, or liquid at intensified pressure. The intensified pressure level will be inlet pressure times the ratio

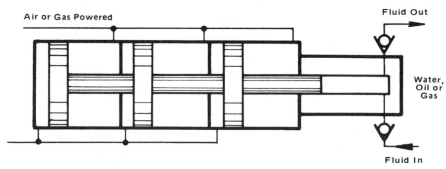

FIGURE 11-5. Three-Section, Tandem-Type Intensifier.

of the combined areas of all pistons in the input section to the output piston area.

Some advantages of the tandem type are that it can operate on lower driving pressure to produce the same hydraulic pressure, or it can produce a higher level of intensified pressure because of the larger ratio of areas possible with this type. It is not recommended for hydraulic operation on the input side because of the pumping time lost on the return stroke.

Double-Ended. Figure 11-6. This type has one input piston alternately driving two output pistons. The input section is designed for hydraulic drive from a low pressure rotary pump, for example a gear pump. Since the input section has the same displacement on both sides of the piston, the piston will travel the same speed in both directions. And since the size of both output pistons is the same, the output flow will remain steady. There will be very brief inter-

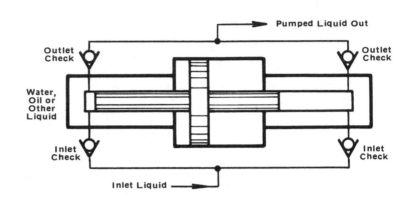

FIGURE 11-6. Double-Ended Intensifier for Liquids.

ruptions in output flow as the pistons reverse direction, but in some circuits the oil compressibility will smooth this out. Or, a small accumulator can be floated on the output to smooth out pulsations.

On most compressed air drive applications there is no significant advantage in using a double ended rather than a single ended intensifier, although the momentary interruptions in flow during piston reversal will be shorter. This type, with hydraulic drive, can also be used to pump high pressure gas.

Two-Stage. Figure 11-7. For moderate ratios of intensification of air or gas up to 3:1 ratio, the single-ended model of Figure 11-4 would normally be used. For higher ratios of compression, it is much better to intensify the pressure in two or more stages. If each stage has a 3:1 ratio, the overall intensification will theoretically be 9:1 if the gas is cooled between stages.

The input piston is in the center. It can be powered with any fluid medium, oil, water, or gas. The output pistons, one at each end,

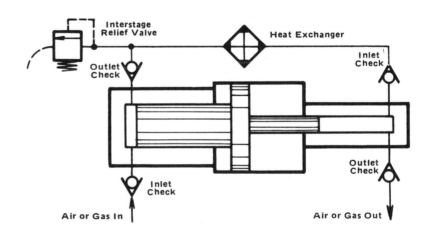

FIGURE 11-7. Two-Stage, Series-Type Intensifier, for Gas Only.

can only be used on gas because of the difference in areas. The two pistons are connected in series with a heat exchanger between stages. The first output stage is on the left. Compressed gas from this piston is delivered through the heat exchanger to the inlet of the second output stage. The displacement volume in the second stage is made smaller because the same quantity of gas, after being compressed in the first stage, occupies a much smaller volume. The area ratio between the first and second output stage pistons is made approximately the same as the compression ratio of the first stage.

The heat exchanger is an important and necessary part of a 2-stage gas compressor. The temperature of the compressed gas delivered from the first output stage is quite high. If its temperature is not reduced before it enters the second stage, the efficiency and compression ratio of the second stage will be greatly reduced. Sometimes finned tubing connecting the two stages will offer sufficient cooling. If necessary, a water cooled heat exchanger or aftercooler can be used.

The interstage pressure relief valve can be omitted if the area ratio of the two output stage pistons is designed correctly.

APPLICATIONS FOR PISTON-TYPE PRESSURE INTENSIFIERS

Pressure Testing. Figure 11-8. Water, rather than oil is ordinarily used for pressure testing of vessels, pipe, and production parts such as castings, forgings, etc. Water is economical, readily available, evaporates without leaving the vessel dirty, does not create a mess if spilled, and can be discarded after use. Most pressure intensifiers will pump water as well as oil, or can be purchased for this service. A small, air-driven, reciprocating type intensifier is suitable for most testing applications. It can be purchased in ratings up to 20,000 PSI, and in some cases to more than 100,000 PSI. It will operate from ordinary shop air lines and is the most economical means for generating high liquid or gas pressure in small volumes.

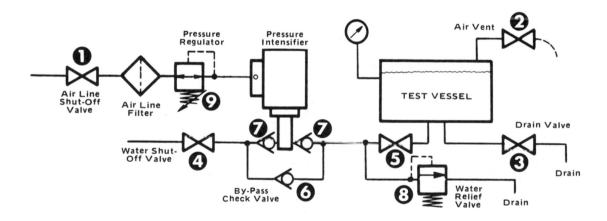

FIGURE 11-8. Example of Pressure Testing With an Intensifier.

To save testing time, large vessels should be pre-filled at a rapid rate from the city water line or by means of a low pressure centrifugal pump. Then when all air has been expelled, the intensifier is started and continues to pump until the pressure is raised to the desired test level. Figure 11-8 is typical of a pressure testing application.

The first step in pressure testing is to pre-fill the vessel at a rapid rate while venting the trapped air. This is done by opening vent Valve 2, closing drain Valve 3, opening series Valve 5, then turning on the water supply with Valve 4. Initial water flow can freely pass through intensifier check Valves 7. If these valves will not carry the pre-fill water at a rapid rate, an auxiliary check Valve 6 of large size can be connected with large pipe or tubing to by-pass the intensifier.

When the vessel has been completely filled with water, and all air has been expelled, air vent Valve 2 should be closed and the intensifier started by opening the air line shut-off Valve 1.

Maximum pressure of the water is controlled by the pressure setting of the regulator, Item 9. When water pressure builds to a level equal to air line pressure times the area ratio in the intensifier, the intensifier automatically ceases to pump and will maintain this pressure level indefinitely. If there is leakage in the test circuit, the intensifier will continue to pump at a rate just sufficient to make up the leakage and keep the vessel at the test pressure.

Water relief Valve 8 is for additional safety, and may be installed for one of two purposes: (1), as a safety valve to limit the water pressure in case of failure of the air pressure regulator or if the regulator were accidentally set too high. It should have a cracking pressure about 5% higher than the water test pressure. Normally it should not be active; or, (2), the relief valve may be set to crack at the exact test pressure and used instead of the air regulator to regulate maximum test pressure. Used in this way, the time to pump to water test pressure is reduced because the intensifier remains pumping at a relatively faster rate right up to relief valve pressure instead of slowing to a very slow stroking rate as the test pressure is approached.

To test for leaks, Valve 5 can be closed after final test pressure has been reached. This isolates the vessel from associated equipment, especially the relief valve, which might leak and give a false indication. Assuming Valves 2, 3, and 5 are leaktight, any drop in gauge pressure after Valve 5 has been closed indicates a leak. Note: If the water temperature is allowed to change during the leak test, a change in gauge reading can be expected.

Caution! Do not allow the water to become heated in the vessel while Valves 2, 3, and 5 are closed. Heating will cause the water to expand and its pressure to increase possibly to a dangerous level.

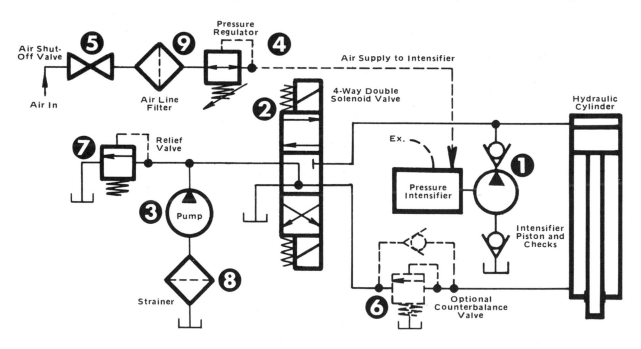

FIGURE 11-9. Intensifier Maintains Holding Pressure on Press While the Pump is Unloaded.

Maintaining Holding Pressure on Presses. Figure 11-9. Presses on certain applications must hold static pressure against a load for a period of time, called the curing period, as described on Page 208.

A pressure intensifier added to the normal press circuit is an ideal way to hold static pressure for an indefinite length of time without wasting power or generating heat. It can maintain the pressure level to a high degree of accuracy during the curing period.

In Figure 11-9, the main press circuit includes the rotary pump, 3, the 4-way directional valve, 2, and the press cylinder itself. The intensifier, 1, is added to bring the cylinder up to full force and to provide make-up oil lost from the circuit by component leakage during the curing period when Valve 2 is shifted to neutral to unload the pump. The intensifier receives its power from the shop air line through shut-off Valve 5, Filter 9, and pressure Regulator 4. Whether or not a lubricator should be used in the air line depends on the brand of intensifier used, as some manufacturers forbid the use of a lubricator in the air line because it will wash out the permanent lubricant.

To operate the press the 4-way valve is shifted to its upper block. The cylinder extends and stalls against the work. The intensifier is put into operation by opening Valve 5 and the 4-way valve is shifted to neutral. The intensifier is allowed to run until pressure comes up to the desired level. It will stall out at this pressure if the regulator, 4, has been set properly. At the end of the curing period the intensifier is stopped and the 4-way valve is shifted to the lower block to retract the cylinder.

Directional Valve 2 should have a "cylinder vent" spool in which the rod end of the cylinder is vented to tank in neutral. This cylinder port should not be blocked in neutral. If it were, any leakage past the cylinder piston during the curing period would cause a counter-pressure to build up on the rod side of the piston and neutralize a part of the force produced on the top side of the piston.

Cylinders working downward and carrying a heavy weight load will need to be counterbalanced to prevent free fall when the 4-way valve is shifted to the lower block. If needed, a counterbalance valve may be installed as shown at Position 6. This valve should be adjusted to the minimum pressure setting which will control the cylinder. Setting it too high will rob some of the press capacity.

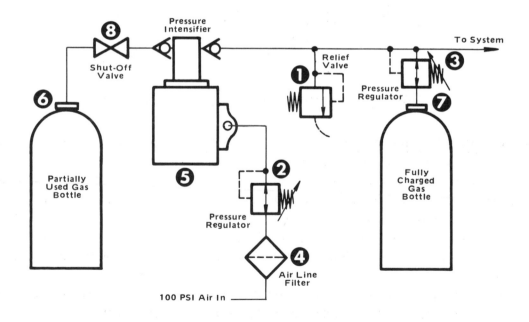

FIGURE 11-10. Salvaging Valuable Gas From a Partially Used Bottle by Means of Cascading.

Cascading. Figure 11-10. This circuit shows how an intensifier may be used to salvage the remaining gas in a pressure bottle when its terminal pressure falls below a usable level. The original bottle pressure would be 2250 to 3000 PSI. In a process requiring a minimum of 2000 PSI pressure, when the bottle terminal pressure dropped to 2000 PSI the bottle would become unusable and ordinarily would be discarded even though most of the gas remained in it. This would be costly if the bottle contained an expensive gas.

213

In Figure 11-10, an intensifier pumps the remaining gas from the discarded bottle into another storage bottle at high pressure. An air-to-air intensifier is used which has an internal area ratio in the range of 15:1 to 30:1. It receives power from a shop air line through a filter and pressure regulator. A high pressure relief valve, 1, should be used to prevent accidental over-pumping to a dangerously high pressure level if someone should inadvertently set the shop air input pressure too high on the intensifier inlet port. The relief valve should have a leaktight poppet and should be adjusted to a pressure about 5% higher than the maximum pressure to be pumped.

The maximum intensified pressure level can be controlled with the pressure level in the shop air line. If possible, the adjustment of the regulator should be locked so no one can tamper with it.

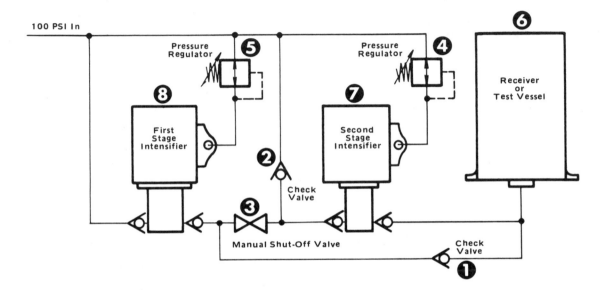

FIGURE 11-11. Two-Stage "High-Low" Pressure Intensification.

Two-Stage Intensification. Figure 11-11. This application is for air or gas, and particularly for reducing the filling time of large vessels or bottles.

Two air-to-air intensifiers, driven from a shop air supply, are used. The first one, Item 8, has a low piston area ratio and will pump high volume but at only moderate pressure. The second one, Item 7, has a higher piston area ratio and will pump to a high pressure but at a low rate of flow. On first starting, both intensifiers work together to produce a high flow until a moderate pressure level is achieved. Then, the first intensifier is stopped and the second one continues to pump vessel pressure to the final level.

When starting, Valve 3 is closed and the shop air supply to the intensifiers is turned on. Both intensifiers combine their flows until a moderate pressure level is reached. At this point the first intensifier stalls and ceases to pump. Valve 3 can now be opened. The first intensifier supplies supercharge pressure to the inlet of the second, allowing it to pump the vessel to the final test pressure.

Valve 3 is shown as a manual type. It can be replaced with an air sequence valve if one can be found to work at the maximum pressure level of the system. This would give automatic changeover at a selected pressure. The changeover can also be made automatic by using a 2-way solenoid valve at Position 3 and working it with a pressure switch.

The circuit, as shown, is designed for shop air on both sections of both intensifiers. Other gases can be pumped by disconnecting the output sections of both intensifiers from shop air and connecting them to a source of the gas to be pumped.

214

ONE-SHOT PISTON-TYPE PRESSURE INTENSIFIERS

Figure 11-12. A 1-shot intensifier is built to deliver a single shot of high pressure oil when activated. The input section is built like a square head industrial cylinder. The output section, with a piston of smaller diameter, is attached to the rod end cap of the input section. The intensifier is not built to automatically reciprocate. It is controlled with an external 4-way valve. Each time the valve is shifted, the intensifier will deliver a measured quantity of high pressure oil through check Valve 2. If more oil is required, the operator must reverse the 4-way valve to retract the pistons. Oil is drawn by suction into the output section through check Valve 1 as the pistons retract. The 4-way valve is again actuated to deliver another shot of oil.

The input section may be driven either from shop air pressure or from a hydraulic oil supply, according to the model used. The output section will pump liquids which are compatible with materials of construction. A 1-shot intensifier is not usually suitable for pumping gas because the displacement of the output section, on a single shot, is too low.

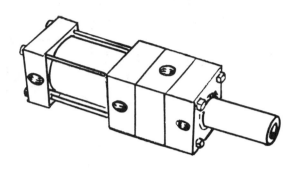

Typical One-Shot Pressure Intensifier

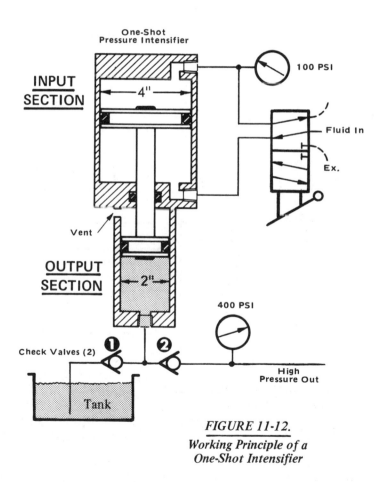

APPLICATIONS

Single-Acting Cylinder. Figure 11-13. The bore and stroke of the intensifier is selected to supply, on one shot, the entire volume required by the cylinder, plus a small extra amount for oil compression and to compensate for nominal leakage from the circuit over a period of time.

FIGURE 11-12.
Working Principle of a
One-Shot Intensifier

Shifting 4-way Valve 1 drives the intensifier pistons downward, delivering a shot of oil at intensified pressure to extend the work cylinder. To retract the work cylinder, 4-way Valve 1 is reversed to retract the intensifier pistons. This allows the internal spring in the work cylinder to push the oil back into the output section of the intensifier.

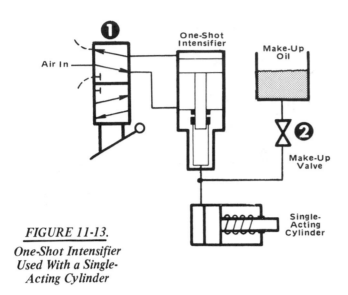

FIGURE 11-13.

One-Shot Intensifier Used With a Single-Acting Cylinder

The work cylinder should have leaktight piston seals. However, over a long period of operation some of the oil may leak from the circuit past the piston or intensifier seals, and will have to be replenished. This will be evident when the work cylinder is unable to make the full forward stroke. To add more oil to the circuit the intensifier pistons should be retracted, then Valve 2 should be opened. The slight vacuum in the intensifier/cylinder line will draw the amount of new oil needed from the make-up oil tank. Be sure to close Valve 2 before operating the intensifier.

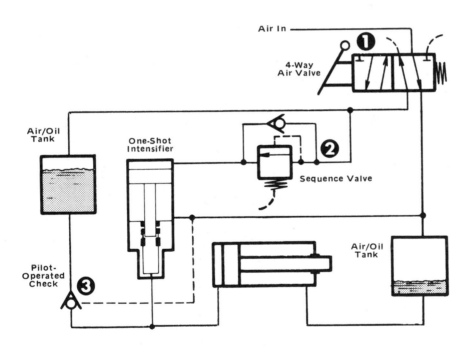

FIGURE 11-14.

One-Shot Intensifier Used With Double-Acting Cylinder. Rapid Advance, Sequence to Feed

Rapid Advance, Sequence to Feed. Figure 11-14. This circuit provides a long travel in the work cylinder in moving up to the work at low pressure. Then, when work resistance builds up, a sequence valve automatically brings the intensifier into action to deliver one shot of oil at intensified pressure.

To start the cycle, 4-way Valve 1 is actuated. This applies shop air pressure to the top of the air-over-oil tank on the left, and this pushes oil at shop air pressure through the pilot-operated check valve and into the cylinder. The cylinder advances rapidly until contacting the work. The speed of advance depends on the shop air pressure level, on the diameter of connecting oil lines, and the size of the pilot-operated check valve. On contact with the work, the oil pressure behind the work cylinder piston rises to the opening pressure setting of sequence Valve 2. Air then flows through Valve 2 into the intensifier, and the intensifier produces a shot of high pressure oil to the cylinder.

After the forward stroke has been completed, the 4-way valve is reversed. This provides shop air pressure to retract the intensifier, to open the pilot-operated check valve, and to pressurize the top of the air-over-oil tank on the right to retract the cylinder.

Pilot-operated check Valve 3 must be able to handle full intensified pressure on its main ports and must be capable of opening on shop air pressure level. Alternatively a 2-way, normally closed hydraulic valve with air pilot actuator could be used, or it could be a manually operated 2-way valve. Sequence Valve 2 must be a true sequence valve; an air pressure relief valve will not work here.

Multiple Clamping. Figure 11-15. Sometimes, due to the nature of the job, several hydraulic clamps must operate from a 1-shot intensifier. The intensifier can be selected with a bore and stroke to supply combined oil to fill all cylinders with one shot, or a smaller size can be used, with the operator actuating the 4-way valve two or three times to supply several shots into the cylinders.

An alternative arrangement is to use a relatively small capacity intensifier and add a rapid advance feature for pre-filling the cylinders until they stall against the work, before starting the intensifier. This is shown in Figure 11-9.

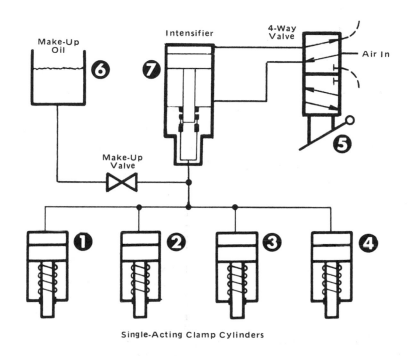

FIGURE 11-15. Multiple Clamping With a One-Shot Intensifier.

Pipe Line Testing. Intensifiers may be used for testing sections of a pipe line, either for leaks or for ability to hold a working pressure. Usually the test medium is either water or the liquid to be carried by the line. The line should first be pre-filled with the liquid from a high volume source until all air has been displaced and purged. Then the intensifier is started and allowed to pump until test pressure is reached. During the pumping, air pockets should be removed by bleeding at high points in the line, or may be flushed with high flow.

ROTARY GEAR-TYPE PRESSURE INTENSIFIER

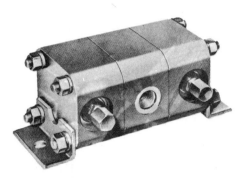

FIGURE 11-16.
Two-Section Rotary Flow Divider.

Figure 11-16. A rotary flow divider consists of two hydraulic pumps or hydraulic motors with shafts coupled so they rotate as one unit. The elements are enclosed in a housing with one inlet port connected to both sections, and with an outlet port from each section. The shafts are not brought out through the housing. The primary purpose is to accept a flow of oil, as from a hydraulic pump, and to divide this flow to two independent branch circuits in the same hydraulic system. If both sections have equal displacement, the inlet oil is split into two equal flows. If one section has a larger displacement than the other, the oil is split in proportion to the relative displacements. Most rotary dividers are constructed from gear-type elements and are available in two, three, or four sections. The model shown in this figure has two sections with a pressure relief valve built into both sections. For a more detailed description of rotary flow dividers, please refer to Volume 3, Chapter 11.

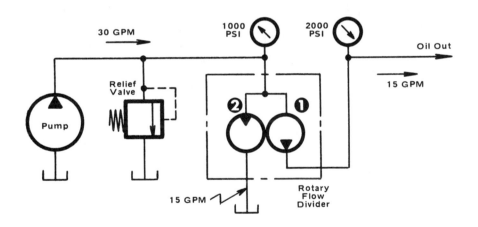

FIGURE 11-17. Pressure Intensification With a Rotary Flow Divider.

Pressure Intensification. Figure 11-17. This circuit illustrates the use of a 2-section rotary flow divider as a pressure intensifier. Both sections have the same displacement.

In this example the system pump is delivering 30 GPM and the pump relief valve is set for 1000 PSI. The divider has a common inlet feeding pump oil to both sections. The inlet flow of 30 GPM is equally divided into two 15 GPM flows. Section 1 delivers 15 GPM to the load circuit. Section 2 outlet is connected directly to tank with minimum flow resistance. Therefore, Section 2 acts as a hydraulic motor, developing torque and transferring it internally to Section 1, adding to the pressure capability of Section 1.

Remember that a flow divider, whether used for flow dividing or for pressure intensification, is always a constant horsepower device. The horsepower available at its inlet will be transferred through to its outlet at the same level (minus flow and friction losses). When the divider is used as a pressure intensifier, the inlet horsepower can be converted into a higher pressure at a lower flow. In the previous example, the power from the pump is at 30 GPM and 1000 PSI. It is converted in the flow divider into an output power of 15 GPM and 2000 PSI, which represents the same horsepower. Section 2 takes its share of the fluid, 15 GPM, and converts it into mechanical power, then transfers this power internally to Section 1 to give that section a higher pressure capability. The losses in this type of flow divider are about the same as in a gear pump, a typical value being about 15% of the input horsepower.

A pressure relief valve should always be used on the pump line, and adjusted so the output section of the flow divider will not be operated above its manufacturers pressure rating. A relief valve is not needed on the outlet when used as an intensifier, but should always be included if it is used in its normal mode as a flow divider.

The usual purpose for a rotary-type pressure intensifier is to deliver a continuous flow of oil at an amplified pressure into one branch of a system in which the other branches are operating at a lower pressure. It is seldom used in a single-branch system. A better plan for such a system would be to select a main pump which has the pressure rating needed.

Rotary intensifiers are used only for oil-to-oil applications; they are not suitable for compressed air. They are ordinarily best suited for different types of applications than the piston-type intensifiers previously described. For example, they can deliver only a moderately intensified pressure which is not sufficiently high for many applications using piston-type intensifiers.

Higher Ratios. Figure 11-18. Greater intensification is possible when using flow dividers of unequal displacement. Section 1, in this figure, has one-half the displacement of Section 2. The ratio of intensification to flow is proportional to displacements in the two sections. Using the same example of 30 GPM pump flow

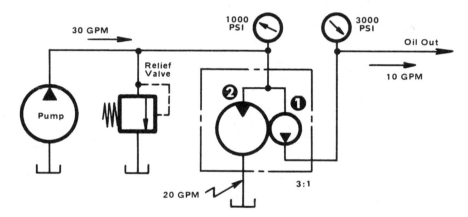

FIGURE 11-18. *Using a Flow Divider With Unequal Sections.*

and a relief valve setting of 1000 PSI, this circuit can deliver up to 3000 PSI and a flow of 10 GPM.

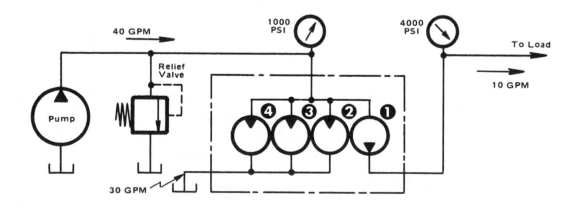

FIGURE 11-19. *Pressure Intensification With a 4-Section Rotary Flow Divider.*

Other Variations. Figure 11-19. Using a 4-section flow divider, an intensification of 4:1 is possible by connecting three of the sections to tank and delivering 1/4th of the pump flow from the remaining section at a pressure four times the pump relief valve setting. By connecting two of the sections to tank and delivering from the remaining two sections, an output of 1/2 the pump GPM at twice pump pressure is obtained.

APPLICATION NOTES ON PRESSURE INTENSIFIERS AND AIR/OIL SYSTEMS

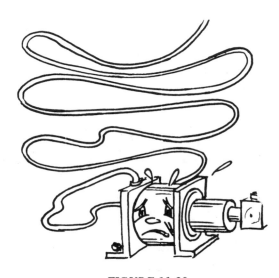

FIGURE 11-20.
The maximum pumping rate is reduced when using small feeder lines of long length.

FIGURE 11-21.
Relieving-type pressure regulators should be used on long-holding applications.

1. Air Supply. On intensifiers driven by a shop air supply the plant air supply including the air compressor must have sufficient capacity to keep up with the air demand where the intensifier will be running for long periods. Full pressure must be maintained on the inlet of the intensifier if maximum pumping rate is to be maintained.

Air line filter-regulator-lubricator units should be of a size at least equal to the intensifier inlet port size. This will usually be 1/2" NPT. For safety the regulator should be the self-relieving type. A non-relieving type with a leaky poppet could allow a slow rise in pressure to a dangerously high level on an intensifier used for a holding application. Be sure to read operating instructions on the intensifier to find whether an air line lubricator is recommended. Some intensifiers are internally pre-lubricated with a moly-type grease and the manufacturer recommends against adding any lubricating oil to the air supply to avoid washing out the permanent lubricant.

Any valves used in the air supply should be generous in size to minimize air flow pressure loss which would reduce maximum pumping speed. Do not use speed control valves in the air supply to an intensifier. If necessary to reduce speed, do it by reducing supply pressure with a pressure regulator.

Air supply lines from the main distribution header into the intensifier should be kept as short as practical and should be large in size, at least as large as the inlet port size on the intensifier.

Exhaust air from the intensifier should be discharged directly to atmosphere through an air muffler. Long or restricted exhaust lines affect the maximum pumping rate even more than small supply lines. Too much back pressure in the exhaust line may cause some intensifiers to stall.

220

2. Intensified Oil Circuit. It almost goes without saying that all components in the high pressure hydraulic side of the system must be rated for the intensified pressure level, including pipe, tubing, and hose. Because of the extra hazard at high pressure, components and piping materials should have a safety factor of at least 6 to 1.

The pipe or tubing diameter to carry a flow of high pressure oil is not critical. Relatively small diameter lines usually work all right; a higher pressure loss can be tolerated than would usually be acceptable in a low pressure system. One exception would be lines which also carry a high flow at low pressure during a pre-fill period.

Seals in all hydraulic components should be as leaktight as possible. Leakage rates which are tolerable in a lower pressure system may be unacceptable in a high pressure system not only because the actual volume of slippage is greater at high pressure but it represents a higher percentage of the total flow. Seals, especially in cylinders which separate driving air pressure from intensified hydraulic pressure must be leaktight to minimize mixing of the two fluids.

At any place in the system where air and oil are in direct contact at 100 PSI or higher, there will be excessive mixing of the fluids. Air will dissolve in the oil and oil will vaporize into the air. The air discharge from such a system should be run through a re-classifier and the oil precipitated before being discharged into the atmosphere. Not only does oil vapor in the atmosphere cause untidiness around the machine, it is a human health hazard if breathed.

Use a minimum of valving in the intensified pressure circuit. Especially avoid the use of spool-type valves at high pressure (over 3000 PSI). Leakage becomes excessive and the spool sticks in the bore if left under full pressure for more than brief periods. Most of the valving can be done in the air circuit.

3. Pumping Oxygen. If any trace of hydrocarbon oil is allowed to mix with oxygen, the heat of compression may cause the mixture to explode. Never pump oxygen with an intensifier not specifically designed for this service. The intensifier must be constructed so that the lubricated rod of the input section can never enter the oxygen area.

4. Oil Compressibility. Table 11-1 shows the per cent reduction in volume of hydraulic oil under pressure at room temperature. Some of the air which dissolves in the oil while it is under pressure will be released as the oil pressure is released and the entrained air will form bubbles in the oil. The system reservoir should be sufficiently large that with the reduction of volume under compression the intensifier inlet will not be cavitated and the bubbles can dissipate before the oil is again compressed.

The compressibility of water is less than that of oil, and it does not entrain air quite to the same extent. As a rule-of-thumb we suggest figuring the reduction in volume of water at 1/3% decrease per 1000 PSI.

Compressibility characterisitcs may affect the choice of a fluid for a pressure test. A fluid which has lower compressibility will have less "explosive" effect in event of rupture of the test vessel. Thus, water would be a better choice than oil from this

TABLE 11-A

Oil Compressibility

Pressure, PSI	% Reduction in Volume
1000	0.5%
2000	1.0%
4000	1.9%
6000	2.7%
8000	3.4%
10,000	4.1%
12,000	4.8%
14,000	5.4%
50,000	12.0%

standpoint. Still better would be glycerine or one of the synthetic fire resistant fluids. Note: Emulsions such as water and oil are not satisfactory for pressure intensification, as the two fluids separate under high pressure. New proprietary fluids are now being marketed which have better characteristics for intensification than water or petroleum oil, and they almost eliminate fluid mist from the exhaust of air/oil systems.

5. Oil Viscosity at High Pressure. The viscosity of liquids increases under pressure and this should be·considered when selecting a liquid for use in a pressure intensifier.

Ordinary 10W motor oil and 150 SSU hydraulic oil will congeal at about 35,000 PSI. MIL-5606 oil will remain liquid up to about 100,000 PSI. This increase in viscosity presents problems in pumping; the liquids will flow with difficulty, or not at all, through small passageways in gauge snubbers, small diameter tubing, check valves, and flow control valves. Pressure gauges may not read accurately or may not read at all.

12

Combination

AIR-OVER-OIL APPLICATIONS

Circuits which use a combination of compressed air and hydraulic oil offer certain advantages on some applications over the use of a single medium. These circuits are often referred to as "air-over-oil" circuits. Compressed air furnishes cylinder force but oil metering is used to control speed.

One or two closed tanks are employed, with compressed air entering at the top to push hydraulic oil out the bottom and to the cylinder. Because of the relatively low pressure, a separator between the two fluids is not ordinarily

required. Although this method does not give an increase in power over straight compressed air, it does give certain advantages where the power of an air cylinder is sufficient but where better control is required. Some of the advantages are:

(1). The cylinder will feed more evenly, with less chatter, at slow rates of travel.

(2). Better "hold-back" control over "lunge" if a tool breaks through the work.

(3). A cylinder can be stopped accurately, and can be held in a mid-stroke position with air-over-oil circuits designed for this purpose. On the other hand, if a straight air cylinder must be stopped in mid-stroke, it cannot be stopped accurately at a desired position, and after stopping it will not hold this position if there is any change in load resistance against the piston rod.

A good example of an air-over-oil application is a machine tool being moved on a slide. For rapid motion an air cylinder, within its limit of power, will move the tool smoothly between positive stops at each end of its stroke. But for moderate or slow speeds an air cylinder may give erratic, jerky, or intermittent motion. Where the greater power of a straight hydraulic system is not required, an air-over-oil system will, in many cases, give a smoothly controlled feed with accurate stops at intermediate points in the cylinder stroke.

The jerky motion of an air cylinder when moving at slow speed is caused by the large area of friction between the slide and the ways on which it moves. To get an air cylinder started, the pressure behind its piston must build up sufficiently to move the slide against static or breakaway friction. As soon as the load starts to move, friction resistance decreases and the slide moves rapidly ahead a short distance and stops, waiting for the air pressure to catch up. This action may occur repetitively at a rapid rate, and the cylinder may move in a series of jerks. Even though the sliding surfaces are well lubricated, this action will usually occur, due to the compressibility of the air, on any mechanism where a sliding surface of large area is involved.

Caution! An air-over-oil arrangement is not a cure-all for every case of jerky cylinder movement. It does give a great deal of increased stability, sufficient in most cases, and is more economical than converting to a straight hydraulic system. If converting from straight air to air-over-oil operation, the air cylinder should have leaktight piston seals to avoid rapid loss of hydraulic oil out to atmosphere.

Although air-over-oil operation gives good results on many cases of slow moving, high friction loads, straight hydraulics must be used in some cases. For example, when the cylinder is pushing a metal cutting tool the greater stability of a straight hydraulic system is usually required.

AIR-OVER-OIL PRINCIPLE

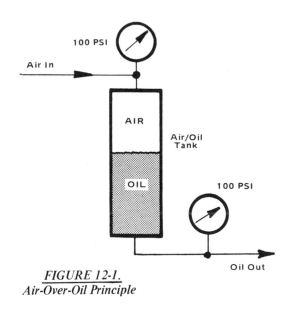

FIGURE 12-1.
Air-Over-Oil Principle

Figure 12-1. Remember that an air-over-oil system is basically a compressed air system, with a trapped volume of oil for smooth metering of travel speed. A straight air system can be converted to an air-over-oil system by interposing a pressure tank in the feed line from 4-way valve to cylinder. A shop air line pressure of 100 PSI can be converted to 100 PSI hydraulic oil pressure which can be metered to control cylinder speed. The working principle is simply that any level of air pressure applied to the top of the tank will force oil out the bottom of the tank with no loss in pressure.

For satisfactory operation the pressure tank must be designed and sized properly, and suitable components selected and mounted in the proper positions. These details will be described in this chapter.

AIR-OVER-OIL WORKING SYSTEMS

Single-Tank System. Figure 12-2. Only one pressure tank is necessary if metered speed is to be in only one direction, with rapid cylinder return. The arrangement shown in this diagram is for metered extension travel, with meter-out flow control Valve 2. For metered speed only on retraction, the pressure tank and flow control valve would be placed in the line to the blind end of the cylinder.

Directional 4-way Valve 1 is a conventional air valve of the same type and size which would normally be used on a straight air circuit for the same size cylinder.

Valve 2, flow control, includes a needle valve internally by-passed with a check valve for free flow in the reverse direction. It is connected in a meter-out mode to meter the flow of oil pushed out of the rod end of the cylinder as the piston advances. Because of the low operating pressure in an air-over-oil system, this valve should be selected one to two sizes larger than the port size on the cylinder, and bushed down to fit the plumbing. If selected the same size as the cylinder ports, the cylinder return speed may be too slow because the internal orifice of the check valve is much smaller than the connection size of the valve. Either a steel or brass body valve may be used.

Important! The pressure tank of an air-over-oil system should be installed at a higher elevation than the cylinder, and all plumbing should slope slightly upward toward the tank to facilitate escape of trapped air. All connecting lines in the oil section should be sized larger than would normally be used for compressed air; they should be as short and direct as possible for maximum return speed of the cylinder.

To operate the system, the operator shifts Valve 1 to start the cylinder forward. This connects air pressure behind the piston. As the piston advances, oil is pushed out the rod port and is metered through Valve 2 for the desired speed.

To retract the cylinder, the operator reverses Valve 1. This puts air pressure on top of the pressure tank, forcing oil out the bottom port of the tank, through the check valve in Valve 2, and into the rod port of the cylinder.

Double-Tank System. Figure 12-3. If oil metering is required in both directions of cylinder travel, two pressure tanks are required, one in each line connecting 4-way valve to cylinder. In other respects the operation is the same as described above, with Valve 2 controlling cylinder extension speed and Valve 3 controlling retraction speed.

The cylinder used in any air-over-oil system should have leaktight piston seals to prevent mixing of the oil and air. However, even with tight seals there may

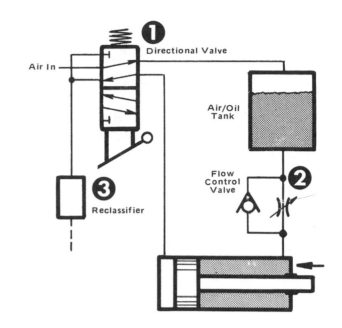

FIGURE 12-2. Single-Tank System.

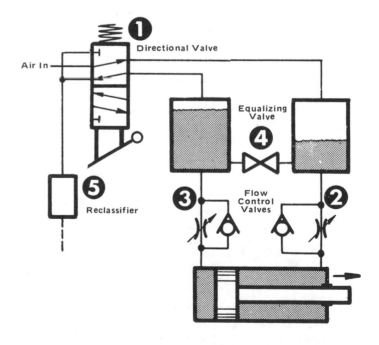

FIGURE 12-3. Double-Tank System.

be a gradual transfer of oil from one tank to the other. If necessary to transfer oil from one tank to the other to bring the system back into balance, shift Valve 1 to pressurize the tank which is too full and open needle Valve 4 until the oil equalizes.

Oil Mist Problem. Where air is in direct contact with hydraulic oil under pressure as in the single and double-tank systems just described, oil evaporates into the air space and is exhausted to atmosphere through the 4-way valve as the cylinder retracts. This is a human health hazard if breathed over a long period of time. The exhaust air should be piped through a muffler/re-classifier to condense and remove the oil vapor. A re-classifier is shown as Item 3 in Figure 12-2. A still better solution to the oil vapor problem is to use a double section cylinder instead of a pressure tank. This is illustrated in the next diagram. There is no oil vapor emitted because the two fluids are kept separated in the two sections of the cylinder.

DOUBLE-SECTION CYLINDER

Figure 12-4. This tandem-type cylinder is available from several manufacturers. It has two sections sealed off from one another. Each section has a piston and two fluid ports. Both pistons are joined to one piston rod. In addition to the special uses described here, the two sections can be operated in parallel from the same fluid source to produce approximately twice the force of a standard cylinder having the same bore. Or, the sections can be operated separately with the same or different fluids in each section.

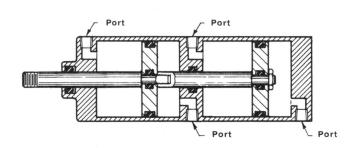

FIGURE 12-4. Double-Section Cylinder.

Air/Oil System With Double-Section Cylinder. Figure 12-5. The rear section is powered from a shop air line through a 4-way directional Valve 1. The front section is unpowered and is used only for oil metering to control cylinder speed. Flow control Valve 2 meters oil from the front to the back side of the oil piston as it advances. On the return stroke, oil can transfer freely back to the front side of the oil piston through the check valve which is a part of flow control Valve 2.

The front section, rather than the rear section, must be used for metering because areas on opposite sides of the piston are equal, and any oil pushed out from in front of the advancing piston will find an equal space opened up behind the piston to receive it.

Over a period of operation a small amount of oil will be lost through rod seal wipe-off, either to atmosphere or into the

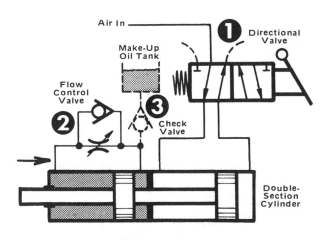

FIGURE 12-5. Air/Oil Circuit Using Double-Section Cylinder.

air section. This will create a slight vacuum in the front section. Make-up oil is then drawn in from the reserve tank through the check valve. The check valve should have a cracking pressure of 1 PSI or less.

SYNCHRONIZATION WITH DOUBLE-SECTION CYLINDERS

Figure 12-6. This is one of the few accurate ways of synchronizing two air cylinders so they will travel in step throughout the length of their stroke. The two cylinders must have the same bore and stroke and the same piston rod diameter. They are the double-section type previously described.

The rear sections operate on compressed air and produce the power. The front sections are filled with oil and provide a means of speed control as well as synchronization. As the cylinders move, the oil is cross ported from one cylinder to the other to keep them traveling together. Flow control Valves 2 and 3 meter the oil flow for accurate fine feed in each direction.

Over a period of time a small amount of oil may transfer from one loop to the other through piston seal leakage or to atmosphere because of rod wipe-off. The small vented reservoir contains a supply of replenishing oil which may be admitted into either cylinder, as needed through needle Valves 4 and 5. The need for replenishment will be evident when one cylinder lags slightly behind the other when first started.

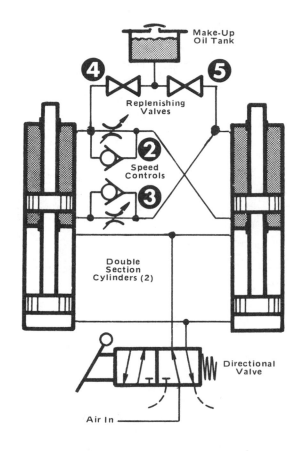

FIGURE 12-6. Cylinder Synchronization.

To replenish, completely retract the cylinders, then open both needle valves for a short time. Then, close the needle valves and extend both cylinders. Again open both needle valve for a short time. The make-up tank should be installed at a higher elevation than the cylinders so oil can flow by gravity into the cylinders.

If the cylinders are carrying unequal loads, pressure intensification, up to twice the air inlet pressure, could develop in the oil loops. Therefore, the pressure rating of the cylinders and other components in the oil loop should be at least twice the inlet pressure.

Figure 12-7(A). This is a series loop between two cylinders which have the same bore and stroke, and which have double piston rods of the same diameter. If the cylinders have leaktight piston seals, synchronization is very accurate. The oil loop from the rod end of one cylinder to the blind end of the other provides a means of equalizing the loads if one cylinder is carrying more than the other.

Air, at shop air pressure, provides the power for the cylinders. However, since the cylinders are connected in series, they must divide the inlet pressure in proportion to the load carried by each. If one cylinder carries twice the load carried by the other, it will receive 2/3rds of the inlet pressure. The

227

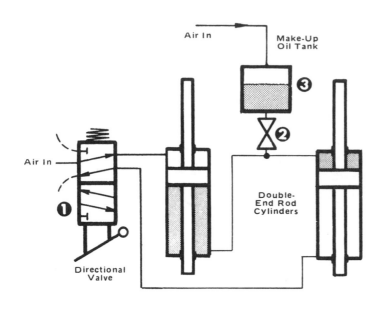

FIGURE 12-7(A). Synchronization for Air Cylinders.

total force produced by **both** cylinders cannot exceed the force which could be produced by only one of them if it were operating by itself on the inlet pressure level.

Reservoir 3 contains make-up oil for replenishing any oil which escapes from the oil loop through seal leakage. Full air pressure is maintained on top of the reservoir but needle Valve 2 is kept closed except for replenishing. The need for make-up oil will be evident when one cylinder does not make a full stroke. For replenishing the oil loop, first run the cylinders to one end of their stroke and open Valve 2 until both cylinders square up against their end caps. Then close Valve 2. For this circuit to be successful, cylinders must be used which have very tight piston seals.

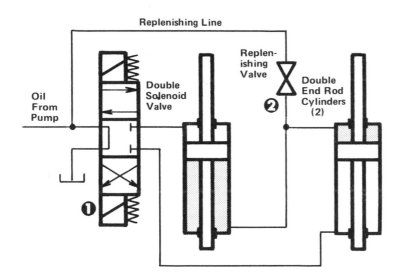

FIGURE 12-7(B). Synchronization for Hydraulic Cylinders.

Figure 12-7(B). For synchronizing hydraulic cylinders, make-up oil, when needed, is obtained from the hydraulic pump line by opening Valve 2 when Valve 1 is in one of its side positions and the cylinders are stalled against positive stops or their own end caps. Replenishing may be indicated when one cylinder fails to make its full stroke.

Using this basic circuit, automatic replenishing valves can be added, for example to be actuated by a limit switch when the cylinders retract, to permit entry of low pressure oil into the oil loop between the cylinders.

AIR POWERED PLATFORM LIFT

Figure 12- 8. Hydraulic rather than air cylinders should be used to operate any lifting platform which must be stopped in mid position for adding or removing any part of the load. An air cylinder would sag if load were added or would rise if load were removed. Lifts operating with air cylinders are suitable on applications where load is never added or removed in mid stroke.

A shop air line provides motive power for the lift in this diagram. Air is not applied directly to the cylinder; it is applied to the top of an air/oil pressure tank and the cylinder is raised by hydraulic oil forced out of the tank by air pressure. A hydraulic valve, 2, must be provided in the oil loop and must be closed before load is added or removed from the platform. Valve 2 is also a means of metering the oil for precise positioning of the cylinder.

Valve 1 is a 4-way, 2-position air valve with one cylinder port plugged. It can be a manual or solenoid type. Usually a 1/4 or 1/2-inch size is sufficient.

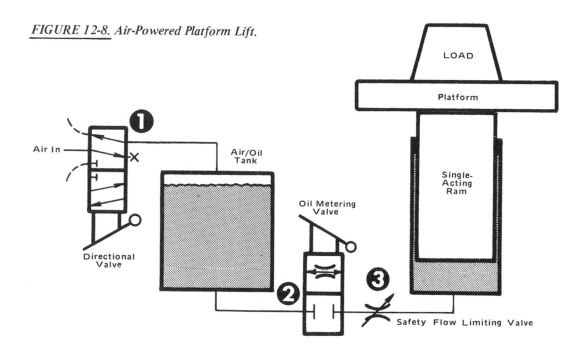

FIGURE 12-8. Air-Powered Platform Lift.

Valve 2 is a 2-way hydraulic valve, manually operated, which has good metering characteristics and which can be completely closed. For example, a low pressure rotary shear seal type gives good results. Valve 3 is a needle valve adjusted, for safety, to a maximum rate of descent to prevent the operator from accidentally allowing the platform to drop too quickly.

With the platform down the load is placed on it. The operator keeps Valve 2 closed and shifts Valve 1 to bring the tank up to full pressure. He should then gradually open Valve 2 and control lifting speed by metering with it. At the desired height he closes Valve 2 and locks oil in the cylinder while load is added or removed. To lower the platform, the operator first shifts Valve 1 to exhaust air pressure from the tank, then uses Valve 2 to control the rate of descent.

STABLE AIR CIRCUIT

Figure 12-9. An air cylinder is usually not stable when stopped at an intermediate point in its stroke. When its air supply is cut off it does not immediately stop; it continues to travel a short distance until a balance is reached between the pressure behind its piston and the mechanical load resistance against its piston rod. This overtravel is indeterminate, depending on load, cylinder speed, and other factors. Then too, if the load against its piston rod should change while it is stopped, it will start up and move to a new position of balance.

This circuit shows a means of holding the cylinder against further movement after it has stopped in mid stroke. An oil loop is used for locking the cylinder piston in position. Motivating pressure for the cylinder is from the shop air line through a pair of air/oil tanks.

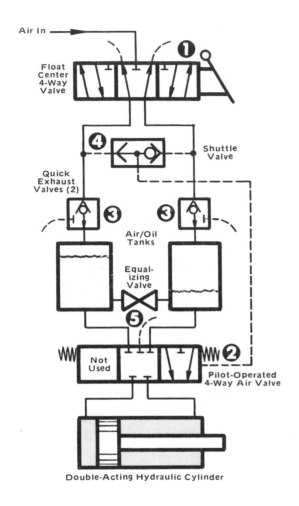

FIGURE 12-9. Stable Air Circuit.

Directional Valve 1 must have a float center spool in which both cylinder ports are vented to atomosphere when the spool returns to neutral.

Valve 2 is a 3-position, double piloted 4-way hydraulic valve, spring centered to a closed neutral. One side position of this valve is not used. Pilot pressure shifts the spool from neutral to a working side position. Loss of pilot pressure allows the spool to go to its closed neutral, locking oil on both sides of the cylinder piston. The opposite pilot port is left open; it must not be plugged.

To move the cylinder piston forward or back, Valve 1 is shifted to a side position. Air pressure connected to the top of one of the tanks produces an equal hydraulic pressure to move the piston. Shuttle Valve 4 receives air pressure when the 4-way valve is shifted to either side position, and connects this pressure to the pilot port of Valve 2. This valve then shifts to open the cylinder ports, allowing the cylinder piston to move.

If speed controls should be needed they can be installed between the cylinder ports and Valve 2. Needle Valve 5 is provided to enable the operator to re-adjust the oil levels between the tanks if necessary.

The quick exhaust valves vent the air/oil tanks to atmosphere very quickly when the main 4-way valve is centered. This shortens the time it takes for the cylinder to come to a stop.

CONSTRUCTION OF AIR/OIL TANKS

Figure 12-10. Pressure tanks can be made in almost any shop using a suitable length of standard pipe, 2″, 3″, 4″, or larger in diameter, threading both ends and using standard pipe caps as closures. The top cap should have a removable pipe plug for insertion of a dip stick to check oil level. An oil level gauge attached to the side of the pipe is a preferred way of keeping a check on oil level, if one is available which is rated for the system pressure. Sometimes an air cylinder, with piston and rod removed, can be converted to a pressure tank, or ready built tanks are available in the smaller sizes.

Tanks should be carefully calculated for volume. If excessively large, air consumption will be unnecessarily high and system response will be slow. If sized too small, they will have to be replenished more often, and oil may get into the air system. We recommend a tank volume of 1½ times the volume displaced by the cylinder to be operated.

Tanks should be constructed with sufficient strength to withstand full air line pressure with a safety factor of 4:1 or 6:1. The following points of construction should be followed:

(1). The top must be removable for installing and servicing the internal diffusers. It should have a removable plug for filling and for dipsticking the oil level without removing the tank cover.

230

(2). The air port inside the tank should be covered with a diffuser to prevent a high velocity jet air stream from causing excessive turbulence in the oil. A standard air exhaust muffler should serve quite well if it is open on the sides and closed on the bottom.

(3). The tank should be relatively tall with respect to its diameter. This provides a greater distance separation between air port and oil port, and reduces transfer of oil into the air stream.

(4). The oil connection should also be protected with a diffuser to prevent the return oil from jetting up into the air port. A suitable diffuser can be made by drilling a series of radial holes in a standard pipe cap as shown in the figure, and screwing this on to a short pipe nipple welded into the bottom cover of the tank.

(5). To reduce turbulence inside the tank, enlarge the oil connecting pipe for a foot or so before it connects to the tank. This will reduce the velocity of the oil as it returns to the tank.

(6). There will inevitably be mixing of the two media inside the tank. Some of the air will dissolve in the oil and some of the oil will vaporize into the air and will be carried to atmosphere through the 4-way valve exhaust. It will condense into oil mist, and this is a health hazard if breathed. A muffler/reclassifier should be connected to the valve exhaust port(s) to condense the oil vapor and remove it.

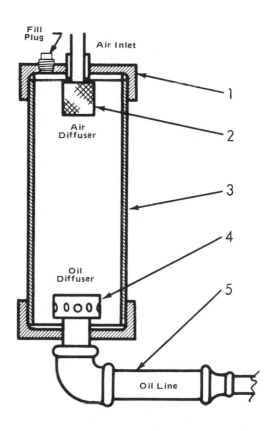

FIGURE 12-10. Construction of Air/Oil Tank.

DESIGN NOTES FOR AIR/OIL SYSTEMS AND PRESSURE INTENSIFIERS

Information in this section applies to systems which use both hydraulic oil (or water) and compressed air. For additional information refer to Chapter 11.

(1). Mounting Position for Components. Figure 12-11. Components such as cylinders and intensifiers which have air on one side of a piston and oil on the other side, should be mounted in such a position that air which may leak across piston seals and get into the oil can escape to the air/oil tank or other high point where it can be purged from the system. For example, a cylinder should be mounted vertically, if possible, with the air end up. If it must be mounted horizontally it should be mounted with ports up. The point is to eliminate pockets in the plumbing where air could accumulate and from which it could not be purged except by opening the plumbing.

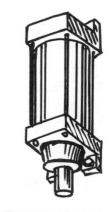

FIGURE 12-11. Preferred Position for Work Cylinder.

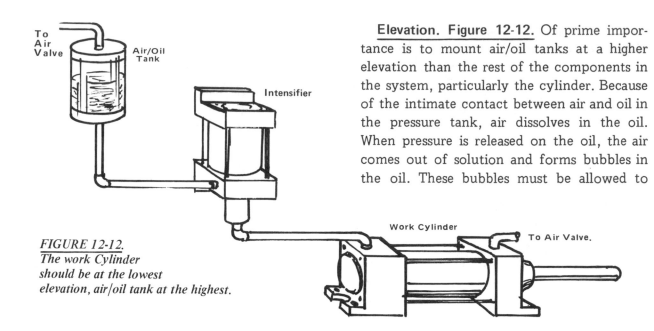

Elevation. Figure 12-12. Of prime importance is to mount air/oil tanks at a higher elevation than the rest of the components in the system, particularly the cylinder. Because of the intimate contact between air and oil in the pressure tank, air dissolves in the oil. When pressure is released on the oil, the air comes out of solution and forms bubbles in the oil. These bubbles must be allowed to

FIGURE 12-12.
The work Cylinder
should be at the lowest
elevation, air/oil tank at the highest.

escape to atmosphere through the air/oil tank. If the system is not laid out so the air can return by gravity to the tank, the system will perform poorly and will have to be purged of air manually by periodically opening shut-off valves located at one or more points in the plumbing. Plumbing runs should be sloped slightly upward toward the air/oil tank, and downward loops must be particularly avoided. On existing systems which have been improperly laid out and where air accumulation is a problem, the use of special synthetic fluids in place of hydraulic oil may reduce the amount of air which can dissolve in the liquid.

(3). Leaktight Piston Seals. As previously emphasized, the choice of a cylinder having leaktight piston seals is one of the more important factors in successful operation. Usually cup packings are best. O-ring seals which are leaktight at higher pressures do not seal as tightly at pressures of 100 PSI and below. Any seepage across the seals will necessitate frequent replenishment or re-adjustment of the oil level and will produce an excessive amount of oil vapor discharged from the system.

(4). Plumbing Sizes. In low pressure oil lines such as air-over-oil circuits, use oversize lines and components. Because of the low pressure, an air-over-oil system normally operates much slower than an air system at the same pressure. To obtain a reasonable speed, restrictions in the oil lines must be reduced to a minimum. Flow control valves, check valves, and plumbing runs should be one to two sizes larger than the cylinder port size. The return line to the pressure tank, especially, should be large in size to reduce velocity of return oil and reduce turbulence in the tank.

(5). Oil Mist. In an air-over-oil system where there is intimate contact between air and oil at pressures up to 125 PSI, a considerable amount of oil vapor is produced which condenses to mist when discharged to atmosphere. Exhaust air should be run through a muffler/reclassifier to condense and remove it.

(6). Air Bleeding. To reduce the amount of manual bleeding required to keep air out of the oil circuit these points are important: (a), Physical placement of pressure tank higher than other components; (b), Upward slope of plumbing toward the pressure tank; (c), Proper mounting of the cylinder with air end up; (d), Large return line to pressure tank to reduce tank turbulence; (e), Adequate volume in the air/oil tank; (f), Leaktight piston seals in cylinder; (g), The use of a glycol fluid instead of oil to reduce amount of air dissolved in liquid.

13

Application Ideas for Cylinders

In addition to the ideas for cylinder applications in this chapter, many other typical applications will be found throughout the book, and especially in Chapter 1.

MATERIAL HANDLING WITH CYLINDERS

Gravity Conveyor. Figure 13-1. Heavy material packaged in rigid containers can be transported quite a distance on unpowered conveyors, sloped in the direction of movement. Operation can be made automatic with limit switches located to start the elevating cylinder when a container approaches.

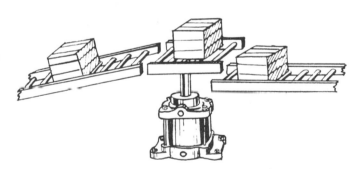

FIGURE 13-1. Gravity Conveyor.

Turn-Over. Figure 13-2. Rotary actuators are ideal for turning parts over 180°. Cylinders can also be employed for the same purpose by using two or more of them. Rather than straight air operation, air-over-oil circuitry as in Chapter 12 may give better results because it can provide the "hold-back" which may be needed on heavy parts as they are turned over center. One cylinder will turn the parts 90° or two cylinders, as shown, may be needed for 180° turn-over. In practice, a guard (not shown) should be placed to prevent a second object from entering the turn-over area until both cylinders retract.

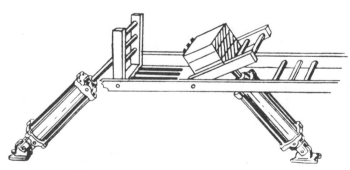

FIGURE 13-2. Turn-Over With Cylinders.

FIGURE 13-3. Sorting Parts.

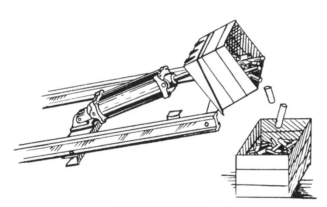

FIGURE 13-4. Dumping.

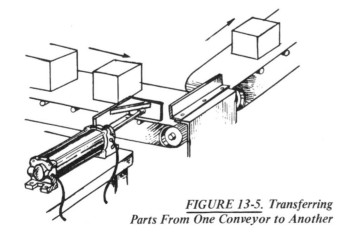

FIGURE 13-5. Transferring Parts From One Conveyor to Another

Sorting. Figure 13-3. A cylinder, controlled with a solenoid 4-way valve can be used for sorting parts according to weight, size, color, or content. For example, a weighing scale can check proper weight; an electric eye can check for proper filling of bottles; a photoelectric system with color filters can sort according to colors. The signal from any one of these devices can be applied to the solenoid valve which controls cylinder action.

Dumping. Figure 13-4. There are many jobs handling small parts, granular material, or even liquids, where dumping can be done with cylinders. A manual control valve can be used if the operator must have fine control over cylinder speed. Cylinder speed can be varied by throttling. On quick dump applications, a solenoid 4-way valve could be used.

Automatic dumping can be iniated when a certain weight has been reached as signalled by a scale-mounted switch, or when a certain count has been reached on an electric counter. Suitable safety precautions should be taken to protect personnel on automatic dumping operations. On most dumping jobs the cylinder should dwell before retracting to allow all parts to dump. Electric timers may be designed into the electric circuit to control dwell time. Pneumatic timers may also be employed on pneumatic applications.

Transferring. Figure 13-5. A cylinder, usually an air cylinder, can transfer moving objects from one conveyor to another operating at right angles to the first.

The transfer operation can start in response to a limit switch, proximity switch, electric eye, sonic or fluidic sensor according to the nature of the product and the fluid medium used.

If the pusher attached to the cylinder is properly shaped, the moving object can be turned at an angle, even up to 90º, while being transferred. A guard should be placed to prevent a second object from entering the transfer area until the cylinder has retracted.

Pumping. Figure 13-6. The cylinder is the power source for a single-piston reciprocating pump. It can be manually stroked one cycle at a time by an operator, using a 4-way manual valve, or it can be continuously stroked automatically with one of the reciprocating circuits shown in Chapter 4. This arrangement is useful for pumping fluids which are corrosive to ordinary pumps.

The pump body and piston can be shop built from suitable materials such as stainless steel, Teflon, Nylon, or others. Inlet and outlet check valves, also of suitable materials, are installed as in the diagram, with their free flow arrows pointing toward the pump discharge.

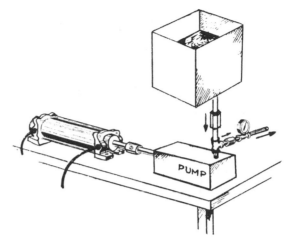

FIGURE 13-6. Single-Piston Reciprocating Pump.

Metering. Figure 13-7. This is also a specially built, single-stroke pump for dispensing an exact volume of liquid on each stroke. The pump body can be constructed of materials compatible with the liquid to be handled. The check valves must also be compatible.

An air cylinder, controlled with a 4-way air valve, manual or solenoid type, furnishes power. A very precise adjustment of cylinder stroke, and of the volume dispensed, can be made with an adjustable collar mounted on the cylinder piston rod to control the length of the return stroke. This method has been successfully used in packaging paint, food products, and various chemicals.

Note: The inlet check valve must have a very low cracking pressure, or the liquid must be fed into the pump under a very low pressure to avoid cavitating the pump.

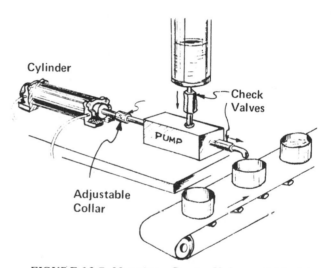

FIGURE 13-7. Metering a Precise Volume of Liquid.

Stacking. Figure 13-8. Each of these two cylinders is arranged to stroke only after a stack of (so many) pieces accumulates. Counting the number of pieces may be done electrically with a stepping relay or by "feeling" with a limit switch to actuate when a certain stack width has been

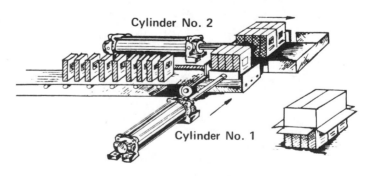

FIGURE 13-8. Stacking or Multiple Counting.

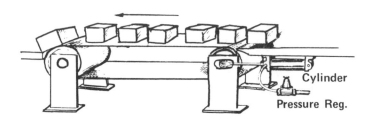

FIGURE 13-9. *Tensioning a Conveyor Belt.*

reached. The desired number of stacked pieces is pushed out by Cylinder 2 to a position where an operator can slip an open shipping container over the entire group.

Belt Tensioner. Figure 13-9. One or more cylinders (usually air cylinders) may be used to maintain a constant tension on conveyor belts or other parts of a machine where a heavy spring might normally be used.

ADDING CYLINDERS TO SHOP MACHINERY

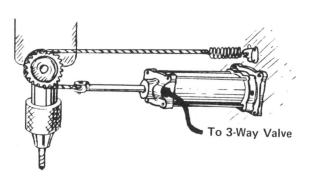

FIGURE 13-10. *Power Feed of a Drill Press.*

Drill Press Feed. Figure 13-10. On production drilling, operator fatigue can be reduced by adding an air cylinder for "crowding" the drill. The regular manual feed lever can be removed, and replaced with a sprocket.

A 3-way valve directs air to the rod end of the cylinder for crowding, and the other cylinder port is vented to atmosphere.

For drilling through-holes, a heavy spring may be needed under a depth stop nut to prevent "grab" when the drill breaks through the work.

A guard should be provided over the sprocket to prevent clothing or fingers getting caught.

Metal Shear. Figure 13-11. One of many ways to power a sheet metal shear is shown here.

In sizing the cylinder, about a 300-lb. force should be provided, since this is approximately the amount an average man could develop by "bouncing" on the treadle.

If the cylinder is mounted elsewhere on the machine, the difference in leverage must be considered, to provide a force of 300 lbs.

Safety guards should be provided to keep the operator's hands and feet out of the machine.

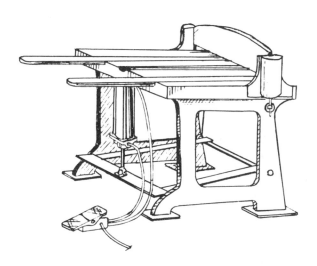

FIGURE 13-11. *Converting a Metal Shear to Power Operation*

Punch Press. Figure 13-12. On mechanically operated punch presses, a small air cylinder can be substituted for the built-in mechanical trip, and can be activated with a solenoid valve or miniature 3-way air valves. Then, the press can

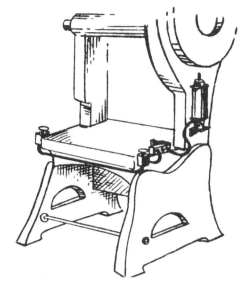

FIGURE 13-12. Electrical Trip for Punch Press.

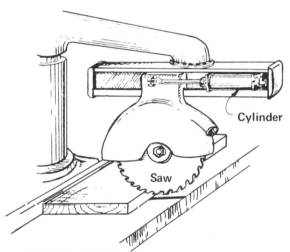

FIGURE 13-13. Power Feed on Radial Arm Saw.

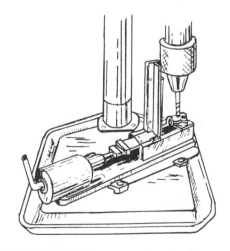

FIGURE 13-14. Cylinder-Operated Slide Feeder.

be tripped with a foot switch, with pushbuttons, or can be arranged to trip from a two-hand safety circuit to protect the operator. Details of the safety circuit are given in the Womack book "Electrical Control of Fluid Power".

Other applications for cylinders around punch presses include material clamps, ejection of finished work pieces, or moving material through the press.

Saw Feeding. Figure 13-13.

A straight air circuit can be used for saw feeding if the saw is on a hinged arm (radial mounting). This is a low friction application, and movement can be controlled with an auxiliary hydraulic checking cylinder for better control of feed and to prevent the saw from being jammed into the material at too fast a speed.

If the saw is mounted on a machine slide, with a relatively large area of frictional contact (breakaway friction), it can be fed with an air-over-oil circuit to eliminate any tendency for erratic or jumpy movement.

A cable cylinder or rodless cylinder has a shorter length for a given stroke, and can be used to good advantage on this application.

Parts Feeding. Figure 13-14.

The cylinder shown in this figure is operating the slide on a device which feeds parts under the drill. In an automatic operation, the feeder cylinder should make a stroke every time the drill retracts. The mechanism should be designed to eject a finished piece on the same stroke which feeds a new part.

The shape of the parts holder should be designed to accommodate the particular part to be machined.

To carry the automation still further, the parts chute could be fed automatically from some type of vibratory feeder.

Other Shop Machinery.

Applications are too numerous to mention, such as moving sheet material or parts in and out of machines, fast or slow feed of grinding or milling tables, clamping parts to be machined, pressure pads or fluid springs on presses.

BENDING AND FORMING

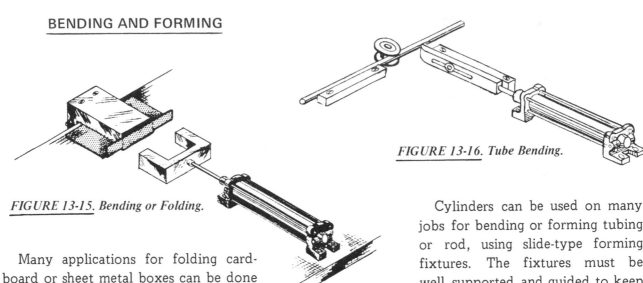

FIGURE 13-16. Tube Bending.

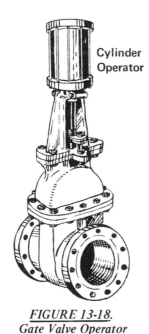

FIGURE 13-15. Bending or Folding.

Many applications for folding cardboard or sheet metal boxes can be done with simple fixtures powered with small air or hydraulic cylinders.

Figure 13-17. Radius-type benders are often used to form tubing and rod. They can be powered with cylinders. The cylinder must be hinge mounted so it can swing with the bender.

Cylinders can be used on many jobs for bending or forming tubing or rod, using slide-type forming fixtures. The fixtures must be well supported and guided to keep side load off of the cylinder rod.

FIGURE 13-17. Radius Bending.

VALVE OPERATORS

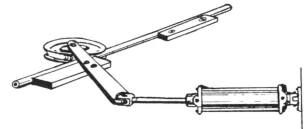

Operator for Rising Stem Gate Valves. Figure 13-18. Due to the wedging action of the gate valve when it closes, it must be closed with considerably less force than is available for breaking it loose to open. Either an air or hydraulic cylinder can be used, depending on the thrust required, for driving the stem. A circuit is shown in Chapter 4 of Volume 1, using a 5-way valve, to develop high force in one direction and a lower force in the other direction.

Cylinder Operator

Plug Valve Operator. Figure 13-19. By means of a simple lever, fabricated to fit the valve plug stem, hydraulic or pneumatic cylinders can be mounted to operate 1/4-turn plug valves of any size, and can be remotely controlled.

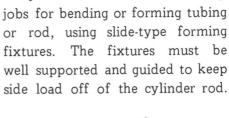

FIGURE 13-18.
Gate Valve Operator

FIGURE 13-19.
Plug Valve Operator

CYLINDERS ON DOORS AND GATES

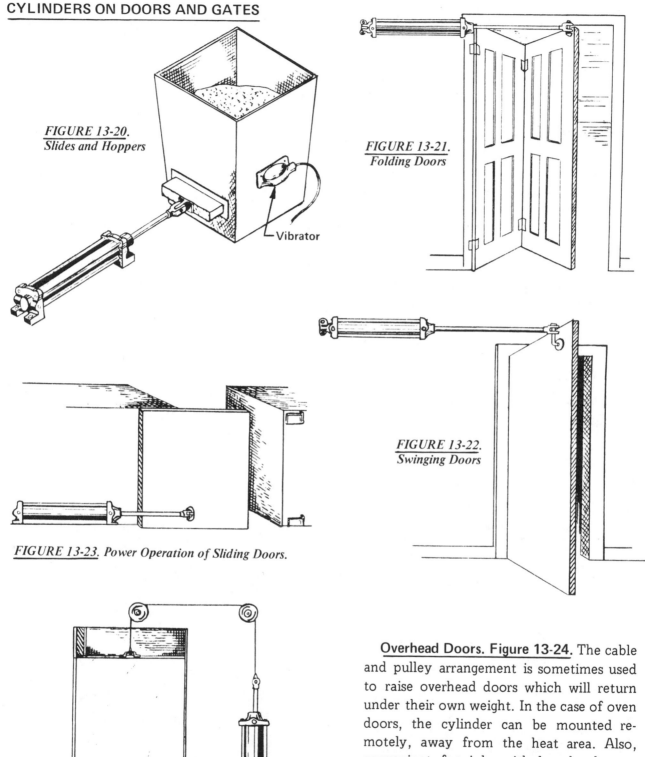

FIGURE 13-20.
Slides and Hoppers

Vibrator

FIGURE 13-21.
Folding Doors

FIGURE 13-23. *Power Operation of Sliding Doors.*

FIGURE 13-22.
Swinging Doors

FIGURE 13-24. Overhead Door Operation.

Overhead Doors. Figure 13-24. The cable and pulley arrangement is sometimes used to raise overhead doors which will return under their own weight. In the case of oven doors, the cylinder can be mounted remotely, away from the heat area. Also, convenient for jobs with low headroom, where there is not enough space to mount a cylinder for a direct lift.

The door can be counter-weighted in the same way as an overhead hoist, by keeping a constant low pressure on the rod end of the cylinder at all times.

ROTARY MOTION FROM CYLINDERS

Lever Action. Figure 13-25. A simple method of producing a limited rotary movement is to connect a cylinder to the end of a lever arm. Angular motion up to 180° could theoretically be obtained, but in practice a practical maximum is about 90°. The cylinder must be hinged on its rear end so it can swing with the lever motion. Speed of rotation will vary with the angle between lever and cylinder.

Maximum torque is obtained when cylinder axis is at right angles to the lever, so the cylinder should be positioned to obtain maximum torque where it is most needed, at the start, at the end, or in the middle of its stroke.

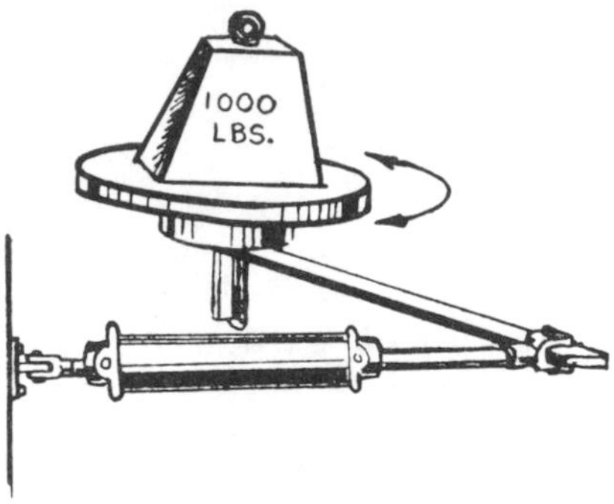

FIGURE 13-25.
Limited Rotation From a Cylinder

The length of cylinder stroke for a given angular swing can be calculated by geometry, or a scale layout can be made. The retracted and extended lengths of the cylinder can be measured. The necessary stroke will be the difference between these lengths. See Chapter 1 for calculating force when lever axis is not perpendicular to cylinder axis.

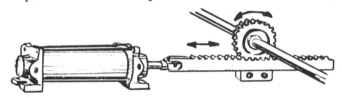

FIGURE 13-26. Rack and Pinion.

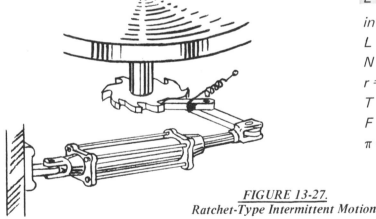

FIGURE 13-27.
Ratchet-Type Intermittent Motion

Rack and Pinion. Figure 13-26. Up to several revolutions are possible, with uniform torque, although torque will be less on retraction due to less piston area.

These formulae may be used for figuring cylinder stroke, pinion turns, and torque:

$$L = N \times 2\pi r, \qquad N = L \div 2\pi r, \qquad T = r \times F$$

in which:
L = Length of cylinder stroke, inches
N = Number of pinion turns
r = Pitch radius of pinion
T = Torque on pinion, inch-lbs.
F = Cylinder force, pounds
π = 3.14

Figure 13-27. Ratchet motion can be obtained with a cylinder and toothed wheel with spring-loaded pawl. It may be necessary to add a friction brake to prevent overrun on high speed indexing.

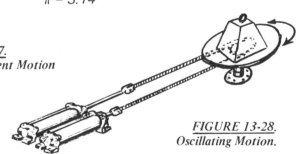

FIGURE 13-28.
Oscillating Motion.

Figure 13-28. Calculations for torque, number of turns on the sprocket, and cylinder stroke can be made with the formulae of Figure 13-26. Each cylinder should have its blind end port vented to atmosphere (or to tank). One 4-way valve connected to their rod ends will operate both cylinders. On massive loads a friction brake may be needed, particularly if cylinders must be stopped in mid stroke.

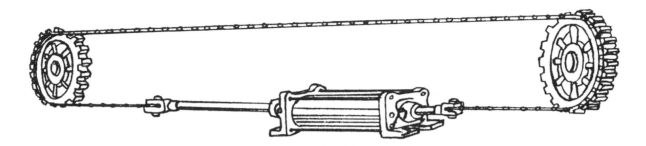

FIGURE 13-29. Rotary Oscillating Motion With Chain and Sprockets.

Chain and Sprocket. Figure 13-29. Several complete turns of the sprocket may be made by using a double-end-rod cylinder as shown. The sprockets may be the same or different diameters, and power can be taken from either or both of them. Select smallest rod offered in chosen bore size to keep net area on piston as large as possible. Use formulae on preceding page to make calculations for torque, thrust, stroke, and number of turns on sprockets.

CYLINDERS FOR WORK CLAMPING

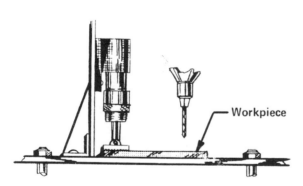

FIGURE 13-30. Drill Press Clamp.

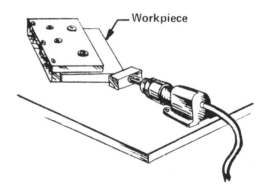

FIGURE 13-31. Drill Jig Clamp.

Cylinders are used in dozens of ways, limited only by the ingenuity of the designer, for holding parts while they are being machined. Automatic "clamp and work" circuits are given in Chapter 3.

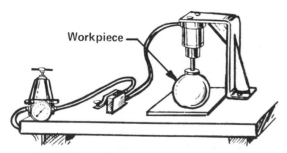

FIGURE 13-32. Clamping Delicate Parts.

Use a pressure regulator or relief valve to keep from crushing delicate parts.

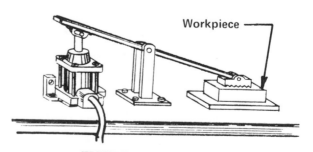

FIGURE 13-33. Offset Clamping.

Sometimes a lever arrangement is more convenient than a direct-acting clamp.

MULTIPLE POSITION INDEXING

FIGURE 13-34.
Positive Indexing at
Four Positions.

Four-Position Indexing. Figure 13-34. This is one method of positive indexing to any one of four pre-determined positions using two air (or hydraulic) cylinders mounted back to back. Each is operated with its own 4-way control valve, usually solenoid type. Please refer to Page 83 for more details of this application.

To obtain four different extensions of the overall assembly, each cylinder must have a different length of stroke. Maximum and minimum extended positions are with both cylinders extended or retracted. The two intermediate positions are with one cylinder extended, the other retracted.

This arrangement is especially useful in air circuitry because a single air cylinder cannot be stopped accurately at a mid-point in its stroke, and when stopped it will not hold this position firmly.

For more than four fixed positions, additional cylinders can be added to the string. On long strings, to prevent cumulative errors due to slight inaccuracies in the precise stroke of individual cylinders, a stroke adjustment collar can be fitted to each cylinder. Total force developed by the entire string, with air pressure on all cylinders at the same time, is that of only one cylinder acting alone since the cylinders are in a series mechanical arrangement.

Double Cylinder. Figure 13-35. Four-position indexing can be done more compactly and at less cost with this semi-standard double cylinder assembly offered by several manufacturers.

Multiple cylinder assemblies like this and the one above are for operating low power mechanisms. The series arrangement of the cylinders gives a total power output equal to only one cylinder acting alone. To obtain higher power, the parallel cylinder arrangement shown below may be preferred.

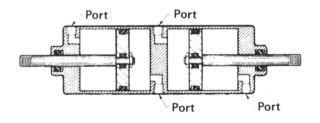

FIGURE 13-35.
Double Cylinder for Four-Position Indexing.

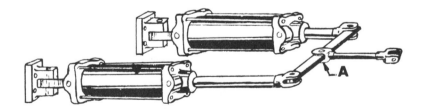

FIGURE 13-36.
Yoked Cylinders for
Multiple Position Indexing.

Yoked Cylinders. Figure 13-36. Four rigidly fixed index positions can be obtained with two cylinders of equal or different strokes. Each cylinder must be controlled with a separate 4-way valve. The force output at Point A depends on the location of Point A along the crossbar, the lever advantage obtained by the crossbar, and the sum of the force outputs of the two cylinders. The output link should be guided at Point A, and the load to be operated should also run in guides. This system must be calculated by mathematics or by making a full scale layout.

CYLINDERS ON PRESSES

Arbor Press. Figure 13-37. Small presses with up to a ton or more of capacity can be assembled from standard air cylinders and valves, and operated at 80 to 100 PSI shop air pressure. They are useful for many small jobs as pressing bushings or sleeves on or off a shaft, punching holes or slots, notching sheet metal, riveting, driving nails or staples, trimming, swaging, marking and the like.

Small air presses are marketed by many manufacturers and can be purchased ready to go, or can be custom built in the shop to suit a highly special application. Hydraulic presses, operated with a hand pump or with a rotary power driven pump are also readily available for immediate use.

The illustration shows a smalll arbor press set up to use two manual control air valves for two-hand operation. Note: This simple circuit does not provide the degree of safety usually required by OSHA. For true safety two-hand operation, see electrical circuit in the Womack textbook "Electrical Control of Fluid Power".

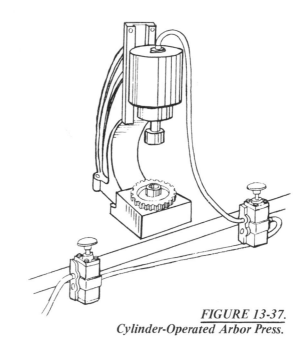

FIGURE 13-37.
Cylinder-Operated Arbor Press.

Shop-Built Hydraulic Press. Figure 13-38. Special-purpose presses can be assembled in most shops using standard hydraulic cylinders and valves. Hydraulic power units for driving the cylinders can either be purchased as a standard assembly of electric motor, pump, relief valve, reservoir, and accessories, or this too can be shop-built from standard components.

Standard die sets having built-in guides are very convenient for small tooling jobs and they simplify construction of a press. Caution! Tools should seldom or never be mounted directly to a cylinder piston rod. The side thrust developed when contacting the load is harmful to the rod bearings and seals of any cylinder. If guided die sets cannot be used, guides should be provided on the press.

Hydraulic Press Break. Figure 13-39. Cylinders may be located at widely separated points as shown on this machine where they may be 10 to 20 feet apart. To keep the cylinders in step with

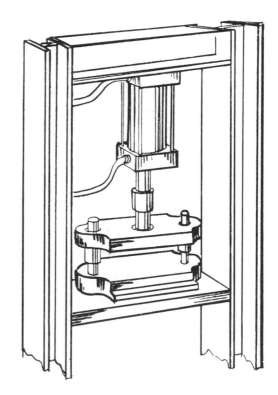

FIGURE 13-38. Shop-Built Hydraulic Press.

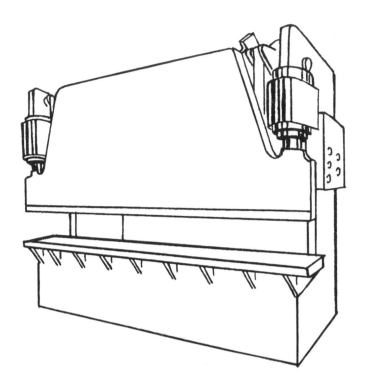

FIGURE 13-39. Hydraulic Press Brake With Synchronized Cylinders.

each other they must be mechanically yoked in some way or an equal division of the hydraulic fluid must be made in the circuit. Or, a separate pump may be used for each cylinder.

On a hydraulic press break, the tonnage rating of the press is based on having the work centered between the two cylinders. The tonnage rating is reduced in proportion as the work is performed off center. If entirely under one cylinder, maximum capacity will be reduced to one-half the nominal press rating.

Cylinder Economics. The amount of force produced by a cylinder is determined by its piston area and the applied fluid pressure. Therefore, there are a number of combinations of cylinder bore and fluid pressure which could be used. Other factors being equal, we can purchase more force per dollar by choosing a bore size in the 4 to 8-inch diameter range. For example, on a 1½" bore cylinder, each pound of force may cost 4 to 5 times the cost on a 6" bore cylinder. And on a 14" bore cylinder, each pound of force may cost 2 to 3 times as much as on a 6" bore size. Of course there are other things, such as valving, plumbing, etc. which must be considered. A wise designer will look at all component costs before deciding on the best bore and pressure combination.

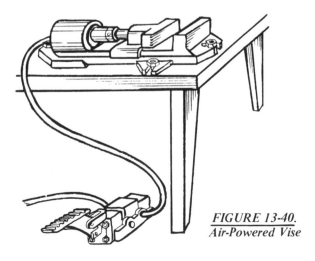

FIGURE 13-40.
Air-Powered Vise

Air-Powered Vise. Figure 13-40. These power vises can be purchased ready-made, or can be assembled from standard air cylinders and valves. They can speed up production work while reducing operator fatigue.

They are better suited for light duty holding (due to air compressibility). They may not hold rigidly enough for parts being milled or otherwise machined. For those applications, hydraulic clamps or toggle clamps should be used.

To protect the operator, a stop should be installed to prevent closure of the jaws more than a short distance beyond normal work closure.

Choice of Cylinder Mounting Style. Figure 13-41. When building a hydraulic press or when selecting a cylinder for other applications, a wise choice between front and rear flange mounting styles can contribute a great deal to the ruggedness and reliability of the press.

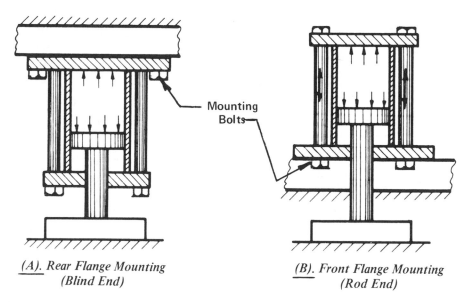

(A). Rear Flange Mounting
(Blind End)

(B). Front Flange Mounting
(Rod End)

FIGURE 13-41. Favorable and Unfavorable Mounting of a Hydraulic Press Cylinder.

If the cylinder will be required to exert its greatest force on its extension (push) stroke, as in Part (A) of the above figure, a rear flange mounting will have greater strength than a front flange mounting. In the figure, the cylinder is backed up against a heavy structural member of the press frame. Reaction force from the load is transmitted directly to the press frame. Thus, there is no longitudinal stress on the tie rods. The only stress in the cylinder is the radial stress from the hydraulic fluid against the side walls. Most cylinders can accept a great deal more radial stress than longitudinal stress, and will safely operate at higher pressure when mounted in this way.

However, to take advantage of this favorable method of mounting, the press should be arranged and the cylinder should have sufficient extra stroke so the piston can never bottom out against its front end cap. If this should happen while load was removed from the press, the internal hydraulic pressure would put full stress against the front end cap, and would subject the tie rods and/or cylinder barrel to full stress. If the piston should also impact against the front end, this would add to the stress. If necessary, an external stop should be provided for stopping the piston.

If a front flange mounting were to be used on this application (see Part B of the figure), the cylinder barrel and/or tie rods would carry the entire longitudinal stress due to load reaction. Add to this the radial stress in the barrel due to fluid pressure and the cylinder would be more likely to fail than if a rear flange mounting were used.

If the greatest force is required during retraction, then a front flange mount would be preferred for the same reason, and an external positive stop should be provided to prevent the piston from bottoming on its back end cap when it retracts. This would subject the tie rods to the full stress.

Favorable mounting gives a greater factor of safety which is always desirable and especially so for cylinders which might be subjected to pressure spikes much higher than their pressure rating. By reducing longitudinal stress, barrel end gaskets or O-ring seals are less likely to leak or blow out.

Note: In Part (A) there is no stress on the mounting bolts while in Part (B) the entire load is causing stress in the mounting bolts.

MISCELLANEOUS CYLINDER APPLICATIONS

FIGURE 13-41. Ideal Application for Telescoping Cylinder(s).

Telescoping Cylinder Application. Figure 13-41. These cylinders are specialty items which, because of their relatively high cost are impractical except where their features of short collapsed length, high force on the first stage and high speed on the last stage will justify their cost which is from three to five times that of a standard cylinder.

The application shown here is ideal because the load requirement very closely matches the cylinder characteristic. Standard cylinders would be impractical because of space limitations.

Telescoping cylinders can be purchased with 2 to 6 or more sleeves. Most of them are custom designed for a specific job. Double-acting models are available but have very low force on the return stroke because of the small net area around the sleeves.

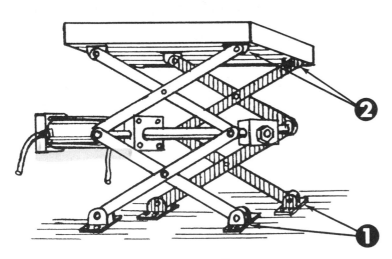

FIGURE 13-42. Scissor Lift.

Scissor Lift. Figure 13-42. To solve the problem of low headroom requirement while having a long stroke, levers can sometimes be used with standard cylinders to save the high cost of a telescoping cylinder.

A scissor is one form of double lever mechanism, although many other mechanical layouts are possible to suit a particular job.

This illustration is intended to show the possibilities of a scissor mechanism and does not represent actual design practice in which the platform and some of the arms should travel in guides. The cylinder weight should also be supported to prevent excessive packing wear and possible bending of the piston rod when extended.

A feature of the scissor mechanism is that its mechanical characteristics are in contrast to those of a telescoping cylinder. Starting force is very low, but increases to a maximum when the lift is fully extended. These characteristics may not be the best for a dump truck application where the maximum force is needed on starting.

Note: To solve lever and angle problems, the table of power factors shown in connection with Figure 1-00 should prove helpful.

Chute Selector. Figure 13-43. Rotary motion produced by a cylinder has many applications for controlling chutes and diverters in feed mills, cement and gravel plants, cotton gins, mining operations, coal and coke, and others too numerous to mention.

In addition to flappers used to divert granular material, cylinders are also used to operate sliding gates to handle the flow of the same materials, for loading into delivery trucks, freight cars, or into other chutes.

FIGURE 13-43.
Cylinder-Powered Chute Selector.

Hydraulic Check. Figure 13-44. Any cylinder can be used in reverse to provide a resistance to a mechanical load applied to its piston rod.

The throttled flow of fluid, usually oil, from one port to the other provides the fluid resistance. A cylinder can be pressed into service for this application if a factory-built hydraulic check cannot be obtained with a suitable bore and stroke.

With the piston rod extended, the cylinder and reservoir are completely filled with oil. All air should be excluded. In the illustration, the flow control valve is piped in a direction to restrict oil discharged from the cylinder rod port. Therefore, the piston rod offers checking resistance when being pulled out, but can be pushed in freely as the circulating oil passes through the check valve in the flow control valve.

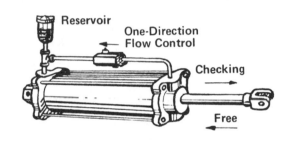

FIGURE 13-44.
Shop-Built Hydraulic Checking Device.

To reverse direction of checking resistance, the flow control valve must be turned around with its controlled flow arrow pointing toward the rod end of the cylinder, and the reservoir must be moved to the rod end port of the cylinder.

Note: A cylinder used for hydraulic checking cannot duplicate the smooth deceleration provided with a factory-built hydraulic check which has variable orifice construction for uniform deceleration.

Design Data Section

Design data in this section is only a part of the overall data available in Womack publications. It has been selected as being appropriate to subjects covered in this book, "Volume 2 – Industrial Fluid Power". Additional data will be found in the apendices of other Womack textbooks.

For example, troubleshooting information on open and closed loop hydraulic systems and on hydraulic pumps and motors will be found in the Womack book "Fluid Power in Plant and Field".

Still further data on many subjects related to fluid power, covering electrical and mechanical as well as fluid power subjects is published in Womack Bulletin W-144. This bulletin is a loose-leaf folder of 160 pages (80 individual sheets printed both sides). Some of the data in this section have been adapted from subjects covered in the bulletin.

Please request latest price list in which major Womack publications on fluid power are listed. The publisher or the author of this book can be contacted at the address on the title page.

CYLINDER PERFORMANCE IN THE REGENERATIVE MODE

Operation of hydraulic cylinders in a regenerative mode has been covered on Pages 124 through 130. Instructions were given for calculating the force, speed, and oil flow through the cylinder connecting lines while the cylinder is extending regeneratively. The kind of applications in which regeneration could be used to advantage were also briefly covered along with typical circuits. The chart on this page is a quick and easy way to determine, without calculations, the force output, the piston travel speed, and the oil flow rate in cylinder lines.

Section 1 of the chart shows standard bore sizes in the left column and the recommended rod size in the 3rd column. These are the maximum diameter rod sizes available for each bore diameter, and are the only rods recommended for regenerative use. Smaller rods would produce excessive speed, and a force too limited to be practical. Full piston area, rod area, and net area are shown for each bore. All three of these areas are used in regenerative cylinder calculations.

Section 2 shows the regenerative force for each 100 PSI of applied pressure on the blind end. Retraction force (for each 100 PSI) was calculated by using the net area.

Section 3 shows the regenerative travel speed and the normal retraction speed for each 1 GPM oil flow from the pump.

In Section 4, Points B and C refer to Figure 7-48 on Page 126. They show the rate of oil flow in each cylinder line while the cylinder is in regeneration. These flow rates are useful for sizing valves placed in these lines. Figures in the chart are GPM flows for each 1 GPM supplied from the pump.

| | | | | | ❶ | | ❷ | | ❸ | | ❹ | |
| | | | | | **For Every 100 PSI of System Pressure** | | **For Every 1 GPM of Pump Oil Flow** | | **For Every 1 GPM of Pump Oil Flow** | |
Bore, Inches	Piston Area, Sq. Ins	Rod Diam., Inches	Rod Area, Sq. Ins	Net Area, Sq. Ins	Regen. Forward Force, Lbs.	Retract Force, Lbs.	Regen. Forward Speed Ins/min	Retract Speed Ins/min	Regen. GPM @ Point B*	Regen. GPM @ Point C*
1-1/2	1.77	1	.7854	0.985	78.5	98.5	294	234	2.25	1.25
2	3.14	1-3/8	1.485	1.66	148	165	156	139	2.72	1.72
2-1/2	4.91	1-3/4	2.405	2.50	240	250	96	92.4	2.04	1.04
3-1/4	8.30	2	3.142	5.16	314	516	73	44.8	2.62	1.62
4	12.57	2-1/2	4.909	7.66	490	766	47	30.2	2.56	1.56
5	19.64	3-1/2	9.621	10.02	962	1002	24	23.1	2.04	1.04
6	28.27	4	12.57	15.70	1257	1570	18.4	14.7	2.25	1.25
7	38.49	5	19.64	18.56	1964	1856	11.7	12.5	1.96	0.96
8	50.27	5-1/2	23.76	26.51	2376	2651	9.7	8.7	2.12	1.12
10	78.54	7	38.49	40.06	3849	4006	6.0	5.8	2.04	1.04
12	113.1	8-1/2	56.75	56.35	5675	5635	4.1	4.1	1.99	0.99
14	153.9	10	78.54	75.40	7854	7540	2.94	3.1	1.96	0.96

Rod sizes in the 3rd column are the only ones recommended for use with the standard cylinder bore sizes in the left column.

*These points refer to the diagram of Figure 7-48 on Page 126

GRAPHIC SYMBOLS FOR FLUID POWER DIAGRAMS

Common symbols are shown here. A complete list of approved ANSI (American National Standards Institute) graphic symbols can be obtained from the National Fluid Power Association. Ask for Publication ANSI Y32.10-1967.

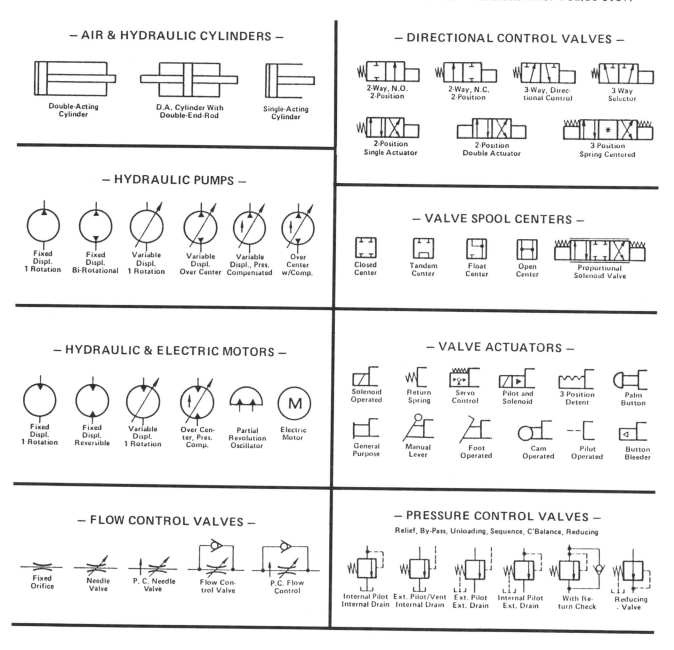

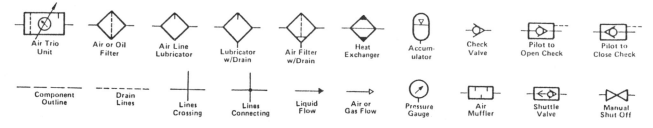

CYLINDER DRIFT PROBLEMS AND THEIR SOLUTIONS

There are certain conditions in some hydraulic cylinder circuits which could cause a cylinder to slowly drift (creep) when its 4-way control valve is in its center neutral position. The cause (in the fluid circuit) for cylinder drift is the unbalance between areas on opposite sides of the piston. Oil which leaks across the spool of the 4-way valve, under pressure, acts on unequal areas of the piston to develop a force unbalance which could cause unwanted piston movement unless there happens to be sufficient dead load or reactionary load against the piston rod to prevent piston drift. Oil leaking internally across the piston seals can also cause drift as discussed later.

Cylinders which have an oversize or 2:1 ratio piston rod have a greater tendency to drift than those cylinders with smallest (standard) diameter rod. Drifting force is equal to system pressure multiplied times rod area. Therefore, on circuits where drifting may be a problem, the cylinder should have the smallest diameter rod which has sufficient column strength.

In this section we will consider those circuits which may have a tendency to drift and will attempt to offer solutions.

Reactionary Load . . .

Figure 1. If the cylinder is expected to support a heavy load or will be subjected to a reactionary force while stopped, internal spool leakage of the 4-way valve may permit the cylinder to drift. Leakage paths in the valve are shown in dotted lines.

Figure 2. Cylinder drift from valve spool leakage can usually be minimized to an acceptable level by placing a pilot-operated check valve in the cylinder line to prevent reverse flow caused by load reaction. This check valve must be piloted from the opposite cylinder line, and this line must be vented to tank through the spool of the 4-way valve in neutral position to prevent pressure build-up on the pilot (from valve spool leakage) which would cause the check valve to open and permit cylinder drift. Note that neither a pilot-operated check valve nor any other external means can prevent cylinder drift caused by leaking piston seals.

If the reactionary load (while the cylinder is stopped) is pulling instead of pushing on the piston rod, the pilot-operated check valve should be located in the rod end line, and the opposite cylinder port on the 4-way valve should be vented to tank in neutral. Since pressure can intensify in the rod end of the cylinder when the rod port is blocked (as by a pilot-operated check valve), the application must be evaluated to be sure pressure intensification will not cause damage.

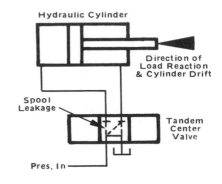

Figure 1. Drift due to valve spool leakage and a reactionary load.

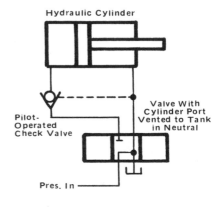

Figure 2. A pilot-operated check valve eliminates valve leakage.

Tandem Center System . . .

Figure 3. If two or more branch cylinder circuits are operated from one pump, the upstream cylinder (Cylinder 1 in Figure 3) may drift when the downstream cylinder is operating at high pressure.

Leakage across Valve 1 spool, entering the blind end will cause pressure intensification in the rod end of the cylinder. This intensified pressure, being higher than pump pressure, may leak across the valve spool into the pump line, causing forward drift of the cylinder, unless reaction or gravity load is sufficient to prevent it. To calculate drifting force, multiply rod area times inlet pressure.

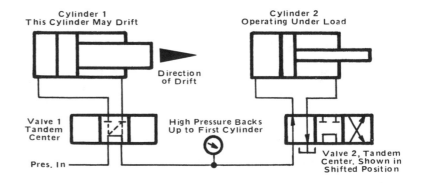

Figure 3. On an open center system, the upstream cylinder may drift due to valve leakage when the downstream cylinder is operated.

250

Closed Center System . . .

Figure 4. This is a system with two or more branch circuits operating in parallel from one pump. The inlet port on all 4-way valves, whether individual valves or one bank valve is used, is closed to oil flow when the valve spool is centered. Pump pressure builds up to relief valve or compensator level and is maintained at this level during the time all valve spools are centered.

To avoid power waste and heat build-up, a variable displacement pump with a pressure compensator is usually employed in a closed center system. When all valve spools are centered and system pressure holds at maximum level, leakage across the 4-way valve spool is at a maximum and may cause one or more of the cylinders to drift, especially those with large diameter rods and those with little reactionary load against them.

In Figure 4 the possible leakage paths across Valve 1 spool are shown in dotted lines. Leakage into Cylinder 1 blind end will cause pressure intensification in the rod end. The intensified pressure, being higher than pump pressure, may leak across the valve spool and into the pressure line, thus causing the cylinder to drift forward. However, if there is a dead load or reactionary load against the piston rod greater than the drifting force of rod square inch area times pump pressure, the cylinder cannot drift.

A suggested solution for closed center systems is shown in Figure 4 for Cylinder 2. A lock valve can be placed in the lines to the cylinder. A lock valve is a double section pilot-operated check valve. In addition to the lock valve, Valve 2 spool must be a float center type, venting both cylinder ports to tank when the spool is centered. The lock valve will prevent cylinder drift in either direction. The vented ports on the 4-way valve will prevent spool leakage from building up on the pilots of the check valve, holding them open. Spool leakage will simply go to tank.

Leakage in the Piston Seals . . .

If piston seals are not leaktight, this can also cause a cylinder to drift either under the influence of valve spool leakage or reactionary load against the piston rod. This type of drift can sometimes be reduced under certain conditions of operation.

Figure 5. If the reaction or gravity load is tending to pull the rod out as in this figure, there is no way to arrange the external circuit to guarantee there will be no drift in either a closed or open center system. Apparently the only remedy is to select a cylinder with leaktight piston seals. Even this may not eliminate drift due to leakage across the valve spool. If drift develops after a period of usage, the piston seals should be replaced.

Figure 6. However, if the gravity or reactionary load is pushing against the piston rod, there is a chance of minimizing drift by using a 4-way valve with blocked cylinder

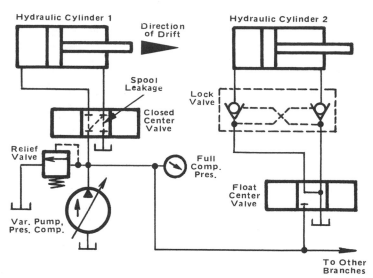

Figure 4. Closed Center System With Two Branch Circuits.

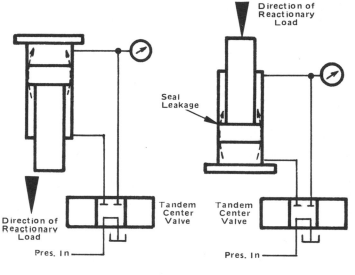

Figure 5
Load Pulling Against Rod.

Figure 6.
Load Pushing Against Rod.

ports in neutral. If leakage across the valve spool could be eliminated, it would be impossible for the cylinder to drift inward regardless of the volume of piston seal leakage. For each 1 cubic inch of leakage from the blind end, there is less than 1 cubic inch of space on the rod end for it to go into.

To further minimize drift due to internal leakage through the piston seals, the 4-way valve itself should be a high quality model rated for low internal leakage. Note that in Figure 6 a pilot-operated check valve or lock valve would not help in a piston seal problem. Since there is no pressure intensification in the cylinder rod end of Figure 6, pilot pressure obtained from from the blind end of the cylinder would hold the check valve open for free flow in both directions and would not allow it to close.

Piston seal leakage can be a very serious problem in closed center systems, and sometimes it may be virtually impossible to eliminate cylinder drift. A lock pin or mechaincal brake can sometimes solve the problem.

FORCE REQUIRED TO ACCELERATE A LOAD

The load on a hydraulic cylinder (or motor) consists of these three components:

(1). Normal load resistance, where fluid power energy is converted into mechanical energy applied to the load.

(2). Friction resistance, where part of the fluid power is expended in overcoming fluid or mechanical friction.

(3). Inertia, where fluid power is needed to get a massive load into motion, sometimes very quickly. This energy sometimes cannot be recovered.

As far as Items (1) and (2) are concerned, acceleration to final velocity is instantaneous as soon as the fluid power is applied to the cylinder (or motor).

If the load has high inertia (due to high mass), as per Item (3), then an *additional* amount of power must be supplied to accelerate the load from a standstill to its final velocity in a required time interval. This extra power is carried in the load as kinetic energy while the load is moving at a constant velocity, and will come back into the system as a pressure surge when the load is stopped,

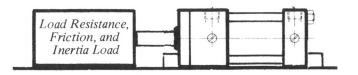

Cylinder Load Consists of Three Components

unless it can be absorbed by the load in the form of work.

The purpose of this section is to show how to calculate the extra PSI or torque needed in a hydraulic system to accelerate an inertia load, Item (3), from standstill or from a lower to a higher velocity in a given time, assuming the PSI for Items (1) and (2), the work load and friction resistance have already been calculated or assumed.

Calculating for Inertia Load . . .

Before calculating the extra PSI needed to rapidly accelerate this vertically moving cylinder, the normal PSI needed to move the load at a constant speed must be calculated by the usual means: Load weight ÷ Piston Area. Allowance should also be made for friction in ways or guides if this is significant.

Use this formula to calculate the EXTRA PSI for acceleration to a final velocity in a specified time:

(a) $F = (V \times W) \div (g \times t)$ **Lbs.**, in which:

F is the accelerating force, in pounds, that will be needed
V is the final velocity, in feet per second, starting from a standstill
W is the weight of the load in pounds
g is acceleration of gravity to convert weight into mass; always 32.16
t is the time, in seconds, during which acceleration takes place

If the cylinder bore is known, the accelerating PSI for its piston can be found directly with Formula *(b)* below:

(b) $PSI = (V \times W) \div (A \times g \times t)$, in which:

A is piston area in square inches. Other symbols are the same as above.

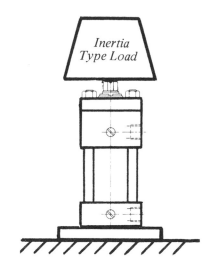

Inertia Type Load

PROBLEM DATA — VERTICAL CYLINDER WITH INERTIA LOAD

Steady Load = 35,000 lbs.
Cylinder Bore = 4" (Piston Area = 12.57 sq.ins.)

Initial Velocity = 0; Final Velocity = 12 ft/sec.
Time Required to Accelerate = 2 Seconds.

Example of Inertia Calculation . . .

Use the problem data in the box to solve for total PSI needed on the vertically moving cylinder not only to lift the given load, but to accelerate it to its final velocity in the specified time.

Note: Remember that the PSI to meet load and acceleration requirements is pressure difference **across the cylinder ports**. An indeterminate amount of additional pressure will be required from the pump to make up flow losses through plumbing and valving.

PSI for Steady Movement . . . 35,000 lbs. (load weight)

÷ 12.57 (piston area) = 2784 PSI to balance the load.

PSI for Acceleration . . . PSI = (12 × 35,000) ÷ (12.57 × 32.16 × 2) = 520 PSI.

Total PSI . . . The cylinder must be provided with 2784 + 520 = 3304 PSI to meet all conditions of the problem.

Non-Inertia Loads . . .

No extra PSI is needed to accelerate loads which consist entirely of frictional resistance and which have a negligible amount of mass (weight).

Rotating Loads — Driven With Hydraulic Motors . . .

Centrifugal pipe casting, in which a high mass must be rapidly accelerated to a high rotational speed, is a typical application in which a significant amount of extra torque, in addition to that required to keep the pipe spinning at a constant speed, must be supplied by the hydraulic motor. Incidentally, many rotating loads driven at high speeds *do* require a considerable amount of extra torque to get them up to full speed very quickly.

Before any calculations can be made, the moment of inertia, Symbol J, must be calculated for the shape and physical dimensions of the work piece. Information in the opposite column shows how to calculate the moment of inertia for the three most common shapes. Formulae for many other shapes can be found in a machinery or mechanics handbook.

After the moment of inertia, J, has been calculated, the *extra* torque required for the rate of acceleration can be found with the basic formula:

(c) $T = J \times \pi \times RPM \div 360 \times t$ inch-lbs., in which:

T is the extra torque, inch-lbs.
J is the moment of inertia of the rotating load which must be calculated with one of the formulae in the opposite column.
π is always 3.14.
RPM is the change in rotational velocity, revolutions per minute, from standstill or from a lower velocity up to a higher velocity.
t is time, in seconds, allowed for the velocity change.

After finding torque, T, the additional PSI on the hydraulic motor to produce this torque can be determined from catalog information on the particular motor model used.

Example of Pipe Spinning . . .

Find the extra torque required to accelerate a 500-lb. pipe from standstill to 700 RPM in 5 seconds. Pipe diameter is 10" O.D. x 7" I.D.

First, solve for J, the moment of inertia, using the formula in the box to the right.

J (moment of inertia) = 500 lbs x $(5^2 + 3.5^2) \div 2 \times 32.16$, or J = 500 x (25 + 12.25) ÷ 64.32 = 289.6

Next, use this calculated value of J in Formula *(c)* above to determine torque required for acceleration.

T = 289.6 x 3.14 x 700 ÷ 360 x 5 = 353.6 in-lbs.

Remember, this calculated torque is in addition to steady torque required to keep the pipe spinning at a constant RPM. Also, remember that the GPM oil flow to the motor must be sufficient to produce the required 700 RPM.

Accelerating From One Speed to Another . . .

When accelerating from a lower to a higher speed, subtract the difference between the two speeds and use it as RPM in the basic formula *(c)* at top of page.

Example: 250 RPM up to 1000 RPM. Use the difference, 750 RPM, in basic formula *(c)*.

CALCULATION OF MOMENT OF INERTIA

Moment of inertia must be calculated before making other calculations. Three common shapes are shown. Refer to machinery handbook for other shapes.

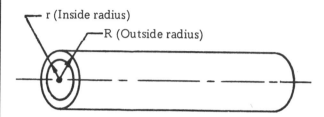

HOLLOW CYLINDER (Pipe) . . .

J (moment of inertia) of a pipe about an axis running lengthwise is:

$J = W \times (R^2 + r^2) \div 2g$ **Inch-Lbs-Secs**^{2}, in which:

W is total weight of pipe in pounds
R is outside radius of pipe in inches
r is inside radius of pipe in inches
g is acceleration of gravity, always 32.16

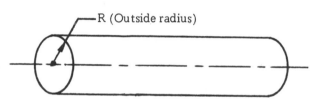

SOLID CYLINDER . . .

J (moment of inertia) of a solid cylinder about an axis running lengthwise is:

$J = W \times R^2 \div 2g$ **Inch-Lbs-Secs**^{2}, in which:

W is total weight of cylinder in pounds
R is outside radius of cylinder in inches
g is acceleration of gravity, always 32.16

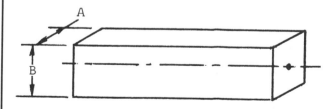

PRISM . . .

J (moment of inertia) of a prism of uniform cross section about the axis shown is:

$J = W \times (A^2 + B^2) \div 12g$ **Inch-Lbs-Secs**^{2}, in which:

W is total weight of prism in pounds
A and B are cross section dimensions in inches
g is acceleration of gravity, always 32.16

HYDRAULIC CYLINDER FORCE — EXTENSION AND RETRACTION

Lines in bold type show extension force, in pounds; italic lines show retraction force, in pounds with various rod diameters. Force values are theoretical; allow at least 5% for cylinder friction. Pressures along top of chart are pressure differentials across cylinder ports, not pressure in the pump line. Allow sufficient extra pump pressure to take care of flow losses through lines and valves as well as back pressure on cylinder exhaust line to tank.

For pressures not shown, use effective area of piston and multiply times pressure differential across ports.

Bore Dia., Ins.	Rod Dia., Ins.	Effec. Area, Sq.Ins.	Pressure Differential Across Cylinder Ports								
			500 PSI	750 PSI	1000 PSI	1250 PSI	1500 PSI	2000 PSI	2500 PSI	3000 PSI	5000 PSI
1-1/2	None	1.77	**885**	**1328**	**1770**	**2213**	**2655**	**3540**	**4425**	**5310**	**8850**
	5/8	*1.46*	*732*	*1097*	*1463*	*1829*	*2195*	*2926*	*3658*	*4390*	*7316*
	1	*.985*	*492*	*738*	*985*	*1231*	*1477*	*1969*	*2462*	*2954*	*4923*
2	None	3.14	**1570**	**2355**	**3140**	**3925**	**4710**	**6280**	**7850**	**9420**	**15,700**
	1	*2.35*	*1177*	*1766*	*2355*	*2943*	*3532*	*4709*	*5887*	*7064*	*11,773*
	1-3/8	*1.66*	*828*	*1241*	*1655*	*2069*	*2483*	*3310*	*4138*	*4965*	*8276*
2-1/2	None	4.91	**2455**	**3683**	**4910**	**6138**	**7365**	**9820**	**12,275**	**14,730**	**24,550**
	1	*4.12*	*2062*	*3093*	*4125*	*5156*	*6187*	*8249*	*10,312*	*12,374*	*20,623*
	1-3/8	*3.43*	*1713*	*2569*	*3425*	*4281*	*5138*	*6850*	*8563*	*10,275*	*17,126*
	1-3/4	*2.50*	*1252*	*1879*	*2505*	*3131*	*3757*	*5009*	*6262*	*7514*	*12,524*
3-1/4	None	8.30	**4150**	**6225**	**8300**	**10,375**	**12,450**	**16,600**	**20,750**	**24,900**	**41,500**
	1-3/8	*6.82*	*3408*	*5111*	*6815*	*8519*	*10,223*	*13,630*	*17,038*	*20,445*	*34,076*
	1-3/4	*5.89*	*2947*	*4421*	*5895*	*7368*	*8842*	*11,789*	*14,737*	*17,684*	*29,474*
	2	*5.16*	*2579*	*3869*	*5158*	*6448*	*7738*	*10,317*	*12,896*	*15,475*	*25,792*
4	None	12.57	**6285**	**9428**	**12,570**	**15,713**	**18,855**	**25,140**	**31,425**	**37,710**	**62,850**
	1-3/4	*10.16*	*5082*	*7624*	*10,165*	*12,706*	*15,247*	*20,329*	*25,412*	*30,494*	*50,824*
	2	*9.43*	*4714*	*7071*	*9428*	*11,786*	*14,143*	*18,857*	*23,571*	*28,285*	*47,142*
	2-1/2	*7.66*	*3831*	*5746*	*7661*	*9577*	*11,492*	*15,323*	*19,153*	*22,984*	*33,307*
5	None	19.64	**9820**	**14,730**	**19,640**	**24,550**	**29,460**	**39,280**	**49,100**	**58,920**	**98,200**
	2	*16.50*	*8249*	*12,374*	*16,498*	*20,623*	*24,748*	*32,997*	*41,246*	*49,495*	*82,492*
	2-1/2	*14.73*	*7366*	*11,048*	*14,731*	*18,414*	*22,097*	*29,463*	*36,828*	*44,194*	*73,656*
	3	*12.57*	*6286*	*9429*	*12,571*	*15,714*	*18,857*	*25,143*	*31,429*	*37,714*	*62,857*
	3-1/2	*10.02*	*5009*	*7514*	*10,019*	*12,524*	*15,028*	*20,038*	*25,047*	*30,057*	*50,095*
6	None	28.27	**14,135**	**21,203**	**28,270**	**35,338**	**42,405**	**56,540**	**70,675**	**84,810**	**141,350**
	2-1/2	*23.36*	*11,681*	*17,521*	*23,361*	*29,202*	*35,042*	*46,723*	*58,403*	*70,084*	*116,807*
	3	*21.20*	*10,601*	*15,901*	*21,201*	*26,502*	*31,802*	*42,403*	*53,004*	*63,604*	*106,007*
	3-1/2	*18.65*	*9324*	*13,987*	*18,649*	*23,311*	*27,973*	*37,298*	*46,622*	*55,947*	*93,245*
	4	*15.70*	*7850*	*11,775*	*15,700*	*19,625*	*23,550*	*31,400*	*39,250*	*47,100*	*78,500*
7	None	38.49	**19,245**	**28,868**	**38,490**	**48,113**	**57,735**	**76,980**	**96,225**	**115,470**	**192,450**
	3	*31.42*	*15,711*	*23,567*	*31,421*	*39,277*	*47,132*	*62,843*	*78,554*	*94,264*	*157,107*
	3-1/2	*28.87*	*14,434*	*21,652*	*28,869*	*36,086*	*43,303*	*57,738*	*72,172*	*86,607*	*144,344*
	4	*25.92*	*12,960*	*19,440*	*25,920*	*32,400*	*38,880*	*51,840*	*64,800*	*77,760*	*129,600*
	4-1/2	*22.59*	*11,293*	*16,940*	*22,586*	*28,233*	*33,879*	*45,172*	*56,465*	*67,758*	*112,930*
	5	*18.56*	*9428*	*14,141*	*18,855*	*23,569*	*28,283*	*37,710*	*47,138*	*56,565*	*94,275*
8	None	50.27	**25,135**	**37,703**	**50,270**	**62,838**	**75,405**	**100,540**	**125,675**	**150,810**	**251,350**
	3-1/2	*40.65*	*20,324*	*30,487*	*40,649*	*50,811*	*60,973*	*81,298*	*101,622*	*121,947*	*203,245*
	4	*37.70*	*18,850*	*28,275*	*37,700*	*47,125*	*56,550*	*75,400*	*94,250*	*131,100*	*188,500*
	4-1/2	*34.37*	*17,183*	*25,775*	*34,366*	*42,958*	*51,549*	*68,732*	*85,915*	*103,098*	*171,830*
	5	*30.64*	*15,318*	*22,976*	*30,635*	*38,294*	*45,953*	*61,270*	*76,588*	*91,905*	*153,175*
	5-1/2	*26.51*	*13,256*	*19,884*	*26,512*	*33,140*	*39,768*	*53,024*	*66,280*	*79,536*	*132,560*
10	None	78.54	**39,270**	**58,905**	**78,540**	**98,175**	**117,810**	**157,080**	**196,350**	**235,620**	**392,700**
	4-1/2	*62.64*	*31,318*	*46,977*	*62,636*	*78,295*	*93,954*	*125,272*	*156,590*	*187,908*	*313,180*
	5	*58.91*	*29,453*	*44,179*	*58,905*	*73,631*	*88,358*	*117,810*	*147,263*	*176,715*	*294,525*
	5-1/2	*54.78*	*27,391*	*41,087*	*54,782*	*68,478*	*82,173*	*109,564*	*136,955*	*164,346*	*273,910*
	7	*40.06*	*20,028*	*30,041*	*40,055*	*50,069*	*60,083*	*80,110*	*100,138*	*120,165*	*200,275*
12	None	113.1	**56,550**	**84,825**	**113,100**	**141,375**	**169,650**	**226,200**	**282,750**	**339,300**	**565,500**
	5-1/2	*89.34*	*44,671*	*67,006*	*89,342*	*111,678*	*134,013*	*178,684*	*223,355*	*268,026*	*446,710*
14	None	153.9	**76,950**	**115,425**	**153,900**	**192,375**	**230,850**	**307,800**	**384,750**	**461,700**	**769,500**
	7	*115.4*	*57,708*	*86,561*	*115,415*	*144,269*	*173,123*	*230,830*	*288,538*	*346,245*	*577,075*

HYDRAULIC CYLINDER SPEEDS – EXTENSION AND RETRACTION

Lines with bold type show cylinder piston extension speeds in "Inches per Minute". Lines with italic type show piston retraction speeds in "Inches per Minute" with various size piston rods. The largest diameter rod in each bore size is a "2:1" rod in which piston area is approximately twice rod area and net area. Since speed is directly proportional to GPM, for flows not shown, multiply speed in the "1 GPM" column times the desired GPM.

Piston Diam.	Rod Diam.	1 GPM	3 GPM	5 GPM	8 GPM	12 GPM	15 GPM	20 GPM	25 GPM	30 GPM	40 GPM	50 GPM	60 GPM	75 GPM	100 GPM
1½"	None*	**131**	**392**	**654**	----	----	----	----	----	----	----	----	----	----	----
	5/8"	*158*	*475*	*791*	----	----	----	----	----	----	----	----	----	----	----
	1	*235*	*706*	----	----	----	----	----	----	----	----	----	----	----	----
2	None*	**74**	**221**	**368**	**588**	**882**	----	----	----	----	----	----	----	----	----
	3/4"	*86*	*257*	*428*	*684*	----	----	----	----	----	----	----	----	----	----
	1	*98*	*295*	*490*	*784*	----	----	----	----	----	----	----	----	----	----
	1-3/8	*139*	*418*	*697*	----	----	----	----	----	----	----	----	----	----	----
2½	None*	**47**	**141**	**235**	**376**	**565**	**706**	**941**	----	----	----	----	----	----	----
	1"	*56*	*168*	*280*	*448*	*672*	*840*	----	----	----	----	----	----	----	----
	1-3/8	*67*	*202*	*337*	*540*	*810*	----	----	----	----	----	----	----	----	----
	1-3/4	*92*	*277*	*461*	*738*	----	----	----	----	----	----	----	----	----	----
3¼	None*	**28**	**84**	**139**	**223**	**334**	**418**	**557**	**696**	**835**	----	----	----	----	----
	1-3/8"	*34*	*102*	*170*	*271*	*407*	*509*	*678*	*848*	----	----	----	----	----	----
	1-3/4	*39*	*118*	*196*	*314*	*471*	*588*	*784*	*980*	----	----	----	----	----	----
	2	*45*	*134*	*224*	*359*	*538*	*672*	*896*	----	----	----	----	----	----	----
	2-1/4	*53*	*160*	*267*	*428*	*642*	*802*	----	----	----	----	----	----	----	----
4	None*	**18**	**55**	**92**	**147**	**221**	**276**	**368**	**460**	**551**	**735**	**919**	----	----	----
	1-1/4"	*20*	*61*	*102*	*163*	*244*	*306*	*407*	*509*	*611*	*815*	----	----	----	----
	1-3/4	*23*	*68*	*114*	*182*	*273*	*341*	*455*	*568*	*682*	*909*	----	----	----	----
	2	*25*	*74*	*123*	*196*	*294*	*368*	*490*	*613*	*735*	*980*	----	----	----	----
	2-1/2	*30*	*91*	*151*	*241*	*362*	*453*	*603*	*754*	*905*	----	----	----	----	----
	2-3/4	*35*	*105*	*174*	*279*	*418*	*523*	*697*	*872*	----	----	----	----	----	----
5	None*	**12**	**35**	**59**	**94**	**141**	**176**	**235**	**294**	**353**	**471**	**598**	**706**	**882**	----
	1-1/2"	*13*	*39*	*65*	*103*	*155*	*194*	*259*	*323*	*388*	*517*	*646*	*776*	*970*	----
	2	*14*	*42*	*70*	*112*	*168*	*210*	*280*	*350*	*420*	*560*	*700*	*840*	----	----
	2-1/2	*16*	*47*	*78*	*125*	*188*	*235*	*314*	*392*	*471*	*627*	*784*	*941*	----	----
	3	*18*	*55*	*92*	*147*	*221*	*276*	*368*	*460*	*551*	*735*	*919*	----	----	----
	3-1/2	*23*	*69*	*115*	*185*	*277*	*346*	*461*	*577*	*692*	*923*	----	----	----	----
6	None*	**8**	**25**	**41**	**65**	**98**	**123**	**163**	**204**	**245**	**327**	**409**	**490**	**613**	**817**
	1-3/4"	*9*	*27*	*45*	*71*	*107*	*134*	*179*	*223*	*268*	*357*	*446*	*538*	*670*	*893*
	2-1/2	*10*	*30*	*49*	*79*	*119*	*148*	*198*	*247*	*297*	*395*	*494*	*593*	*741*	*989*
	3	*11*	*33*	*54*	*87*	*131*	*163*	*218*	*272*	*327*	*436*	*545*	*654*	*817*	----
	3-1/2	*12*	*37*	*62*	*99*	*149*	*186*	*248*	*310*	*372*	*495*	*619*	*743*	*929*	----
	4	*15*	*44*	*74*	*118*	*176*	*221*	*294*	*368*	*441*	*588*	*735*	*882*	----	----
7	None*	**6**	**18**	**30**	**48**	**72**	**90**	**120**	**150**	**180**	**240**	**300**	**360**	**450**	**600**
	3"	*7*	*22*	*37*	*59*	*88*	*110*	*147*	*184*	*221*	*294*	*368*	*441*	*551*	*735*
	3-1/2	*8*	*24*	*40*	*64*	*96*	*120*	*160*	*200*	*240*	*320*	*400*	*480*	*600*	*800*
	4	*9*	*27*	*45*	*71*	*107*	*134*	*178*	*223*	*267*	*356*	*446*	*535*	*668*	*891*
	4-1/2	*10*	*31*	*51*	*82*	*123*	*153*	*205*	*256*	*307*	*409*	*512*	*614*	*767*	---
	5	*12*	*37*	*61*	*98*	*147*	*184*	*245*	*306*	*368*	*490*	*613*	*735*	*919*	----
8	None*	**4**	**14**	**23**	**36**	**55**	**69**	**92**	**115**	**135**	**185**	**230**	**275**	**345**	**460**
	3-1/2"	*5½*	*17*	*28*	*45*	*68*	*85*	*115*	*140*	*170*	*230*	*285*	*340*	*425*	*570*
	4	*6*	*18*	*30*	*49*	*73*	*90*	*122*	*150*	*180*	*240*	*305*	*365*	*460*	*610*
	4-1/2	*6½*	*20*	*33*	*53*	*80*	*100*	*135*	*165*	*200*	*265*	*335*	*405*	*505*	*670*
	5	*7½*	*22*	*38*	*60*	*90*	*114*	*150*	*185*	*225*	*300*	*375*	*450*	*565*	*755*
	5-1/2	*8½*	*26*	*43*	*70*	*104*	*129*	*172*	*215*	*255*	*345*	*430*	*520*	*650*	*870*
10	None*	**3**	**9**	**15**	**23**	**35**	**44**	**60**	**73**	**88**	**115**	**145**	**175**	**220**	**295**
	4-1/2"	*3½*	*11*	*18*	*29*	*44*	*55*	*75*	*92*	*111*	*150*	*185*	*220*	*275*	*370*
	5	*4*	*12*	*20*	*31*	*47*	*60*	*80*	*100*	*120*	*155*	*195*	*235*	*295*	*390*
	5-1/2	*4½*	*13*	*21*	*34*	*50*	*63*	*84*	*105*	*132*	*165*	*210*	*250*	*315*	*420*
	7	*5½*	*17*	*29*	*46*	*69*	*87*	*115*	*145*	*174*	*230*	*285*	*345*	*430*	*575*

Figures in the body of this chart are piston speeds in "Inches per Minute".

These lines are for extension speeds. No piston rod area is involved.

AIR CYLINDER FORCE — EXTENSION AND RETRACTION

Lines in bold type show extension force, in pounds; italic lines show retraction force, in pounds, with various rod diameters. Force values are theoretical; allow about 5% for cylinder friction. Pressures across top of chart are differentials across cylinder ports and not necessarily air line gauge pressures, because of back pressure in the exhaust line to atmosphere. When designing an air circuit, remember to allow a feed line air pressure at least 25% higher than required by the load to make up for back pressure and flow losses while the cylinder is moving.

For pressures not shown, use effective area of piston and multiply times pressure differential across ports.

Piston Dia., Ins.	Rod Dia., Ins.	Travel Direction	Effec. Area, Sq.Ins.	Pressure Differential Across Cylinder Ports								
				60 PSI	70 PSI	80 PSI	90 PSI	100 PSI	110 PSI	120 PSI	130 PSI	150 PSI
1-1/2	None	Extend	1.77	106	124	142	159	177	195	212	230	266
	5/8	Retract	1.46	88	102	117	132	146	161	176	190	205
	1	Retract	.985	59	69	79	89	98	108	118	128	148
1-3/4	None	Extend	2.41	144	168	192	216	241	265	289	313	361
	5/8	Retract	2.10	126	147	168	189	210	231	252	273	315
	1-1/4	Retract	1.18	71	83	95	106	118	130	142	154	177
2	None	Extend	3.14	188	220	251	283	314	345	377	408	471
	5/8	Retract	2.83	170	198	227	255	283	312	340	368	425
	1	Retract	2.35	141	165	188	212	235	259	283	306	353
2-1/2	None	Extend	4.91	295	344	393	442	491	540	589	638	737
	5/8	Retract	4.60	276	322	368	414	460	506	552	598	690
	1	Retract	4.12	247	289	330	371	412	454	495	536	619
	1-3/8	Retract	3.43	206	240	274	308	343	377	411	445	514
3	None	Extend	7.07	424	495	565	636	707	778	848	919	1060
	1	Retract	6.28	377	440	503	565	628	691	754	817	942
	1-3/4	Retract	4.66	280	326	373	420	466	513	560	606	699
3-1/4	None	Extend	8.30	498	581	664	747	830	913	996	1079	1245
	1	Retract	7.51	451	526	601	676	751	827	902	977	1127
	1-3/8	Retract	6.82	409	477	545	613	681	750	818	886	1022
	1-3/4	Retract	5.89	354	413	472	531	589	648	707	766	884
3-1/2	None	Extend	9.62	577	674	770	866	962	1058	1155	1251	1443
	1	Retract	8.84	530	618	707	795	884	972	1060	1149	1325
4	None	Extend	12.57	754	880	1006	1131	1257	1283	1508	1634	1886
	1	Retract	11.78	707	825	943	1061	1178	1296	1414	1532	1768
	1-3/8	Retract	11.09	665	776	887	998	1109	1219	1330	1441	1663
	1-3/4	Retract	10.16	610	712	813	915	1016	1118	1220	1321	1525
5	None	Extend	19.64	1178	1375	1571	1768	1964	2160	2357	2553	2946
	1	Retract	18.85	1131	1320	1508	1697	1885	2074	2263	2451	2828
	1-3/8	Retract	18.16	1089	1271	1452	1634	1816	1997	2179	2360	2723
6	None	Extend	28.27	1696	1979	2262	2544	2827	3110	3392	3675	4241
	1-3/8	Retract	26.79	1607	1875	2143	2411	2679	2946	3214	3482	4018
	1-3/4	Retract	25.9	1552	1811	2069	2328	2586	2845	3104	3362	3880
7	None	Extend	38.49	2309	2694	3079	3464	3849	4234	4619	5004	5774
	1-3/8	Retract	37.01	2220	2590	2960	3331	3701	4071	4441	4811	5551
8	None	Extend	50.27	3016	3519	4022	4524	5027	5530	6032	6535	7541
	1-3/8	Retract	48.79	2927	3415	3903	4391	4879	5366	5854	6342	7318
	1-3/4	Retract	47.90	2872	3351	3829	4308	4786	5265	5744	6222	7180
10	None	Extend	78.54	4712	5498	6283	7069	7854	8639	9425	10,210	11,781
	1-3/4	Retract	76.14	4568	5329	6091	6852	7614	8375	9136	9898	11,420
	2	Retract	75.40	4524	5278	6032	6786	7540	8294	9048	9802	11,310
12	None	Extend	113.1	6786	7917	9048	10,179	11,310	12,441	13,572	14,703	16,965
	2	Retract	110.0	6598	7697	8797	9896	10,996	12,095	13,195	14,295	16,494
	2-1/2	Retract	108.2	6491	7573	8655	9737	10,819	11,901	12,983	14,065	16,229
14	None	Extend	153.9	9234	10,773	12,312	13,851	15,390	16,929	18,468	20,007	23,085
	2-1/2	Retract	149.0	8939	10,429	11,919	13,409	14,899	16,389	17,879	19,369	22,349
	3	Retract	146.8	8810	10,278	11,747	13,215	14,683	16,151	17,620	19,088	22,025

AIR CONSUMPTION OF CYLINDERS

When designing a machine using air cylinders which operate on a fast, continuously reciprocating program, it may be necessary to estimate the compressor HP capacity to keep the cylinders in continuous operation.

To estimate compressor load, first calculate air consumption of the cylinder(s), then convert this into compressor HP.

Air consumption should be calculated at total pressure required to balance the load plus additional pressure to make up for circuit losses in the piping, directional valve, flow control valves, mufflers, etc. In the chart the pressure values across the top are those set on the pressure regulator serving the circuit. This will automatically include circuit losses in the calculation. The system pressure regulator should, of course, be set no higher than necessary to produce the travel speed required. The total pressure required for load plus losses is often about 25% more than load balance pressure.

<u>How to Use the Chart.</u> The chart at the foot of this page is set up for standard bore sizes with standard (smallest) rod. The use of a piston rod larger than standard size for a given bore will cause the air consumption to be slightly less than shown, but the difference will be so slight that it can be disregarded.

To use the chart, find cylinder bore in the left column. Follow this line to the column with your expected pressure regulator setting. Figures in the body of the chart are SCF (standard cubic feet) of free air to operate a 1-inch stroke cylinder through one cycle, forward and return. Multiply this figure times your actual cylinder stroke, then times the number of complete cycles per minute. The final result will be the SCFM (standard cubic feet per minute) which must be supplied by the air compressor on a continuous basis.

This chart was calculated using compression ratios from the chart at the right. Consumption was calculated assuming the cylinder piston will be allowed to stall, at least momentarily, at each end of its stroke, giving it time to fill with air to full pressure regulator setting. If reversed at either end before reaching full stall, air consumption will be slightly less than shown.

<u>Calculation Example.</u> Estimate the SCFM consumption of a 4-inch bore air cylinder having a 28-inch stroke, to cycle 11 times a minute. Assume the load balance pressure for this cylinder is 65 PSI.

<u>Solution:</u> The pressure regulator will probably have to be set for a pressure of 80 PSI to take care of circuit flow losses in addition to load balance.

Use the 80 PSI column and the 4-inch bore line. The chart shows .091 SCF for a 1-inch stroke. Then SCFM air flow required will be:

$$SCFM = .091 \times 28 \times 11 = 17.9$$

<u>Converting SCFM into Compressor HP.</u> Rule-of-thumb: 6 SCFM = 1 HP, or use the chart in the Appendix of Volume 1 for more accurate conversion.

PRESSURE CONVERSIONS			
Pres., PSIG	Pres., PSIA	No. of Atmos*	No. of Bars
50	64.7	4.40	4.46
55	69.7	4.74	4.81
60	74.7	5.08	5.15
65	79.7	5.42	5.50
70	84.7	5.76	5.84
75	89.7	6.10	6.19
80	94.7	6.44	6.53
85	99.7	6.78	6.88
90	104.7	7.12	7.22
95	109.7	7.46	7.57
100	114.7	7.80	7.91
105	119.7	8.14	8.26
110	124.7	8.48	8.60
115	129.7	8.82	8.94
120	134.7	9.16	9.29
125	139.7	9.50	9.63
130	144.7	9.84	9.98
135	149.7	10.2	10.3
140	154.7	10.5	10.7
145	159.7	10.9	11.0
150	164.7	11.2	11.4
155	169.7	11.5	11.7
160	174.7	11.9	12.0
165	179.7	12.2	12.4
170	184.7	12.6	12.7

*Atmospheres are used to determine the quantity of free air contained in a given SCF volume when under compression, and were used to calculate the chart below.

Cylinder Air Consumption per 1-Inch Stroke, Forward and Return

Figures in body of chart are SCF air consumption of cylinders with 1-inch stroke, at various air pressures.

Cylin. Bore, Ins.	Rod Dia., Ins.	PSI Pressure Setting on Regulator Outlet											
		50 PSI	60 PSI	70 PSI	80 PSI	90 PSI	100 PSI	110 PSI	120 PSI	130 PSI	140 PSI	150 PSI	160 PSI
1.50	.625	.008	.010	.011	.012	.013	.015	.016	.017	.018	.020	.021	.022
2.00	.625	.015	.018	.020	.022	.025	.027	.029	.032	.034	.036	.039	.041
2.50	.625	.024	.028	.032	.035	.039	.043	.047	.050	.054	.058	.062	.065
3.00	1.00	.035	.040	.046	.051	.056	.062	.067	.071	.076	.081	.087	.092
3.25	1.00	.040	.047	.053	.059	.065	.071	.077	.084	.090	.096	.102	.109
4.00	1.00	.062	.072	.081	.091	.100	.110	.120	.129	.139	.148	.158	.168
5.00	1.00	.098	.113	.128	.143	.159	.174	.189	.204	.219	.233	.249	.265
6.00	1.38	.140	.162	.184	.205	.227	.249	.270	.292	.313	.334	.357	.379
8.00	1.38	.252	.291	.330	.369	.408	.447	.486	.525	.564	.602	.642	.682
10.00	1.75	.394	.455	.516	.576	.637	.698	.759	.820	.881	.940	1.00	1.07
12.00	2.00	.568	.656	.744	.831	.919	1.01	1.09	1.18	1.27	1.36	1.45	1.54
14.00	2.50	.771	.891	1.01	1.13	1.25	1.37	1.49	1.61	1.74	1.85	1.98	2.10

STEEL TUBING — PRESSURE AND FLOW RATINGS FOR HYDRAULIC PLUMBING

Pressure ratings are for annealed steel tubing "hydraulic grade" having a tensile strength of 55,000 PSI (the kind most often used for plumbing hydraulic systems). See notes at foot of next page for adjustment to other steels.

Pressure ratings are shown for S.F. (safety factors) from 2 to 6. A factor of 4 is recommended for general service. On shockless systems a smaller factor is sometimes used.

GPM ratings are shown for flow velocities from 10 through 30 f/s (feet per second). A general guideline is to use 10 f/s on systems operating at a maximum pressure of 1000 PSI; 15 f/s on systems having maximum pressures from 1000 to 2000 PSI; 20 f/s if system pressure is in the range of 2000 to 3500 PSI; and a velocity of 30 f/s on systems designed for pressures higher than 3500 PSI.

Wall, Inches	.025	.032	.035	.042	.049	.058	.065	.072	.083	.095	.109
5/16" O.D. TUBING — 55,000 PSI Tensile											
PSI @ S.F. = 6	1470	1880	2050	2460	2880	3400	3800	----	----	----	----
PSI @ S.F. = 5	1760	2250	2460	2960	3450	4080	4580	----	----	----	----
PSI @ S.F. = 4	2200	2820	3080	3700	4310	5100	5720	----	----	----	----
PSI @ S.F. = 3	2930	3750	4100	4930	5750	6800	7630	----	----	----	----
PSI @ S.F. = 2	4400	5630	6160	7390	8630	10,200	11,440	----	----	----	----
GPM @ 10 f/s	1.69	1.51	1.44	1.28	1.13	0.95	0.82	----	----	----	----
GPM @ 15 f/s	2.53	2.27	2.16	1.92	1.69	1.42	1.23	----	----	----	----
GPM @ 20 f/s	3.37	3.02	2.88	2.56	2.25	2.25	1.89	----	----	----	----
GPM @ 30 f/s	5.06	4.54	4.32	3.83	3.38	2.83	2.45	----	----	----	----
3/8" O.D. TUBING — 55,000 PSI Tensile											
PSI @ S.F. = 6	1220	1560	1710	2050	2400	2840	3180	3520	----	----	----
PSI @ S.F. = 5	1470	1880	2050	2460	2880	3400	3800	4220	----	----	----
PSI @ S.F. = 4	1830	2350	2570	3080	3590	4250	4770	5280	----	----	----
PSI @ S.F. = 3	2440	3130	3420	4100	4790	5670	6350	7040	----	----	----
PSI @ S.F. = 2	3670	4690	5130	6160	7190	8500	9530	10,550	----	----	----
GPM @ 10 f/s	2.59	2.37	2.28	2.07	1.88	1.64	1.47	1.31	----	----	----
GPM @ 15 f/s	3.88	3.55	3.42	3.11	2.82	2.46	2.20	1.96	----	----	----
GPM @ 20 f/s	5.17	4.74	4.56	4.15	3.76	3.29	2.94	2.61	----	----	----
GPM @ 30 f/s	7.76	7.11	6.84	6.22	5.64	4.93	4.40	3.92	----	----	----
1/2" O.D. TUBING — 55,000 PSI Tensile											
PSI @ S.F. = 6	920	1170	1280	1540	1800	2130	2380	2640	3040	----	----
PSI @ S.F. = 5	1100	1400	1540	1850	2150	2550	2860	3170	3650	----	----
PSI @ S.F. = 4	1380	1760	1920	2310	2700	3190	3580	3960	4560	----	----
PSI @ S.F. = 3	1830	2350	2570	3080	3590	4250	4770	5280	6090	----	----
PSI @ S.F. = 2	2750	3520	3850	4620	5390	6380	7150	7920	9130	----	----
GPM @ 10 f/s	4.96	4.65	4.53	4.24	3.96	3.61	3.35	3.10	2.73	----	----
GPM @ 15 f/s	7.43	6.98	6.79	6.35	5.93	5.41	5.03	4.65	4.10	----	----
GPM @ 20 f/s	9.91	9.31	9.05	8.47	7.91	7.22	6.70	6.20	5.46	----	----
GPM @ 30 f/s	14.9	14.0	13.6	12.7	11.9	10.8	10.1	9.30	8.19	----	----
5/8" O.D. TUBING — 55,000 PSI Tensile											
PSI @ S.F. = 6	730	940	1030	1230	1440	1700	1910	2110	2435	2790	3200
PSI @ S.F. = 5	880	1130	1230	1480	1720	2040	2290	2530	2920	3340	3840
PSI @ S.F. = 4	1100	1400	1540	1850	2160	2550	2860	3170	3650	4180	4800
PSI @ S.F. = 3	1470	1880	2050	2460	2870	3400	3810	4220	4870	5570	6390
PSI @ S.F. = 2	2200	2800	3080	3700	4320	5100	5720	6340	7300	4360	9600
GPM @ 10 f/s	8.10	7.71	7.54	7.17	6.80	6.34	6.00	5.66	5.16	4.63	4.06
GPM @ 15 f/s	12.1	11.6	11.3	10.7	10.2	9.52	9.00	8.50	7.38	6.95	6.08
GPM @ 20 f/s	16.2	15.4	15.1	14.3	13.6	12.7	12.0	11.3	10.3	9.26	8.11
GPM @ 30 f/s	24.3	23.1	22.6	21.5	20.4	19.0	18.0	17.0	15.5	13.9	12.2
3/4" O.D. TUBING — 55,000 PSI Tensile											
PSI @ S.F. = 6	610	780	860	1030	1200	1420	1590	1760	2030	2320	2660
PSI @ S.F. = 5	730	940	1030	1230	1440	1700	1900	2110	2430	2790	3200
PSI @ S.F. = 4	920	1170	1280	1540	1800	2130	2380	2640	3040	3480	4000
PSI @ S.F. = 3	1220	1560	1710	2050	2400	2840	3180	3520	4060	4640	5330
PSI @ S.F. = 2	1840	2340	2560	3080	3600	4260	4760	5280	6080	6960	8000
GPM @ 10 f/s	12.0	11.5	11.3	10.9	10.4	9.84	9.41	8.99	8.35	7.68	6.93
GPM @ 15 f/s	18.0	17.3	17.0	16.3	15.6	14.8	14.1	13.5	12.5	11.5	10.4
GPM @ 20 f/s	24.0	23.0	22.6	21.7	20.8	19.7	18.8	18.0	16.7	15.4	13.9
GPM @ 30 f/s	36.0	34.6	34.0	32.6	31.2	29.5	28.2	27.0	25.1	23.0	20.8

(This Table is Continued on the Next Page)

(This Table is Continued From the Preceding Page)

Wall, Inches	.025	.032	.035	.042	.049	.058	.065	.072	.083	.095	.109
7/8" O.D. TUBING, 55,000 PSI Tensile											
Inside Area	.5346	.5166	.5090	.4914	.4742	.4525	.4359	.4197	.3948	.3685	.3390
PSI @ S.F. = 6	520	670	730	880	1030	1210	1360	1500	1740	1990	2280
PSI @ S.F. = 5	630	800	880	1060	1230	1460	1630	1810	2090	2390	2740
PSI @ S.F. = 4	790	1000	1100	1320	1540	1820	2040	2260	2600	2990	3420
PSI @ S.F. = 3	1050	1340	1470	1760	2050	2430	2720	3020	3480	3980	4570
PSI @ S.F. = 2	1580	2000	2200	2640	3080	3640	4080	4520	5200	5980	6840
GPM @ 10 f/s	16.7	16.1	15.9	15.3	14.8	14.1	13.6	13.1	12.3	11.5	10.6
GPM @ 15 f/s	25.0	24.2	23.8	23.0	22.2	21.2	20.4	19.6	18.5	17.2	15.9
GPM @ 20 f/s	33.3	32.2	31.7	30.6	29.6	28.2	27.2	26.2	24.6	23.0	21.1
GPM @ 30 f/s	50.0	48.3	47.6	46.0	44.3	42.3	40.8	39.2	36.9	34.5	31.7
1" O.D. TUBING, 55,000 PSI Tensile											
Inside Area	.7088	.6881	.6793	.6590	.6390	.6138	.5945	.5755	.5463	.5153	.4803
PSI @ S.F. = 6	460	590	640	770	900	1060	1190	1320	1520	1740	2000
PSI @ S.F. = 5	550	700	770	920	1080	1280	1430	1580	1830	2090	2400
PSI @ S.F. = 4	690	880	960	1150	1350	1590	1790	1980	2280	2610	3000
PSI @ S.F. = 3	920	1170	1280	1540	1800	2130	2380	2640	3040	3480	4000
PSI @ S.F. = 2	1380	1760	1920	2300	2700	3180	3580	3960	4560	5220	6000
GPM @ 10 f/s	22.1	21.4	21.2	20.5	19.9	19.1	18.5	17.9	17.0	16.1	15.0
GPM @ 15 f/s	33.1	32.2	31.8	30.8	29.9	28.7	27.8	26.9	25.5	24.1	22.5
GPM @ 20 f/s	44.2	42.9	42.4	41.1	39.8	38.3	37.1	35.9	34.1	32.1	29.9
GPM @ 30 f/s	66.3	64.3	63.5	61.6	59.8	57.4	55.6	53.8	51.1	48.2	44.9
1¼" O.D. TUBING, 55,000 PSI Tensile											
Inside Area	1.131	1.105	1.094	1.068	1.042	1.010	.9852	.9607	.9229	.8825	.8365
PSI @ S.F. = 6	365	470	515	615	720	850	950	1060	1220	1390	1600
PSI @ S.F. = 5	440	560	610	740	860	1020	1140	1270	1460	1670	1920
PSI @ S.F. = 4	550	700	770	920	1080	1280	1430	1580	1820	2090	2400
PSI @ S.F. = 3	730	940	1030	1230	1440	1700	1900	2110	2430	2790	3200
PSI @ S.F. = 2	1100	1400	1540	1840	2160	2560	2860	3160	3640	4180	4800
GPM @ 10 f/s	35.3	34.4	34.1	33.3	32.5	31.5	30.7	29.9	28.8	27.5	26.1
GPM @ 15 f/s	52.9	51.7	51.2	49.9	48.7	47.2	46.1	44.9	43.2	41.3	39.1
GPM @ 20 f/s	70.5	68.9	68.2	66.6	65.0	63.0	61.4	59.9	57.5	55.0	52.2
GPM @ 30 f/s	106	103	102	100	97.4	94.5	92.1	89.8	86.3	85.5	78.2
1½" O.D. TUBING, 55,000 PSI Tensile											
Inside Area	1.651	1.612	1.606	1.575	1.544	1.504	1.474	1.444	1.398	1.348	1.291
PSI @ S.F. = 6	305	390	430	515	600	710	795	880	1014	1161	1332
PSI @ S.F. = 5	370	470	510	610	720	850	950	1050	1210	1390	1600
PSI @ S.F. = 4	460	590	640	770	900	1060	1190	1320	1520	1740	2000
PSI @ S.F. = 3	610	780	850	1030	1200	1420	1590	1760	2030	2320	2660
PSI @ S.F. = 2	920	1180	1280	1540	1800	2120	2380	2640	3040	3480	4000
GPM @ 10 f/s	51.5	50.2	50.1	49.1	48.1	46.9	45.9	45.0	43.6	42.0	40.2
GPM @ 15 f/s	77.2	75.4	75.1	73.6	72.2	70.3	68.9	67.5	65.4	63.0	60.4
GPM @ 20 f/s	103	101	100	98.2	96.3	93.8	91.9	90.0	87.2	84.0	80.5
GPM @ 30 f/s	154	151	150	147	144	141	138	135	131	126	121
2" O.D. TUBING, 55,000 PSI Tensile											
Inside Area	2.986	2.944	2.926	2.883	2.841	2.788	2.746	2.705	2.642	2.573	2.490
PSI @ S.F. = 6	230	295	320	385	450	530	595	660	760	870	1000
PSI @ S.F. = 5	270	350	390	460	540	640	710	790	910	1040	1200
PSI @ S.F. = 4	340	440	480	580	670	800	890	990	1140	1300	1500
PSI @ S.F. = 3	460	590	640	770	900	1060	1190	1320	1520	1740	2000
PSI @ S.F. = 2	680	880	960	1160	1340	1600	1780	1980	2280	2600	3000
GPM @ 10 f/s	93.1	91.8	91.2	89.4	88.6	86.9	85.6	84.3	82.4	80.2	77.6
GPM @ 15 f/s	139	138	137	135	133	130	128	126	123	120	116
GPM @ 20 f/s	186	184	182	180	177	174	171	169	165	160	233
GPM @ 30 f/s	279	275	274	270	266	261	257	253	247	241	233

For steels with tensile strength other than 55,000 PSI, pressure rating is proportional to tensile strength.

Barlow's Formula *(P = 2t x S ÷ O)* was used for pressure calculations, in which P = burst pressure, PSI; t = wall thickness, inches; S = tensile strength of material, in PSI; and O = outside diameter of tube, in inches. Working pressure = burst pressure ÷ safety factor.

The formula: *GPM = V x A ÷ 0.3208* was used for calculating flow velocity, in which V = velocity in feet per second, and A = inside area, in square inches.

PRESSURE AND FLOW RATINGS OF IRON PIPE

Pressure ratings are for wrought steel, butt welded pipe, the kind used most often in hydraulic plumbing. It is a low carbon steel with 40,000 PSI tensile strength. High carbon steel pipe, with tensile ratings up to 60,000 PSI is also available. Its pressure rating is higher in proportion to the increase in its tensile rating.

Schedule 40 is "standard weight" pipe. Schedules 80 and 160 have the same O. D. for a given size, but have heavier walls and a smaller internal flow area. Double extra strength pipe is also available in sizes of 1/2" and larger where higher pressure ratings are necessary.

Taper pipe threads, if used, should be NPTF (dryseal type) to minimize leakage around the crest of the threads.

Abbreviations used in the table are: S.F. is safety factor on the pressure ratings; f/s is flow velocity in feet per second.

1/4" NPT PIPE — 40,000 PSI Tensile

Pipe Schedule ➤	40	80	160
PSI @ S.F. = 6	2170	2940	- - - -
PSI @ S.F. = 5	2610	3530	- - - -
PSI @ S.F. = 4	3260	4410	- - - -
PSI @ S.F. = 3	4340	5880	- - - -
PSI @ S.F. = 2	6520	8820	- - - -
GPM @ 10 f/s	3.00	2.40	- - - -
GPM @ 15 f/s	4.50	3.60	- - - -
GPM @ 20 f/s	6.00	4.80	- - - -
GPM @ 30 f/s	9.00	7.20	- - - -

3/8" NPT PIPE — 40,000 PSI Tensile

Pipe Schedule ➤	40	80	160
PSI @ S.F. = 6	1800	2490	- - - -
PSI @ S.F. = 5	2160	2990	- - - -
PSI @ S.F. = 4	2700	3730	- - - -
PSI @ S.F. = 3	3590	4980	- - - -
PSI @ S.F. = 2	5390	7470	- - - -
GPM @ 10 f/s	6.00	4.20	- - - -
GPM @ 15 f/s	9.00	5.30	- - - -
GPM @ 20 f/s	12.0	8.40	- - - -
GPM @ 30 f/s	18.0	11.0	- - - -

1/2" NPT PIPE — 40,000 PSI Tensile

Pipe Schedule ➤	40	80	160
PSI @ S.F. = 6	1730	2330	2980
PSI @ S.F. = 5	2080	2800	3580
PSI @ S.F. = 4	2600	3500	4480
PSI @ S.F. = 3	3460	4670	5970
PSI @ S.F. = 2	5190	7000	8950
GPM @ 10 f/s	9.50	7.20	5.33
GPM @ 15 f/s	14.0	11.0	8.00
GPM @ 20 f/s	19.0	14.0	10.6
GPM @ 30 f/s	29.0	22.0	16.0

3/4" NPT PIPE — 40,000 PSI Tensile

Pipe Schedule ➤	40	80	160
PSI @ S.F. = 6	1430	1950	2780
PSI @ S.F. = 5	1720	2350	3340
PSI @ S.F. = 4	2150	2930	4170
PSI @ S.F. = 3	2870	3910	5560
PSI @ S.F. = 2	4300	5870	8340
GPM @ 10 f/s	16.0	15.0	9.33
GPM @ 15 f/s	25.0	22.0	14.0
GPM @ 20 f/s	33.0	30.0	18.6
GPM @ 30 f/s	50.0	44.0	28.0

1" NPT PIPE — 40,000 PSI Tensile

Pipe Schedule ➤	40	80	160
PSI @ S.F. = 6	1350	1810	2530
PSI @ S.F. = 5	1620	2180	3040
PSI @ S.F. = 4	2020	2720	3800
PSI @ S.F. = 3	2700	3630	5070
PSI @ S.F. = 2	4040	5440	7600
GPM @ 10 f/s	27.0	22.2	16.0
GPM @ 15 f/s	41.0	33.3	24.0
GPM @ 20 f/s	55.0	44.0	32.0
GPM @ 30 f/s	83.0	66.0	48.0

1¼" NPT PIPE — 40,000 PSI Tensile

Pipe Schedule ➤	40	80	160
PSI @ S.F. = 6	1120	1530	2000
PSI @ S.F. = 5	1350	1840	2410
PSI @ S.F. = 4	1690	2300	3010
PSI @ S.F. = 3	2250	3070	4020
PSI @ S.F. = 2	3370	4600	6020
GPM @ 10 f/s	47.0	40.2	32.6
GPM @ 15 f/s	70.0	60.0	49.0
GPM @ 20 f/s	94.0	68.0	65.0
GPM @ 30 f/s	140	120	98.0

1½" NPT PIPE — 40,000 PSI Tensile

Pipe Schedule ➤	40	80	160
PSI @ S.F. = 6	1020	1400	1970
PSI @ S.F. = 5	1220	1680	2370
PSI @ S.F. = 4	1530	2100	2960
PSI @ S.F. = 3	2030	2810	3940
PSI @ S.F. = 2	3050	4210	5920
GPM @ 10 f/s	65.0	55.2	43.3
GPM @ 15 f/s	95.0	83.0	65.0
GPM @ 20 f/s	130	110	87.0
GPM @ 30 f/s	190	166	130

2" NPT PIPE — 40,000 PSI Tensile

Pipe Schedule ➤	40	80	160
PSI @ S.F. = 6	865	1220	1930
PSI @ S.F. = 5	1040	1470	2320
PSI @ S.F. = 4	1300	1840	2900
PSI @ S.F. = 3	1730	2450	3860
PSI @ S.F. = 2	2590	3670	5790
GPM @ 10 f/s	105	91.2	68.6
GPM @ 15 f/s	156	138	103
GPM @ 20 f/s	210	185	137
GPM @ 30 f/s	312	276	206

2½" NPT PIPE — 40,000 PSI Tensile

Pipe Schedule ➤	40	80	160
PSI @ S.F. = 6	940	1280	1740
PSI @ S.F. = 5	1130	1540	2090
PSI @ S.F. = 4	1410	1920	2610
PSI @ S.F. = 3	1880	2560	3480
PSI @ S.F. = 2	2820	3840	5220
GPM @ 10 f/s	120	132	111
GPM @ 15 f/s	222	198	166
GPM @ 20 f/s	300	265	220
GPM @ 30 f/s	444	396	332

3" NPT PIPE — 40,000 PSI Tensile

Pipe Schedule ➤	40	80	160
PSI @ S.F. = 6	820	1140	1670
PSI @ S.F. = 5	985	1370	2000
PSI @ S.F. = 4	1230	1710	2500
PSI @ S.F. = 3	1650	2290	3340
PSI @ S.F. = 2	2470	3430	5000
GPM @ 10 f/s	225	206	167
GPM @ 15 f/s	345	310	250
GPM @ 20 f/s	450	412	334
GPM @ 30 f/s	690	620	500

DIMENSIONS AND FLOW AREAS OF PIPES

Nom. Pipe Size	Outside Diam., Inches	Circumference, Inches	Schedule 40		Schedule 80		Schedule 160		Schedule XXS	
			Inside Diam., Inches	Inside Area, Sq. Ins.	Inside Diam., Inches	Inside Area, Sq. In.	Inside Diam., Inches	Inside Area, Sq. In.	Inside Diam., Inches	Inside Area, Sq. Ins.
1/8 NPTF	.405	1.27	.269	.057	.215	.036	- - - -	- - - -	- - - -	- - - -
1/4 NPTF	.540	1.70	.364	.104	.302	.072	- - - -	- - - -	- - - -	- - - -
3/8 NPTF	.675	2.12	.493	.191	.423	.141	- - - -	- - - -	- - - -	- - - -
1/2 NPTF	.840	2.64	.622	.304	.546	.234	.464	.169	.252	.050
3/4 NPTF	1.05	3.30	.824	.533	.742	.432	.612	.294	.434	.148
1" NPTF	1.32	4.13	1.05	.866	.957	.719	.815	.522	.599	.282
1¼ NPTF	1.66	5.22	1.38	1.50	1.28	1.29	1.16	1.06	.896	.631
1½ NPTF	1.90	5.97	1.61	2.04	1.50	1.77	1.34	1.41	1.10	.950
2" NPTF	2.38	7.46	2.07	3.37	1.94	2.96	1.69	2.24	1.50	1.77
2½ NPTF	2.88	9.03	2.47	4.79	2.32	4.23	2.13	3.56	1.77	2.46
3" NPTF	3.50	11.0	3.07	7.40	2.90	6.61	2.62	5.39	2.30	4.15

FLOW AREAS OF STEEL TUBING

Figures in the body of this chart are internal flow areas, in square inches, of steel tubing. When connecting from pipe into steel tubing, use this chart to find equivalent internal flow area to match areas of pipe in the chart above.

Tube O.D.	Wall Thickness, Inches										
	.025	.032	.035	.042	.049	.058	.065	.072	.083	.095	.109
5/16	.0541	.0485	.0462	.0410	.0361	.0303	.0262	- - - -	- - - -	- - - -	- - - -
3/8	.0830	.0760	.0731	.0665	.0603	.0527	.0471	.0419	- - - -	- - - -	- - - -
1/2	.1590	.1493	.1452	.1359	.1269	.1158	.1075	.0995	.0876	- - - -	- - - -
5/8	.2597	.2472	.2419	.2299	.2181	.2035	.1924	.1817	.1655	.1486	.1301
3/4	.3848	.3696	.3632	.3484	.3339	.3157	.3019	.2884	.2679	.2463	.2223
7/8	.5346	.5166	.5090	.4914	.4742	.4525	.4359	.4197	.3948	.3685	.3390
1	.7088	.6881	.6793	.6590	.6390	.6138	.5945	.5755	.5463	.5153	.4803
1¼	1.131	1.105	1.094	1.068	1.042	1.010	.9852	.9607	.9229	.8825	.8365
1½	1.651	1.612	1.606	1.575	1.544	1.504	1.474	1.444	1.398	1.348	1.291
2	2.986	2.944	2.926	2.883	2.841	2.788	2.746	2.705	2.642	2.573	2.490

COPPER TUBING TO SCHEDULE 40 PIPE — EQUIVALENT FLOW CAPACITY

Copper tubing is a good plumbing medium for compressed air but is not recommended for hydraulic oil plumbing. This chart shows the size tubing which should be used to connect into components which have NPTF pipe thread portholes. For example, if the porthole size is 1/4" NPTF, 3/8" O.D. tubing must be used if full flow capacity is to be maintained. Usually brass fittings or braze-type fittings are recommended for permanent installations. The SAE flare angle of 45° is most often used rather than the JIC angle of 37° as used on steel tubing for hydraulics. Ferrule type or plastic fittings are not recommended for permanent plumbing.

If Pipe Size is:	1/8" NPT	1/4" NPT	3/8" NPT	1/2" NPT	3/4" NPT	1" NPT
Nearest Equivalent Tubing Size is:	1/4" or 5/16"	3/8" O.D.	1/2" O.D.	5/8" O.D.	3/4" O.D.	1" O.D.

HOSE — EQUIVALENT PIPE AND TUBING SIZES

Hose is not standarized to the same extent as pipe and tubing. There is a wide variation in dimensions, pressure ratings, and availability between manufacturers.

Hose is specified by its inside diameter, overall length fitting to fitting, type and size of end fittings, pressure rating, composition, and temperature range. This table is limited to showing characteristics of one popular brand of low pressure, small diameter hose. Space does not permit listing the wide variety of sizes and types which are available. Consult catalogs of hose manufacturers.

Hose I.D.	Hose O.D.	Minimum Bend Radius	Working Pressure	Min. Burst Pressure	Equivalent Pipe Size	Nearest Equiv. Tubing Size
1/4"	.50"	4"	400 PSI	1250 PSI	1/8" NPT	5/16" O.D.
3/8"	.63"	4"	300 PSI	1000 PSI	1/4" NPT	3/8" or 1/2"
1/2"	.78"	6"	150 PSI	750 PSI	3/8" NPT	1/2" or 5/8"
5/8"	.91"	6"	140 PSI	700 PSI	1/2" NPT	5/8" or 3/4"

PRESSURE LOSS DUE TO OIL FLOW THROUGH PIPES

This table has been calculated from a formula published by the Crane Company on Page 3-12 of Technical Paper 410. It shows the *approximate* pressure loss per 100 feet of Schedule 40 pipe with hydraulic oil of known specific gravity and known viscosity flowing through it.

The formula used is: $\Delta P = 0.0668\ \mu\nu \div D^2$, in which: ΔP is pressure loss per 100 feet of pipe; μ is viscosity in centipoises (not SSU); ν is the flow velocity in feet per second; D is inside diameter of pipe, in inches.

Note: Absolute viscosity in centipoises must be used in the formula. For any fluid this is kinematic viscosity in centistokes times the specific gravity. An absolute viscosity of 40 centipoises was used for calculating the table. This corresponds approximately to a hydraulic oil with 0.9 specific gravity and a viscosity of 220 SSU (or 44.4 centistokes). See next page for other fluids.

TABLE 1. Pressure Loss per 100 Feet of Schedule 40 Pipe With Oil of 220 SSU and 0.9 Specific Gravity

See Next Page for Adjustment to Other Fluids and Viscosities

GPM	Pipe Size*	Pres. Drop**	Flow Veloc†	GPM	Pipe Size*	Pres. Drop**	Flow Veloc†	GPM	Pipe Size*	Pres. Drop**	Flow Veloc†
3	1/8	624	17	40	3/4	95	24	80	1	75	31
	1/4	187	9.3		1	36	15		1¼	24	17
	3/8	55	5.0		1¼	12	8.6		1½	13	13
	1/2	22	3.2		1½	6.5	6.3		2	4.8	7.7
	3/4	7.1	1.8		2	2.4	3.8		2½	2.3	5.4
6	1/4	373	19	45	3/4	106	27	90	1	80	33
	3/8	111	10		1	41	17		1¼	27	19
	1/2	44	6.3		1¼	14	9.7		1½	15	14
	3/4	14	3.6		1½	4.4	7.1		2	5.4	8.6
	1	5.4	2.2		2	2.7	4.3		2½	2.6	6.0
10	3/8	185	17	50	3/4	122	31	100	1	92	38
	1/2	73	11		1	46	19		1¼	30	22
	3/4	24	6.0		1¼	15	11		1½	16	16
	1	9.0	3.7		1½	8.1	7.9		2	6.0	9.6
	1¼	3.0	2.2		2	3.0	4.8		2½	2.9	6.7
15	1/2	109	16	55	3/4	130	33	125	1	114	47
	3/4	36	9.0		1	50	20		1¼	38	27
	1	14	5.6		1¼	17	12		1½	20	20
	1¼	4.5	3.2		1½	8.9	8.7		2	7.5	12
	1½	2.4	2.4		2	3.3	5.3		2½	9.8	8.4
20	1/2	146	21	60	3/4	142	36	150	1¼	44	31
	3/4	47	12		1	53	22		1½	24	24
	1	18	7.4		1¼	18	13		2	8.9	14
	1¼	6.0	4.3		1½	9.8	9.5		2½	4.4	10
	1½	3.2	3.2		2	3.6	5.7		3	1.8	6.4
25	1/2	180	26	65	3/4	154	39	175	1¼	53	38
	3/4	59	15		1	59	24		1½	29	28
	1	23	9.3		1¼	20	14		2	10	17
	1¼	7.6	5.4		1½	11	10		2½	5.1	12
	1½	4.0	3.9		2	3.9	6.2		3	2.2	7.6
30	1/2	214	31	70	3/4	205	42	200	1¼	60	43
	3/4	71	18		1	63	26		1½	32	31
	1	27	11		1¼	21	15		2	12	19
	1¼	9.0	6.4		1½	11	11		2½	5.9	13
	1½	4.8	4.7		2	4.2	6.7		3	2.5	8.7
35	1/2	249	36	75	1	68	28	225	1¼	69	49
	3/4	83	21		1¼	23	16		1½	37	36
	1	32	13		1½	12	12		2	13	22
	1¼	11	7.5		2	4.5	7.2		2½	6.6	15
	1½	5.7	5.5		2½	2.2	5.0		3	2.8	9.8

*Standard Schedule 40 pipe.   **Pressure loss per 100 feet of pipe.   †Oil flow velocity in feet per second.

TABLE 2. Factors for Converting Table 1 for Use With Steel Tubing

For pressure loss per 100 feet of steel tubing, use the nearest NPT size shown in this table. Find pressure loss from Table 1 on preceding page. Then multiply this loss times the factor shown in the last column of this table.

Example: For flow of 50 GPM through a 1½" O.D. tube with .095 wall, use the 1¼" pipe size under 50 GPM in Table 1. This shows a 265 PSI loss per 100 feet. Multiply this times 1.11 from Table 2 = 295 PSI per 100 feet loss.

Tube O.D.	Wall Thick.	Tube I.D.	Use NPT	Mult. by	Tube O.D.	Wall Thick.	Tube I.D.	Use NPT	Mult. by	Tube O.D.	Wall Thick.	Tube I.D.	Use NPT	Mult. by
3/16	.032	.124	1/4	8.69	3/4	.049	.652	1/2	.910	1¼	.072	1.106	1	.901
1/4	.035	.180	1/4	4.09		.058	.634	1/2	.962		.083	1.084	1	.938
	.042	.166	1/4	4.81		.065	.620	1/2	1.01		.095	1.060	1	.981
	.049	.152	1/4	5.73		.072	.606	1/2	1.08		.109	1.032	1	1.03
	.058	.134	1/4	7.38		.083	.584	1/2	1.13		.120	1.010	1	1.08
	.065	.120	1/4	9.20		.095	.560	1/2	1.23	1½	.065	1.370	1¼	1.01
3/8	.035	.305	1/4	1.42		.109	.532	1/2	1.37		.072	1.356	1¼	1.04
	.042	.291	1/4	1.56	1	.049	.902	3/4	.835		.083	1.334	1¼	1.07
	.049	.277	1/4	1.73		.058	.884	3/4	.869		.095	1.310	1¼	1.11
	.058	.259	1/4	1.97		.065	.870	3/4	.897		.109	1.282	1¼	1.16
	.065	.245	1/4	2.21		.072	.856	3/4	.927		.120	1.260	1¼	1.20
1/2	.035	.430	3/8	1.31		.083	.834	3/4	.976	2	.065	1.870	2	1.22
	.042	.416	3/8	1.40		.095	.810	3/4	1.03		.072	1.856	2	1.24
	.049	.402	3/8	1.50		.109	.782	3/4	1.11		.083	1.834	2	1.27
	.058	.384	3/8	1.65		.120	.760	3/4	1.18		.095	1.810	2	1.30
	.065	.370	3/8	1.78	1¼	.049	1.152	1	.830		.109	1.782	2	1.35
	.072	.356	3/8	2.01		.058	1.134	1	.857		.120	1.760	2	1.38
	.083	.334	3/8	2.18		.065	1.120	1	.878		.134	1.732	2	1.42

How to Adjust for Other Fluids and Conditions

First use Table 2, if necessary, to find factor to convert a tube size to equivalent pipe size. Next, use Table 1 to find pressure loss per 100 feet. Next, use Table 3 to adjust for viscosities other than 220 SSU. If using a fluid other than oil, adjust for its gravity as explained elsewhere on this page.

Generally, as shown by the formula at the top of the preceding page, pressure loss increases in direct proportion to a velocity increase. This can also be seen in the velocity columns of Table 1.

Always remember that *centistoke* viscosity defines only the flow resistance to shear in the fluid. *Centipoise* viscosity defines the combined flow resistance including both shear in the fluid and specific gravity. Centipoises = centistokes x specific gravity.

Adjusting for Other Gravities

Pressure loss through a pipe is directly proportional to specific gravity of the fluid. Other hydraulic fluids have a higher specific gravity than petroleum oil and (at the same viscosity) will have a higher pressure loss. Water/oil emulsion will have a 7% higher, water/glycol a 14%, and phosphate ester will have a 22% higher pressure loss than petroleum oil.

TABLE 4. Pressure Loss Through Fittings

Pressure loss through common fittings is shown in terms of the equivalent length of straight pipe of the same size. *Example:* The flow from the side outlet of a 1½" tee suffers approximately the same pressure loss as if it were flowing through a 9-foot straight length of the same size pipe. For pipe sizes less than 1/2", pressure loss through fittings is little more than for a straight section of the same length (The Crane Company).

	NPT Pipe Size							
Tee,	3/4"	1"	1¼"	1½"	2"	2½"	3	3½"
Side →	4½	5½	7½	9	11½	14	16½	20
45° El	1	1¼	1¾	2	2½	3	3¾	4½
90° El	2	2¾	3¼	4¼	5	6	8	9½

Adjusting for Other Viscosities

Pressure loss through a pipe is directly proportional to viscosity in centistokes (for a given specific gravity). This table may be used with Table 1 to adjust pressure loss per 100 feet to oil with viscosity other than 220 SSU (44.4 centistokes).

Example: A hydraulic oil of 500 SSU will have a higher pressure loss than shown in the table by a factor of 2.48 for the same size pipe and the same flow.

In using Table 3, multiply factor in 3rd column times the pressure loss taken from Table 1.

Water is a special case. For plain water, pressure loss will be approximately half the values shown in Table 1.

TABLE 3

SSU Vis.	Centistokes	Factor
80	15.8	.356
100	20.8	.468
150	33	.743
300	65	1.46
400	87	1.96
500	110	2.48
750	163	3.67
1000	220	4.95
2000	420	9.46
3000	630	14.2
4000	850	19.1

INTERCHANGE BETWEEN SI, METRIC, AND U. S. CUSTOMARY UNITS

International Standard (SI) units are shown in the first column of each chart. Values with exponents can be handled directly on a pocket calculator which has an exponent key.

For manual calculations, remember that the + or – sign in front of an exponent tells whether to move the decimal point to the right (for a + sign), or to the left (for a – sign), and the exponent tells how far to move it. Examples: $2.540 \times 10^{-5} = .0000254$, and $3.048 \times 10^{2} = 304.8$, etc.

Equivalent values of all units are shown on the same horizontal line. Perhaps the easiest way to use the charts is to look down the column of the unit to be converted and find the line on which the "1" appears. Then move to the left or right to the column of the desired new unit. That figure is a multiplier.

Example: Look down the "Inch" column to the "1" line. The chart shows 1 inch = 1.578×10^{-5} mile. Therefore, 627 inches would be: 627 x [1.578×10^{-5}] miles, etc.

TORQUE

Newton-Metres	Kilopond-Mtrs.	Foot-Pounds	Inch-Lbs.
1	1.020×10^{-1}	7.376×10^{-1}	8.851
9.807	1	7.233	86.80
1.356	1.382×10^{-1}	1	12
1.130×10^{-1}	1.152×10^{-2}	8.333×10^{-2}	1

LENGTH (Linear Measurement)

Metre	Centimetre	Millimetre	Kilometre	Mile	Inch	Foot
1	100	1000	1×10^{-3}	6.214×10^{-4}	39.370	3.281
0.01	1	10	1×10^{-5}	6.214×10^{-6}	3.937×10^{-1}	3.281×10^{-2}
1×10^{-3}	0.10	1	1×10^{-6}	6.214×10^{-7}	3.937×10^{-2}	3.281×10^{-3}
1×10^{3}	1×10^{5}	1×10^{6}	1	6.214×10^{-1}	3.937×10^{4}	3.281×10^{3}
1.609×10^{3}	1.609×10^{5}	1.609×10^{6}	1.609	1	6.336×10^{4}	5280
2.540×10^{-2}	2.540	25.40	2.540×10^{-5}	1.578×10^{-5}	1	8.333×10^{-2}
3.048×10^{-1}	30.479	3.048×10^{2}	3.048×10^{-4}	1.894×10^{-4}	12	12

VOLUME — (Cubic)

Cubic Metre	Cu. Decimetre (Litre)	Cu. Centimetre	Imperial Gallon	U.S. Gallon	Cubic Inch	Cubic Foot
1	1×10^{3}	1×10^{6}	2.20×10^{2}	2.642×10^{2}	6.102×10^{4}	35.314
1×10^{-3}	1	1×10^{3}	2.20×10^{-1}	2.642×10^{-1}	61.024	3.531×10^{-2}
1×10^{-6}	1×10^{-3}	1	2.20×10^{-4}	2.642×10^{-4}	6.102×10^{-2}	3.531×10^{-5}
4.546×10^{-3}	4.546	4.546×10^{3}	1	1.200	2.774×10^{2}	1.605×10^{-1}
3.785×10^{-3}	3.785	3.785×10^{3}	8.327×10^{-1}	1	2.310×10^{2}	1.337×10^{-1}
1.639×10^{-5}	1.639×10^{-2}	16.387	3.605×10^{-3}	4.329×10^{-3}	1	5.787×10^{-4}
2.832×10^{-2}	28.317	2.832×10^{4}	6.229	7.481	1.728×10^{3}	1

AREA — (Square measurement)

Square Metre	Sq. Centimetre	Sq. Millimetre	Sq. Kilometre	Square Inch	Square Foot	Square Mile
1	1×10^{4}	1×10^{6}	1×10^{-6}	1.550×10^{3}	10.764	3.861×10^{-7}
1×10^{-4}	1	100	1×10^{-10}	1.550×10^{-1}	1.076×10^{-3}	3.861×10^{-11}
1×10^{-6}	1×10^{-2}	1	1×10^{-12}	1.550×10^{-3}	1.076×10^{-5}	3.861×10^{-13}
1×10^{6}	1×10^{-10}	1×10^{12}	1	1.550×10^{9}	1.076×10^{7}	3.861×10^{-1}
6.452×10^{-4}	6.452	6.452×10^{2}	6.452×10^{-10}	1	6.944×10^{-3}	2.491×10^{-10}
9.290×10^{-2}	9.290×10^{2}	9.290×10^{4}	9.290×10^{-8}	144	1	3.587×10^{-8}
2.590×10^{6}	2.590×10^{10}	2.590×10^{12}	2.590	4.014×10^{9}	2.788×10^{7}	1

FORCE — (Including force due to weight)

Newton	Dyne	Kilopond	Metric Ton (Tonne)	Long Ton	U.S. Ton	Pound
1	1×10^{5}	1.020×10^{-1}	1.020×10^{-4}	1.004×10^{-4}	1.124×10^{-4}	2.248×10^{-1}
1×10^{-5}	1	1.020×10^{-6}	1.020×10^{-9}	1.004×10^{-9}	1.124×10^{-9}	2.248×10^{-6}
9.807	9.807×10^{5}	1	1×10^{-3}	9.842×10^{-4}	1.102×10^{-3}	2.205
9.807×10^{3}	9.807×10^{8}	1000	1	9.842×10^{-1}	1.102	2.205×10^{3}
9.964×10^{3}	9.964×10^{8}	1.016×10^{3}	1.016	1	1.120	2.240×10^{3}
8.896×10^{3}	8.896×10^{8}	9.072×10^{2}	9.072×10^{-1}	8.929×10^{-1}	1	2000
4 448	4.448×10^{5}	4.536×10^{-1}	4.536×10^{-4}	4.464×10^{-4}	5×10^{-4}	1

MASS — (Not Weight)

Kilogram	Gram	Metric Ton (Tonne)	Newton	Pound	Slug	U.S. Ton
1	1000	1×10^{-3}	9.807	2.205	6.853×10^{-2}	1.102×10^{-3}
1×10^{-3}	1	1×10^{-6}	9.807×10^{-3}	2.205×10^{-3}	6.853×10^{-5}	1.102×10^{-6}
1×10^{3}	1×10^{6}	1	9.807×10^{3}	2.205×10^{3}	68.530	1.102
1.020×10^{-1}	1.020×10^{2}	1.020×10^{-4}	1	2.248×10^{-1}	6.988×10^{-3}	1.124×10^{-4}
4.536×10^{-1}	4.536×10^{2}	4.536×10^{-4}	4.448	1	3.108×10^{-2}	5×10^{-4}
14.594	1.459×10^{4}	1.459×10^{-2}	1.431×10^{2}	32.170	1	1.609×10^{-2}
9.072×10^{2}	9.072×10^{5}	9.072×10^{-1}	8.896×10^{3}	2000	62.170	1

VELOCITY

Metres/Sec.	Decimetres/Sec.	Kilometres/Hr.	Miles/Hr.	Feet/Min.	Feet/Sec	Inches/Min.
1	10	3.6	2.237	1.968×10^{2}	3.281	2.362×10^{3}
1×10^{-1}	1	1×10^{-4}	6.214×10^{-5}	5.468×10^{-3}	9.113×10^{-5}	6.562×10^{-2}
2.778×10^{-1}	2.278	1	6.214×10^{-1}	5.468×10^{1}	9.113×10^{-1}	6.562×10^{2}
4.470×10^{-1}	4.470	1.609	1	88	1.467	1.056×10^{3}
5.080×10^{-3}	5.080×10^{-2}	1.829×10^{-2}	1.136×10^{-2}	1	1.667×10^{-2}	12
3.048×10^{-1}	3.048	1.097	6.818×10^{-1}	60	1	7.2×10^{2}
4.233×10^{-4}	4.233×10^{-3}	1.524×10^{-3}	9.470×10^{-4}	8.333×10^{-2}	1.389×10^{-3}	1

UNIT PRESSURE (Either fluid or mechanical)

Bar	Newtons/m² (Pascal)	Kilopond/m²	Kilopond/cm²	Atmosphere	Pounds/Ft²	Pounds/Inch²
1×10^{-5}	1	1.020×10^{-1}	1.020×10^{-5}	9.869×10^{-6}	2.088×10^{-2}	1.45×10^{-4}
1	1×10^{5}	1.020×10^{4}	1.020	9.869×10^{-1}	2.088×10^{3}	14.5
9.807×10^{-5}	9.807	1	1×10^{-4}	9.678×10^{-5}	2.048×10^{-1}	1.422×10^{-3}
9.807×10^{-1}	9.807×10^{4}	1×10^{4}	1	9.678×10^{-1}	2.048×10^{3}	14.220
1.013	1.013×10^{5}	1.033×10^{4}	1.033	1	2.116×10^{3}	14.693
4.789×10^{-4}	47.893	4.884	4.884×10^{-4}	4.726×10^{-4}	1	6.944×10^{-3}
6.897×10^{-2}	6.897×10^{3}	7.033×10^{2}	7.033×10^{-2}	6.806×10^{-2}	1.440×10^{2}	1

POWER — (Fluid, Electrical, or Mechanical)

Kilowatt	Watt, Joules/s and N-m/s	U.S. & U.K Horsepower	Foot-Pounds per Minute	Foot-Pounds per Second	BTU per Hour	BTU per Min.
1	1000	1.340	4.425×10^{4}	7.376×10^{2}	3.412×10^{3}	56.862
1×10^{-3}	1	1.340×10^{-3}	44.254	7.376×10^{-1}	3.412	5.686×10^{-2}
7.461×10^{-1}	746	1	3.300×10^{4}	5.500×10^{2}	2.545×10^{3}	42.44
2.260×10^{-5}	2.260×10^{-2}	3.029×10^{-5}	1	1.667×10^{-2}	7.710×10^{-2}	1.285×10^{-3}
1.356×10^{-3}	1.356	1.817×10^{-3}	60	1	4.626	7.710×10^{-2}
2.931×10^{-4}	2.931×10^{-1}	3.928×10^{-4}	12.971	2.162×10^{-1}	1	1.667×10^{-2}
1.759×10^{-2}	17.586	2.357×10^{-2}	7.783×10^{2}	12.971	60	1

ENERGY OR WORK

Kilowatt-Hour	Watt-second Joule, or N-m	Dyne-Cm. or Erg	Horsepower-Hr.	Foot-Pound	Inch-Pound	BTU
1	3.6×10^{6}	3.6×10^{13}	1.341	2.655×10^{6}	3.187×10^{7}	3.412×10^{3}
2.778×10^{-7}	1	1×10^{7}	3.725×10^{-7}	7.376×10^{-1}	8.851	9.477×10^{-4}
2.778×10^{-14}	1×10^{-7}	1	3.725×10^{-14}	7.376×10^{-8}	8.851×10^{-7}	9.477×10^{-11}
7.457×10^{-1}	2.685×10^{6}	2.685×10^{13}	1	1.980×10^{6}	2.376×10^{7}	2.544×10^{3}
3.766×10^{-7}	1.356	1.356×10^{7}	5.051×10^{-7}	1	12	1.285×10^{-3}
3.138×10^{-8}	1.130×10^{-1}	1.130×10^{6}	4.209×10^{-8}	8.333×10^{-2}	1	1.071×10^{-4}
2.931×10^{-4}	1.055×10^{3}	1.055×10^{10}	3.931×10^{-4}	7.783×10^{2}	9.339×10^{3}	1

WORKING WITH METRIC CYLINDERS

Sometime in the future, cylinders as well as other fluid power components will be built to ISO (International Standards Organization) dimensions in which metric measurements are used. Conversion to international standards has been slow in the United States, and at this writing the availability of metric dimension cylinders is quite limited, but complete conversion will come in due time.

Cylinder charts in this section cover standardized bore and rod combinations from 25mm through 200 mm bore and with minimum and maximum size piston rods. Intermediate size piston rods will no doubt be offered by most manufacturers. ISO standard sizes also include bore sizes of 8, 10, 12, 16, 20, 250, 320, and 400 mm.

Calculations of cylinder force and velocity are not quite as straightforward as in the U. S. system because of extra conversions between units which become necessary. The ISO units which will be used in cylinder calculations are these:

FORCE. Force values are in Newtons (N). One Newton is equal to about ¼ pound (.2248 lb. to be exact), or 1 pound is equal to about 4½ N (4.448 N to be exact). This unit should serve for most cylinder calculations except where very large forces are involved in which the kilo Newton (kN) equal to 1000 N may be used.

PISTON AREA. Piston bore is cataloged in units of millimetres (mm) as shown in the charts. For area, the mm^2 is too small for convenient calculations, so the unit for piston surface area will be the square centimetre (cm^2). To calculate piston area change bore diameter to cm. by dividing by 10. Then find cm^2 piston area with the formula: $A = \pi r^2$ in the usual manner.

PRESSURE. Fluid pressure will usually be expressed in kilo Pascals (kPa) because the Pascal which is defined as one Newton of force per square metre, is such a small unit that it is hard to work with in making calculations.

METRIC AIR CYLINDERS — FORCE CHART — 3 TO 6 BARS PRESSURE

Bars	3	3½	4	4½	5	5½	6
Kilo Pascals	300	350	400	450	500	550	600
PSI	43.5	50.8	58.0	65.3	72.5	79.8	87.0

Bore mm.	Bore cm.	Area sq. cm.	Theoretical Cylinder Force in Newtons						
25	2.5	4.91	147	172	196	221	245	270	294
32	3.2	8.04	241	281	322	362	402	442	483
40	4.0	12.57	377	440	503	565	628	691	754
50	5.0	19.63	589	687	785	884	982	1 080	1 178
63	6.3	31.17	935	1 091	1 247	1 403	1 559	1 714	1 870
80	8.0	50.27	1 508	1 759	2 011	2 262	2 513	2 765	3 016
100	10.0	78.54	2 356	2 749	3 142	3 534	3 927	4 320	4 712
125	12.5	122.72	3 681	4 295	4 909	5 522	6 136	6 749	7 363
160	16.0	201.06	6 032	7 037	8 042	9 048	10 053	11 058	12 064
200	20.0	314.16	9 225	10 996	12 566	14 137	15 708	17 279	18 850

METRIC AIR CYLINDERS — FORCE CHART — 6½ TO 11 BARS PRESSURE

Bars	6½	7	7½	8	9	10	11
Kilo Pascals	650	700	750	800	900	1000	1100
PSI	94.3	102	109	116	131	145	160

Bore mm.	Bore cm.	Area sq. cm.	Theoretical Cylinder Force in Newtons						
25	2.5	4.91	319	343	368	393	442	491	540
32	3.2	8.04	523	563	603	643	724	804	885
40	4.0	12.57	817	880	942	1 005	1 131	1 257	1 382
50	5.0	19.63	1 276	1 374	1 473	1 571	1 767	1 963	2 160
63	6.3	31.17	2 026	2 182	2 338	2 494	2 805	3 117	3 429
80	8.0	50.27	3 267	3 519	3 770	4 021	4 524	5 027	5 529
100	10.0	78.54	5 105	5 498	5 890	6 283	7 069	7 854	8 639
125	12.5	122.72	7 977	8 590	9 204	9 817	11 044	12 272	13 499
160	16.0	201.06	13 069	14 072	15 080	16 085	18 095	20 106	22 117
200	20.0	314.16	20 420	21 991	23 562	25 133	28 274	31 416	34 557

One kPa is equal to 1000 Pa.

The bar is a more convenient unit for fluid pressure and will be allowed, at least for a limited time. The bar is related to the Pascal. One bar = 100,000 Pascals or 100 kPa. It is also equal to 14.5 PSI which is very close to one atmosphere. Pressure values in the following charts are given in three pressure units, bars, kPa, and PSI, to help the student get a "feel" for the way metric pressure units compare with PSI units he has been using.

PUMP FLOW. Oil flow from a hydraulic pump is expressed in litres per minute (l/min). A litre is defined as one cubic decimetre (dm^3), and is roughly ¼ gallon (.2642 gal. to be exact). Or, 1 gal. = 3.785 litres. On very large flows units of litres per second (l/s) can be used.

Cylinder Calculations

FORCE CALCULATION. Cylinder force is calculated by multiplying piston surface area times fluid pressure.

$$F = A \times P \div 10, \text{ in which:}$$

F = force, in Newtons (N).
A = piston area in square centimetres (cm^2).
P = differential pressure across ports in kPa.
10 is a necessary metric conversion constant.

When working with pressure in bars, the formula becomes:

$$F = A \times P \times 10, \text{ in which,}$$

P is pressure differential in bars.

VELOCITY CALCULATIONS. The travel speed of a cylinder piston is calculated with this formula:

$$S = V \times 10 \div A, \text{ in which:}$$

S = travel speed expressed in metres per minute.
V = Pump oil flow in l/min. (dm^3/min).
A = Piston or net area, in cm^2.
10 is a necessary metric conversion between cm and dm.

Refer to Appendix D of Volume 1 for more data on metric cylinders

METRIC HYDRAULIC CYLINDERS — FORCE CHART — 25 TO 175 BARS PRESSURE

Bars			25	50	75	100	125	150	175
Kilo Pascals			2500	5000	7500	10 000	12 500	15 000	17 500
PSI			363	725	1 088	1 450	1 813	2 175	2 538
Bore mm.	*Bore cm.*	*Area sq. cm.*	Theoretical Cylinder Force in Newtons						
25	2.5	4.91	1 227	2 454	3 680	4 907	6 138	7 361	8 587
32	3.2	8.04	2 011	4 021	6 032	8 042	10 053	10 053	12 063
40	4.0	12.57	3 142	6 283	9 425	12 566	15 708	18 849	21 991
50	5.0	19.63	4 909	9 817	14 726	19 634	24 543	29 451	34 360
63	6.3	31.17	7 793	15 585	23 378	31 170	38 963	46 755	54 548
80	8.0	50.27	12 566	25 133	37 699	50 265	62 831	75 398	87 964
100	10.0	78.54	19 635	39 270	58 904	78 539	98 174	117 809	137 443
125	12.5	122.72	30 679	61 358	92 037	122 716	153 395	184 074	214 753
160	16.0	201.06	50 265	100 531	150 796	201 061	251 326	301 592	351 857
200	20.0	314.16	78 540	157 080	235 619	314 159	392 699	471 239	549 778

METRIC HYDRAULIC CYLINDERS — FORCE CHART — 200 TO 350 BARS PRESSURE

Bars			200	225	250	275	300	325	350
Kilo Pascals			20 000	22 500	25 000	27 500	30 000	32 500	35 000
PSI			2 900	3 263	3 625	3 988	4 350	4 713	5 075
Bore mm.	*Bore cm.*	*Area sq. cm.*	Theoretical Cylinder Force in Newtons						
25	2.5	4.91	9 814	11 041	12 268	13 494	14 721	15 948	17 175
32	3.2	8.04	16 084	18 095	20 105	22 116	24 126	26 137	28 147
40	4.0	12.57	25 132	28 274	31 415	34 557	37 698	40 840	43 981
50	5.0	19.63	39 268	44 177	49 085	53 994	58 902	63 811	68 719
63	6.3	31.17	62 340	70 133	77 925	85 718	93 510	101 303	109 095
80	8.0	50.27	100 530	113 096	125 663	138 229	150 795	163 361	175 928
100	10.0	78.54	157 078	176 713	196 348	215 982	235 617	255 252	274 887
125	12.5	122.72	245 432	276 111	306 790	337 469	368 148	398 827	429 506
160	16.0	201.06	402 122	452 387	502 653	552 918	603 183	653 448	703 714
200	20.0	314.16	628 318	706 858	785 398	863 937	942 477	1 021 017	1 099 557

DRIVE HORSEPOWER FOR A HYDRAULIC PUMP

Figures in the body of this table show the horsepower needed to drive a hydraulic pump having an efficiency of 85%. Most positive displacement pumps (gear, vane, piston) fall in the range of 80% to 90% efficiency so this chart should be accurate to within 5% for almost any pump. The table was calculated from the formula:

$$HP = PSI \times GPM \div [1714 \times 0.85]$$

For pumps with other than 85% efficiency, the formula can be used, substituting actual efficiency, in decimals, in place of 0.85.

Using the Table. The range of 500 to 5000 PSI covers most hydraulic systems, but power requirements can be determined for conditions outside the range of the table, or for intermediate values. For example, power at 4000 PSI will be exactly 2 times the figure shown for 2000 PSI. At 77 GPM, power will be the sum of the figures shown in the 75 and 2 GPM lines, etc. For systems operating below 500 PSI, horsepower calculations tend to become inaccurate because mechanical friction losses reduce pump efficiency.

Rules-of-Thumb. Approximate HP requirement can be estimated by our "rule of 1500" which states that 1 HP is required for each multiple of 1500 when multiplying PSI x GPM. For example, a 5 GPM pump at 1500 PSI would require 5 HP, or at 3000 PSI would require 10 HP. A 10 GPM pump at 1000 PSI would require 6-2/3 HP or the same pump at 1500 PSI would require 10 HP, etc.

Another rule-of-thumb states that about 5% of the pump maximum rated HP is required to idle the pump when it is "unloaded" and the full flow is circulating at near 0 PSI. This amount of power is consumed in flow losses plus mechanical friction losses in bearings and pumping elements.

Figures in Body of Table are HP's Required to Drive a Hydraulic Pump (A Pump Efficiency of 85% is Assumed)

GPM	500 PSI	750 PSI	1000 PSI	1250 PSI	1500 PSI	1750 PSI	2000 PSI	2500 PSI	3000 PSI	3500 PSI	4000 PSI	5000 PSI
3	1.03	1.54	2.06	2.57	3.09	3.60	4.12	5.15	6.18	7.21	8.24	10.3
5	1.72	2.57	3.43	4.29	5.15	6.00	6.86	8.58	10.3	12.0	13.7	17.2
7½	2.57	3.86	5.15	6.43	7.72	9.01	10.3	12.9	15.4	18.0	20.6	25.7
10	3.43	5.15	6.86	8.58	10.3	12.0	13.7	17.2	20.6	24.0	27.5	34.3
12½	4.29	6.43	8.58	10.7	12.9	15.0	17.2	21.4	25.7	30.0	34.3	42.9
15	5.15	7.72	10.3	12.9	15.4	18.0	20.6	25.7	30.9	36.0	41.2	51.5
17½	6.01	9.01	12.0	15.0	18.0	21.0	24.0	30.0	36.0	42.0	48.0	60.1
20	6.86	10.3	13.7	17.2	20.6	24.0	27.5	34.3	41.2	48.0	54.9	68.6
22½	7.72	11.6	15.4	19.3	23.2	27.0	30.9	38.6	46.3	54.1	61.8	77.2
25	8.58	12.9	17.2	21.4	25.7	30.0	34.3	42.9	51.5	60.1	68.6	85.8
30	10.3	15.4	20.6	25.7	30.9	36.0	41.2	51.5	61.8	72.1	82.4	103
35	12.0	18.0	24.0	30.0	36.0	42.0	48.0	60.1	72.1	84.1	96.1	120
40	13.7	20.6	27.5	34.3	41.2	48.0	54.9	68.6	82.4	96.1	110	137
45	15.4	23.2	30.9	38.6	46.3	54.1	61.8	77.2	92.7	108	124	154
50	17.2	25.7	34.3	42.9	51.5	60.1	68.6	85.8	103	120	137	172
55	18.9	28.3	37.8	47.2	56.6	66.1	75.5	94.4	113	132	151	189
60	20.6	30.9	41.2	51.5	61.8	72.1	82.4	103	124	144	165	206
65	22.3	33.5	44.6	55.8	66.9	78.1	89.2	112	134	156	178	223
70	24.0	36.0	48.0	60.1	72.1	84.1	96.1	120	144	168	192	240
75	25.7	38.6	51.5	64.3	77.2	90.1	103	129	154	180	206	257
80	27.5	41.2	54.9	68.7	82.4	96.1	110	137	165	192	220	275
85	29.2	43.8	58.3	72.9	87.5	102	117	146	175	204	233	292
90	30.9	46.3	61.8	77.2	92.7	108	124	154	185	216	247	309
100	34.3	51.5	68.6	85.8	103	120	137	172	206	240	275	343

Oversizing or Undersizing of an Induction-Type Electric Motor

Optimum results are obtained if HP rating of an electric motor is neither too far oversize or undersize for the job. Some effects of motor power mismatch are:

Oversize Motor. Using a 20 HP motor to do a 10 HP job, for example, will give good results as far as running the fluid power system is concerned, but it will consume a little more current for the same 10 HP output. It will also cause the power factor of the plant electric system to be poorer, with higher power cost. Idling current will be higher so more power will be wasted during periods in the cycle when the motor is running at idle condition.

Undersize Motor. A 3-phase induction motor can usually be safely overloaded during peak parts of the cycle as explained on Page 195, but during these peak periods, motor current will be all out of proportion to the excess power being produced with a considerable amount of overheating. If the motor is too far undersize it will, of course, burn out in a short time.

HP REQUIRED FOR COMPRESSING AIR

Values in this 3-part table are for single-stage, 2-stage, and 3-stage piston-type air compressors operating at 85% efficiency and working on intake air under approximately standard conditions. Compression conditions intermediate between adiabatic and isothermal have been assumed, which we believe are representative of conditions existing in the shop air supply of most plants. (See explanation of these terms below). The table was prepared from information in Machinery's Handbook. Please refer to your copy of the Handbook for more information on air compression and for formulae from which the tables were calculated.

Of course, if your compressor operates with efficiency greater or less than 85%, an allowance can be made in table values.

Explanation of the Table

The table is useful either in determining the HP for a new application or for checking the capacity of an existing system for addition of more air-operated equipment, especially equipment which requires a large amount of air. Figures in the tables are HP to compress 1 SCFM from 0 PSIG to the gauge pressures shown. Calculate cylinder air consumption from Page 257, then multiply times the figure in the table.

Adiabatic Compression. This is defined as compression taking place without allowing the escape of heat of compression. This is a theoretical condition because cooling starts immediately after compression.

Isothermal Compression. This is compression which takes place over a period of time, allowing escape of all heat of compression. This is also a theoretical condition.

Single-Stage Compressor, 85% Eff.		Two-Stage Compressor, 85% Eff.		Three-Stage Compressor, 85% Eff.	
PSIG	HP*	PSIG	HP*	PSIG	HP*
5	.021	50	.116	100	.159
10	.033	60	.128	150	.190
15	.056	70	.138	200	.212
20	.067	80	.148	250	.230
25	.079	90	.155	300	.240
30	.089	100	.164	350	.258
35	.099	110	.171	400	.269
40	.108	120	.178	450	.279
45	.116	130	.185	500	.289
50	.123	140	.190	550	.297
55	.130	150	.196	600	.305
60	.136	160	.201	650	.311
65	.143	170	.206	700	.317
70	.148	180	.211	750	.323
75	.155	190	.216	800	.329
80	.160	200	.220	850	.335
85	.165	210	.224	900	.340
90	.170	220	.228	950	.345
95	.175	230	.232	1000	.290
100	.179	240	.236	1050	.354
110	.191	250	.239	1100	.358
120	.196	260	.243	1150	.362
130	.204	270	.246	1200	.366
140	.211	280	.250	1250	.370
150	.218	290	.253	1300	.374
160	.225	300	.255	1350	.378
170	.232	350	.269	1400	.380
180	.239	400	.282	1450	.383
190	.244	450	.289	1500	.386
200	.250	500	.303	1550	.390

*HP to compress 1 SCFM from 0 PSIG to the values shown.

HP LOST THROUGH A PRESSURE REGULATOR

Air compressor power is wasted by compressing air to a pressure higher than necessary then reducing it through a regulator. The amount of power lost cannot be accurately calculated because accurate data cannot be obtained, but it can be estimated with sufficient accuracy for practical purposes.

The method used for preparing the chart below was to calculate the compressor power from the chart above for compressing the air to the regulator inlet pressure in the first column. Then calculating power to compress it to the regulator outlet pressure along the top of the chart, then subtracting the two.

Additional Horsepower Required Due to Over-Compression

Figures in the body of this table show the HP lost for every 1 SCFM of air which passes through the regulator, suffering a pressure drop from inlet pressure (left column) to outlet pressure (along top of table) for operation of the air cylinder. Multiply table values times the SCFM flow through the circuit.

Regulator Inlet PSI	Regulator Outlet Pressure, PSI							
	50 PSI	60 PSI	70 PSI	80 PSI	90 PSI	100 PSI	110 PSI	120 PSI
70	0.024	0.011	- - - - -	- - - - -	- - - - -	- - - - - -	- - - - - -	- - - - - -
80	0.035	0.022	0.011	- - - - -	- - - - -	- - - - - -	- - - - - -	- - - - - -
90	0.044	0.031	0.020	0.009	- - - - -	- - - - - -	- - - - - -	- - - - - -
100	0.053	0.040	0.029	0.018	0.009	- - - - - -	- - - - - -	- - - - - -
110	0.062	0.049	0.038	0.027	0.018	0.009	- - - - - -	- - - - - -
120	0.070	0.057	0.046	0.035	0.026	0.017	0.008	- - - - - -
130	0.077	0.064	0.053	0.042	0.033	0.024	0.015	0.007
140	0.084	0.071	0.060	0.049	0.040	0.031	0.022	0.016
150	0.091	0.078	0.067	0.056	0.047	0.038	0.029	0.021
160	0.096	0.083	0.072	0.061	0.052	0.043	0.034	0.029

Index to Volume 2

Other Womack Books
on Industrial Fluid Power

VOLUME 1 — INDUSTRIAL FLUID POWER. The first of a series of three textbooks on fluid power as used in the industrial plant and on mobile equipment. This is the basic textbook, covering the full range of hydraulic, compressed air, and vacuum usage.

This Third Edition has been completely re-written, with more photos of equipment, more circuit diagrams, and about 40 more pages of design calculations, troubleshooting information, and design tables and charts. The new metric S. I. system of units is explained and charts are included for sizing metric bore cylinders.

Subjects covered include laws and terms relating to fluid power; simple layouts for plant hydraulic and air systems; fluid flow through pipes; selection of pipe size; air and hydraulic cylinders — direction control, speed control, and sizing for power; air and hydraulic direction, pressure, and speed control valves; hydraulic pumps — how they operate and how to use them; trio units for compressed air; air dryers; hydraulic filters; hydraulic power units; hydraulic accumulators and heat exchangers.

This book is excellent for home study, with review questions at the end of each chapter. It is now being used by several hundred vocational/technical schools and is the official course textbook in some of these schools. Also used for in-plant employee training by hundreds of nationally known companies in many industries. Recommended as a training manual by the NFPA (National Fluid Power Association) and the FPS (Fluid Power Society). It has been selected as the most suitable training manual by the FPDA (Fluid Power Distributors Association) and is used as a part of their audio/visual program which was prepared for training employees of their member companies.

VOLUME 3 — INDUSTRIAL FLUID POWER. Although covering many advanced applications which require rotary mechanical output, the presentation is simple as it is in all the Womack books, and can easily be understood by anyone with mechanical or electrical aptitude. Mathematics are limited to very simple formulae, and in most cases the charts and tables can be used instead of the formulae.

The format includes first an explanation of each component and how it works, using photos and sectional views. Then circuit diagrams plus design charts show how to design the component into circuits.

Among the components covered in this particular book are common types of hydraulic motors — gear, vane, and piston types. Shows how to match a hydraulic motor to its load; circuits for direction and speed control; closed loop hydrostatic transmissions; motor starting and running torque, efficiency, HP, life expectancy, and installation; air motor operation and circuit design; power steering design; rotary and spool-type flow dividers and how to use them; bootstrapping principles and circuits for saving power; and many other topics.

ELECTRICAL CONTROL OF FLUID POWER. This is a textbook on how to draw electrical circuits and ladder diagrams for air and hydraulic applications. Electrical diagrams are shown alongside the fluid circuits.

First, common electrical components, switches, relays, solenoid valves, timers, counters, etc., are described, then how to draw schematic ladder diagrams for JIC hard wired circuits. How to draw circuits for directional control and sequencing of cylinders using 4-way solenoid valves, how to peck drill, cylinder dwell, deceleration at the end of the stroke, index table operation, automatic clamp and work, feed and drill, one-cycle control, multiple indexing, counting and stacking, turnover of parts, jogging, safety circuits, etc. Electric motors and motor starters are also covered.

Electronic control with programmable controllers shows how controllers work, how to design ladder diagrams for them, how to convert JIC diagrams for programming in a controller, how to enter a program into a controller, how to store and retrieve programs, etc.

Servo valves and systems are explained; what a servo valve is and how to design circuits for electro-hydraulic servo valves. How to select servo valves. Typical applications for servo valves.

Finally, proportional solenoid valves are explained along with the kind of applications which can use this kind of directional, pressure, or speed control.

FLUID POWER IN PLANT AND FIELD. The new Second Edition is a revision with 56 new pages added to include more information on installation and start-up of hydrostatic and open loop systems; more information on troubleshooting of both hydrostatic and open loop hydraulic systems; and more reference data in the Appendix on pipe sizes, pump and motor shafts and flanges, electric motor frame sizes, and much more of interest to installers.

Although this is not one of the regular textbook series, it has much practical information that could not be included in a textbook. Some companies use it as a training manual for mechanics, field service personnel, and installers. Contains many practical ideas on the best way to install cylinders, hydraulic pumps and motors, air and hydraulic valves, hydraulic reservoirs, heat exchangers, accumulators, air trio units, air dryers, hydrostatic systems, and vacuum systems. Unique ideas on increasing speed and efficiency of air and hydraulic cylinders, causes of recurring problems, etc. Material in this book is of a practical nature which is usually not included in an ordinary fluid power design handbook. We recommend this book as a companion reference to students of the 3-volume industrial fluid power textbooks.

HOW TO ORDER. SEE INFORMATION IN THE FRONT OF THIS BOOK